6G Enabling Technologies
New Dimensions to Wireless Communication

RIVER PUBLISHERS SERIES IN COMMUNICATIONS AND NETWORKING

The "River Publishers Series in Communications and Networking" is a series of comprehensive academic and professional books which focus on communication and network systems. Topics range from the theory and use of systems involving all terminals, computers, and information processors to wired and wireless networks and network layouts, protocols, architectures, and implementations. Also covered are developments stemming from new market demands in systems, products, and technologies such as personal communications services, multimedia systems, enterprise networks, and optical communications.

The series includes research monographs, edited volumes, handbooks and textbooks, providing professionals, researchers, educators, and advanced students in the field with an invaluable insight into the latest research and developments.

Topics included in this series include:-

- Communication theory
- Multimedia systems
- Network architecture
- Optical communications
- Personal communication services
- Telecoms networks
- Wifi network protocols

For a list of other books in this series, visit www.riverpublishers.com

6G Enabling Technologies

New Dimensions to Wireless Communication

Editors

Ramjee Prasad

Aarhus University, Denmark

Anand Raghawa Prasad

Deloitte Tohmatsu Cyber, Japan

Albena Mihovska

Aarhus University, Denmark

Nidhi

Aarhus University, Denmark

Published 2023 by River Publishers
River Publishers
Alsbjergvej 10, 9260 Gistrup, Denmark
www.riverpublishers.com

Distributed exclusively by Routledge
605 Third Avenue, New York, NY 10017, USA
4 Park Square, Milton Park, Abingdon, Oxon OX14 4RN

6G Enabling Technologies / by Ramjee Prasad, Anand Raghawa Prasad, Albena Mihovska, Nidhi.

Routledge is an imprint of the Taylor & Francis Group, an informa business

ISBN 978-87-7022-774-2 (print)
ISBN 978-10-0075-029-4 (online)
ISBN 978-10-0336-088-9 (ebook master)

Towards New Horizons of Wireless Communication Era

Contents

Preface

यथैधांसि समिद्धोऽग्निर्भस्मसात्कुरुतेऽर्जुन ||

ज्ञानाग्नि: सर्वकर्माणि भस्मसात्कुरुते तथा ||४- ३७||

yathaidhānsi samiddho 'gnir bhasma-sāt kurute 'rjuna //

jānāgniḥ sarva-karmāṇi bhasma-sāt kurutetathā // 4-37 //

Translation

Just as a burning flame turns fuel into ashes, juna, the fire of knowledge, turns all actions into ashes.

——— Bhagavad Gita Chapter 4, verse 37 ———

यथैधांसि समिद्धोऽग्निर्भस्मसात्कुरुतेऽर्जुन ||

ज्ञानाग्नि: सर्वकर्माणि भस्मसात्कुरुते तथा ||४- ३७||

yuktāhāravihārasya yuktaceṣṭasyakarmasu //

yuktasvapnāvabodhasya yogobhavati du : khahā // 6-5 //

Translation

Elevate yourself through your efforts, and not degrade yourself. For, the mind can be the friend and the self's enemy.

——— Bhagavad Gita Chapter 6, verse 5 ———

"*6G Enabling Technologies: New Dimensions to Wireless Communication*" intends to cater for an exhaustive understanding of the research activities for future networks. It drafts the fundamental system requirements, enabling technologies, eminent drivers, evolved use cases, research requirements, challenges, and open issues expected to drive 6G research. This book is an outcome of the workshop "*6G Knowledge Lab Opening* and *36th GISFI Workshop*" held 21-22 December 2020, organized jointly by the CTIF Global Capsule (CGC) and the Global ICT Standardisation Forum for India (GISFI) and technically sponsored by IEEE.

The workshop was a two-day event featuring industry and academia invited talks on various research challenges related to 6G and enabling technologies. Its theme was 6G as a universal technology and infrastructure to support Further-Enhanced Mobile Broadband (FeMBB) services. The focus was on the current digitalization trends, network intelligence, and the user requirements for accessing and transmitting high-definition data in varied scenarios. The workshop highlighted the importance of digital technologies in day-to-day lives and how it efficiently drives various tasks with ease, more than ever since the global pandemic.

To make this book a fundamental resource, we have invited world-renowned experts for 6G from the industry and academia to pen down their ideas on different aspects of 6G research. The chapters in this book cover a broader scope and various related and unrelated verticals. Specifically, this book covers the following topics:

- 6G Use Cases, Requirements, and Enabling Technologies
- New Spectrums and their challenges for 6G
- Privacy Preservation in 6G Networks
- Aerial Infrastructure for 6G Networks
- Economic Challenges associated with 6G Wireless Networks

The encompassing intent of this book is to explore the evolution of current 5G networks towards the future 6G networks from a service, air interface, and network perspective, thereby laying out a vision for 6G networks. This book not only discusses the potential 6G use cases, requirements, metrics, and enabling technologies but also discusses the emerging technologies and topics such as 6G Physical Layer (PHY) technologies, Reconfigurable Intelligent Surface (RIS), Millimetre-Wave (mmWave), and Terahertz (THz) communications, Visible Light Communications (VLC), transport layer for Tbit/s communications, high-capacity backhaul connectivity, cloud-native approach, Machine-Type Communications (MTC), edge intelligence, and pervasive Artificial Intelligence (AI), network security and blockchain, and the role of the open-source platform in 6G. This book provides a systematic treatment of the state-of-the-art in these emerging topics and their role in supporting a wide variety of verticals in the future. As such, the book provides a comprehensive overview of the expected applications of 6G with a detailed discussion of their requirements and possible enabling technologies. This book also outlines the potential challenges and research directions to facilitate the future research and development of 6G mobile wireless networks.

List of Figures

List of Tables

List of Contributors

Anant, Aaloka, *CGC - Aarhus University, Birk Centerpark 10, DK-7400 Herning Denmark; E-mail: aaloka@anantprayas.org*

Chandra, Ashok, *(Former) Wireless Adviser to the Government of India, New Delhi, India; E-mail: drashokchandra@gmail.com*

Chandramouli, R., *Stevens Institute of Technology; E-mail: mouli@ieee.org*

Cooklev, Todor, *Purdue University Fort Wayne, Indiana, USA; E-mail: cooklevt@pfw.edu*

Craciunescu, Razvan, *Bucharest, Romania; E-mail: razvan.craciunescu@upb.ro*

Ekova, Dessislava, *Technical University of Sofia, Sofia, Bulgaria*

Fratu, Octavian, *Telecommunication Department, University Politehnica of Bucharest, Romania; E-mail: octavian.fratu@upb.ro*

Halunga, Simona, *Telecommunication Department, University Politehnica of Bucharest, Romania; E-mail: simona.halunga@upb.ro*

Henrique, Paulo Sergio Rufino, *CTIF Global Capsule (CGC) / Aarhus University/ Spideo, Paris, France; E-mail: rufino@spideo.tv*

Kloch, Christian, *5G Innovation Architect, TDC Net,; E-mail: chkl@tdcnet.dk*

Konhäuser, Walter, *Managing Director, Oktett64 GmbH and Associated Partner, Management Consulting Kastner GmbH & Co. KG; E-mail: walter.konhaeuser@gmx.de*

Kuipers, Berend W.M., *INOV-INESC Inovação / Lusíada University of Lisbon, Lisbon, Portugal; E-mail: martijn.kuipers@gmail.com*

Kulkani, Nandkumar, P., *Smt Kashibai Navale College of Engineering, Pune, India; E-mail: npkulkarni.pune@gmail.com*

Kumar, Ambuj, *Department of Business Development and Technology, Aarhus University, Aarhus; Denmark, E-mail: ambuj@btech.au.dk*

Kumar, Navin, *ECE Dept., Amrita School of Engineering Bengaluru, Amrita Vishwa Vidyapeetham; E-mail: navinkumar@ieee.org*

Lindgren, Peter, *CGC - Aarhus University, Birk Centerpark 10, DK-7400 Herning Denmark; E-mail: peterli@btech.au.dk*

Mantri, Dnyaneshwar, *S., Sinhgad Institute of Technology, Lonavala, Pune, India; E-mail: dsmantri@gmail.com*

Mihovska, Albena, *CGC Research Lab, Dept. of Business Development and Technology, Aarhus University, Denmark; E-mail: amihovska@btech.au.dk*

Nidhi, *CGC Research Lab, Dept. of Business Development and Technology, Aarhus University, Denmark; E-mail: nidhi@btech.au.dk*

Pawar, Pranav M., *BITS Pillani Dubai;*
E-mail: pranav@dubai.bits-pilani.ac.in

Poulkov, Vladimir, *Technical University of Sofia, Sofia, Bulgaria;*
E-mail: vkp@tu-sofia.bg

Prasad, Anand, R., *Deloitte Tohmatsu Cyber, Tokyo, Japan;*
E-mail: anandraghawa.prasad@tohmatsu.co.jp

Prasad, Neeli, R., *TrustedMobi "VehicleAvatar Inc.", USA;*
E-mail: neeli.prasad@smartavatar.nl

Prasad, Ramjee, *CGC - Aarhus University, Birk Centerpark 10, DK-7400 Herning Denmark; E-mail: ramjee@btech.au.dk*

Raj Dua, Tilak, *Tower and Infrastructure Providers Association, 7, Bhai Vir Singh Marg, Gole Market, New Delhi, India, Vice-Chairman, Global ICT Standardization Forum for India (GISFI), and Chairman, ITU-APT Foundation of India; E-mail: tr.dua@taipa.in*

Rufino Henrique, Paulo S., *CTIF Global Capsule (CGC) / Aarhus University/ Spideo, Paris, France; E-mail: rufino@spideo.tv*

Sagar, Vidya, *Spectronn Inc.; E-mail: vidya@spectronn.com*

Sanyal, Rajarshi, *67, Route De Bouillon, Arlon,6700, Belgium; E-mail: rajarshi.sanyal@proximus.lu*

Srikanth, S., *Nanocell Networks,; E-mail: sribommi@gmail.com*

Subbalakshmi, K.P., *Stevens Institute of Technology,;*
E-mail: ksubbala@stevens.edu

Tripathi, P S M., *Ministry of Communications, Government of India, New Delhi, India; E-mail: psmtripathi@gmail.com*

Uraz, Thomas, *5G Core Partner Lead, Ericsson Teglholmsgade 1, 2450 København C, Denmark; E-mail: thomas.uraz@ericsson.com*

Vlahov, Atanas, *Technical University of Sofia, Sofia, Bulgaria;*
E-mail: avlahov66@gmail.com

List of Abbreviations

1G	First-generation networks
2G	Second-generation networks
3G	Third-generation networks
3GPP	Third generation partnership project
4G	Fourth-generation networks
5G	Fifth-generation
5G EIR	5G equipment identity register
5G NR	5G new radio
6G	Sixth-generation (6G) networks
6GMVNO	6g mobile virtual networks operators
AI	Artificial intelligence
AMF	Access and mobility management function
AMQP	Advanced message queuing protocol
AR	Augmented reality
AUSF	Authentication server function
BBU	Baseband units
Bluetooth LE	Wireless personal area network technology
BSF	Binding support function
CA	Carrier aggregation
CAAS	Cloud-As-A-Service
Capex	Capital expenditure
CHP	combined heat and power unit
CIA	Confidentiality, integrity, and availability
CN	Core network
CoAP	Constrained application protocol
CONASENSE	Communication, navigation, sensing and services
COTS	Custom of the shelf
CPRI	Common public radio interface
CRAN	Cloud RAN
C-RAN	Cloud radio access networks
CSI	Channel state information

D2D	Device-To-Device
D2M	Device-To-Machine
DMZ	Demilitarized zone
DNN	Deep neural networks
DSO	Distribution system operator
E2E	End-To-End
eMBB	Enhanced mobile broadband
emMTC	Enhanced Mmtc
EnOcean	Energy harvesting wireless technology used primarily in building automation systems
eURLLC	Enhanced ultra reliable low latency communication
EV	Electrical vehicle
FEC	Forward error correction
FeMBB	Further-enhanced mobile broadband
FTTH	Fibre to the home
FPS	Frames per second
FRAN	Future-Radio-Access-Networks
FWA	Fixed wireless access
HBC	Human bond communication
HBF	Holographic beam forming
HI	Human intelligence
HVAC	Heating and Air Conditioning
IAB	Integrated access backhaul
IAV	Industrie-Automation Vertriebs – GmbH
ICT	Information communication technologies
IIoT	Industrial IoT
IoB	Internet of beings
IoT	Internet of things
ITU	International telecommunication union
KNX-RF	Open standard for RF-links in buildings
KPI	Knowledge performance index
LEO	Low earth orbit
LISs	Large intelligent surfaces
LoRa	Long Range Radio Technology
M2M	Machine-To-Machine
MEC	Multi-access edge computing
MIMO	Multiple input multiple output
ML	Machine learning

mMTC	Massive machine-type communication
mmWave	Millimeter wave
MNO	Mobile network operators
MQTT	Message queuing telemetry transport
MVNO	Mobile virtual network operator
NAAS	Network-As-A-Service
NB-IoT	Narrowband internet of things
NEF	Network exposure function
NFV	Network function virtualization
NGMN	Next generation mobile network
NMO	Network management and orchestration
NOMA	Non-orthogonal multiple access
NRF	Network repository function
NS	Network slicing
NSSF	Network slicing selection function
Open RAN	Open-radio access network
Opex	Operational expenditure
OWC	Optical wireless communications
P2M	People-To-Machine
P2P	People-To-People
PAN	Personal area network
PAPR	Peak average power ratios
PSM	Power saving mode
PV	Photovoltaic
QML	Quantum machine learning
QoE	Quality of experience
QoL	Quality of life
QoS	Quality of service
RAN	Radio access network
RFID	Radio frequency identification
ROI	Return over investments
RRH	Remote radio head
SBA	Service based architecture
SCP	Service communication proxy
SDG	Societal development goal
SDN	Software defined network
SEPP	Security edge protection proxy
SIM	Subscriber identity modules
SLA	Service level agreement

SMF	Session management function
SM-MIMO	Supermassive multiple-input and multiple-output system
SNR	Signal-to-noise ratio
STI	Science, technology, and innovation
SW	Software updates
TCO	Total cost of ownership
THz	Terahertz frequency
TSP	Telecommunication service provider
TVO	Total value of opportunity
U4SSC	UN project united for smart sustainable cities
UAV	Unmanned aerial vehicle
UDM	User data management
UDR	User data repository
UDSF	User data service function
UE	User equipment
UPF	User plane function
URLLC	Ultra-reliable low latency communications
USN	Underwater-Sensor-Node
V2X	Vehiclw-To-Everything
VLC	Visible light communication
VR	Virtual reality
WUS	Wake up signal
XMPP	Extensible messaging and presence protocol
ZigBee	IEEE 802.15.4-based specification for a suite of high-level communication protocols
Z-Wave	Wireless communications protocol used primarily for home automation

1

Introduction

Nidhi[1], Albena Mihovska[1], Anand R. Prasad[2] and Ramjee Prasad[1]

[1]CGC Research Lab, Dept. of Business Development and Technology, Aarhus University, Denmark
[2]Deloitte Tohmatsu Cyber, Tokyo, Japan
E-mail: nidhi@btech.au.dk; amihovska@btech.au.dk; anandraghawa.prasad@tohmatsu.co.jp; ramjee@btech.au.dk

The best way to predict the future is to create it. Technology innovation connects societies and becomes a part of everyday life.

The sixth generation of wireless communication (6G), succeeding the 5G cellular technology, opens up several possibilities in terms of technology and offered services. 6G is expected to allow usage of available higher frequency spectrums to cater to increased capacity, throughput, and low latency ($< 1\,\mu s$). 6G will witness the unification of various technologies like artificial intelligence (AI), Machine Learning (ML), Augmented/Virtual Reality (AR/VR), etc., to provide an immersive user experience. It is foreseen as the accelerator of transformation and innovation globally. These developments challenge the current capabilities of the enabling wireless communication systems from various aspects, such as delay, data rate, degree of intelligence, coverage, reliability, and capacity. Thus the research should seek breakthroughs from the current network architecture and communication theory to provide novel concepts that can be key for designing a new system such as 6G. At the same time, it is crucial to enable such technology developments to stay 'green' and take into account major environmental concerns, such as climate change, which can be achieved by novel, 'green' digitalized business models and other requirements for sustainability and beyond.

The fifth generation of wireless communication (5G) is in its deployment phase and has set new parameters to speed up the initiatives towards 6G.

The commercial deployment has exposed the shortcomings in the present communication techniques. Thus, it laid out the well-defined expectations from the 6G systems to support enhanced user experience applications ranging from autonomous systems to extended reality (XR) [1]. 6G has been defined with numerous use cases and applications [2], but it lacks its fundamental network architecture and design. It will exploit high-frequency bands and new access techniques to evolve the wireless communication era. Various enabling technologies are identified to have the potential to cater requirements of a new set of services.

1.1 Evolution of Mobile Communication

Communication has been an essential part of our lives since the beginning. The Evolution of Communication is an ongoing process aided by advancing technologies. The journey from the wired telephone to AI-aided real-time communication [3] has marked importance for a better, well-connected and progressing society. The wireless communication sector is considered a "Living" industry as it continues to grow. The growth is directly proportional to demands in society. So far, we have witnessed five generations of mobile communication and are on our way to more upcoming generations. In the First Generation (1G), Second Generation (2G), Third Generation (3G), Fourth Generation (4G) mobile communication, the "G" represents the "Generation". Every decade a generation of mobile communication is succeeded by a new one offering distinct features and services to meet user and network demand. The critical characteristic differences in mobile generations are data traffic, data rate, throughput, latency, coverage, device density, reliability, and security [4].

1.1.1 1G – First Generation of Mobile Communication (Analogue Systems)

1G was launched in 1979, and it marked the first generation of cellular telephony. It was based on analogue technology and was dedicated to voice calling. It uses narrowband frequency channels to transmit voice signals. With 1G, several standards were developed regionally simultaneously [5, 6], such as nordic mobile telephone (NMT), advanced mobile phone system (AMPS) and total access communications system (TACS). The maximum achieved speed for 1G systems was 2.4 Kbps with poor battery life, less security and voice quality.

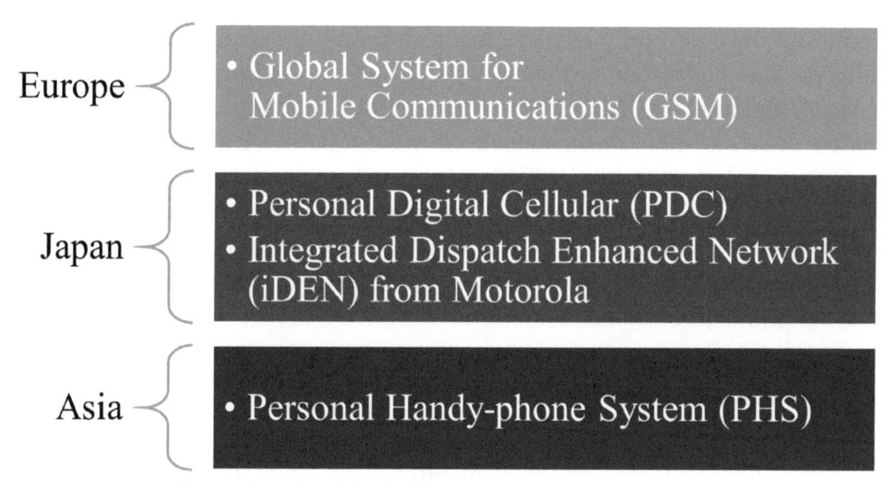

Figure 1.1 Worldwide initiatives for 2G mobile systems

1.1.2 2G – Second Generation (Digital Systems)

The second generation of mobile communication was launched in the early 1990s, and these mobile systems use digital radio signals. The driving force behind 2G was to add security and reliability to the communication channel [4]. 2G standard systems witnessed several independent initiatives regionally. Figure 1.1 illustrates the country-wise initiatives.

GSM was introduced by european telecommunications standards institute (ETSI) in 1987 [7]. It implemented time division multiple access (TDMA) and frequency division duplex (FDD) schemes. Alternatively, code division multiple access (CDMA) was introduced by Qualcomm [8]. These multiplexing schemes allowed 2G systems to incorporate multiple users on a single channel. In addition to the voice calls, 2G systems offered data services, short messaging systems (SMS) and enhanced features of conventional calling like call hold, conferencing, etc. With the general packet radio service (GPRS), 2G systems achieved a speed of 50 Kbps, and with enhanced data rates for GSM evolution (EDGE), the rate was up to 1 Mbps.

1.1.3 3G – Third Generation

2001 witnessed the rising of 3G Mobile Systems. The international telecommunication union (ITU) defined the International Mobile Telecommunications 2000 (IMT-2000) framework [9] for the 3G cellular networks. 3G systems marked the revolutionary changes in mobile communications by offering services like web browsing, email, video downloading, picture sharing and fast internet speed at a rate of 200Kbps. It marked the switch from multimedia phones to Smartphones. It offered a cost-efficient increased data rate along with enhanced voice quality. Universal mobile telecommunications system (UMTS) was introduced with 3G systems. It used wideband CDMA (WCDMA) with achieved data rates up to 2 Mbps (stationary) and 384kbps (mobility). The theoretical maximum data rate for HSPA+ is 21.6 Mbps. High-speed internet made data streaming one of the killing applications of 3G.

1.1.4 4G – Fourth Generation

4G mobile systems were based on the IMT-Advanced requirements [10]. 4G systems' approach was practical to offer users high speed, high quality and high capacity. It promised to achieve 100 Mbps (Mobility) and 1 Gbps (Stationary) with a reduced latency of 300ms to $<$100ms. 4G systems reduced the network congestion significantly. It improved in-built security and introduced the concept of multimedia and the internet over IP. The key enabling technologies include multiple input multiple output (MIMO) and orthogonal frequency division multiplexing (OFDM). It also defined heterogeneous networks (HetNets), small cells (SC) [12] and additional features like carrier aggregation (CA) [11], Coordinated Multipoint (CoMP) and advanced multiple-input multiple-output (MIMO) transmissions.

4G systems also introduced the following positioning methods and enhancements [13];

- *Network-Based Positioning* (Release 10 [14, 15])
- *Radio Frequency Pattern Matching* (Release 9 [16]- Release 12 [16])
- *Positioning Enhancements* (Release 9 [17]- Release 12 [18])

1.1.5 5G – Fifth Generation

5G systems are currently in the commercial deployment phase and intended to improve their predecessor-4G through data rates, connection density, latency, and other enhancements. These systems exploited the 30 GHz – 300 GHz spectrum of millimetre waves (mmWave) to achieve a maximum of 35.46 Gbps. It utilized the concepts of Small Cells to increase coverage and reduce latency [12]. 5G systems also exploited advanced access technologies such as beam division multiple access (BDMA), non-orthogonal multiple access (NOMA), Quasi-Orthogonal Sequences, filter bank multi-carrier (FBMC),

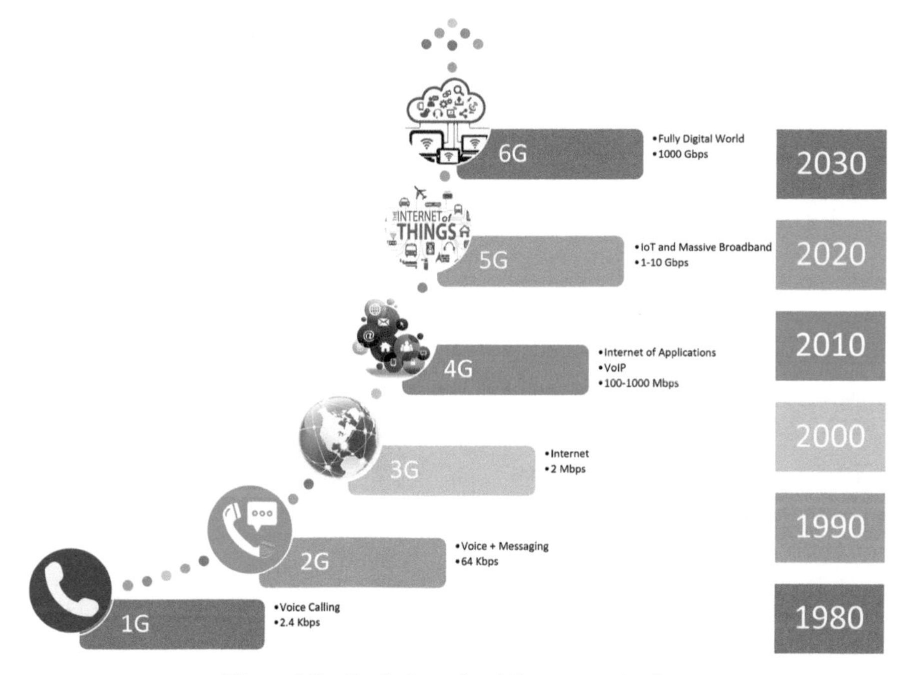

Figure 1.2 Evolution of mobile communication

[19, 20] etc. The BDMA technique provides base stations multiple access by facilitating the antenna beam divided based on the base station location. The scalable orthogonal frequency-division multiplexing (OFDM) is dominant in 5G networks to achieve ultra-low latency. It also incorporated network slicing (NS) [21] to cater customized services from mobile network operator (MNO) side based on the service level agreement (SLA). On the user experience side, implementing AI, AR/VR, and XR opened new use cases, applications, and services. Figure 1.2 illustrates the distinctions between 2G, 3G, 4G, 5G, and 6G.

1.2 6G KPIs and Use Cases

Different studies and research bodies, including the 3rd Generation Partnership Project (3GPP), have specified standards based on bandwidth, data rate, latency, access technologies, energy and spectrum efficiency, connection density, etc. [22, 23] for the upcoming 6G networks. 6G systems have their Key Performance Indicators (KPIs), use cases, and requirements [24] to guide the research activities for academia and industry. Various KPIs are mapped

Figure 1.3 6G KPIs

according to the 6G network requirements, as illustrated in Figure 1.3. Table 1.1 lists the difference between 5G and 6G KPIs.

Use cases are essential to validate the development process. It defines the final results and determines the system's behaviour. Figure 1.4 lists different use cases to support identified 6G features [24].

1.3 6G Networks: Considerations and Requirements

1.3.1 Network Considerations

6G is expected to take the legacy of commercial 5G forward with advancements at a broader scale. 6G deployments are envisioned to provide new customer markets offering a plethora of services and parallel/vertical applications. AI will be imperative for 6G networks and act as a tool to implement other interoperable technologies. In addition to AI, Cloud and Big Data[25] will be critical to addressing enormous data. Immersive media

Table 1.1 Difference between 5G and 6G KPIs

KPIs	5G	6G
Maximum Bandwidth	1 GHz	100 GHz
Peak Data Rate	20 Gb/s	>= 1 Tb/s
Experienced Data Rate	0.1 Gb/s	1 Gb/s
Spectrum Efficiency	Peak: 30 b/s/Hz Experienced: 0.3 b/s/Hz (3 times that of 4G)	Peak: 60 b/s/Hz Experienced: 3 b/s/Hz (5 to 10 times that of 5G)
Network Energy Efficiency	Not Specified	1 pJ/b
Area Traffic Capacity	10 $Mb/s/m^2$	1 $Gb/s/m^2$
Connection Density	10^6 devices/Km^2	10^7 devices/Km^2
Latency	1 ms	10 to 100 μs
Jitter	Not specified	1 μs
Reliability or FER	1×10^{-5}	1×10^{-9}
Mobility	500 Km/h	>= 1000 Km/h
Uniform User Experience	50 Mb/s, 2D everywhere	10 Gb/s, 3D everywhere
Localization Accuracy	10 cm in 2D	1 cm in 3D

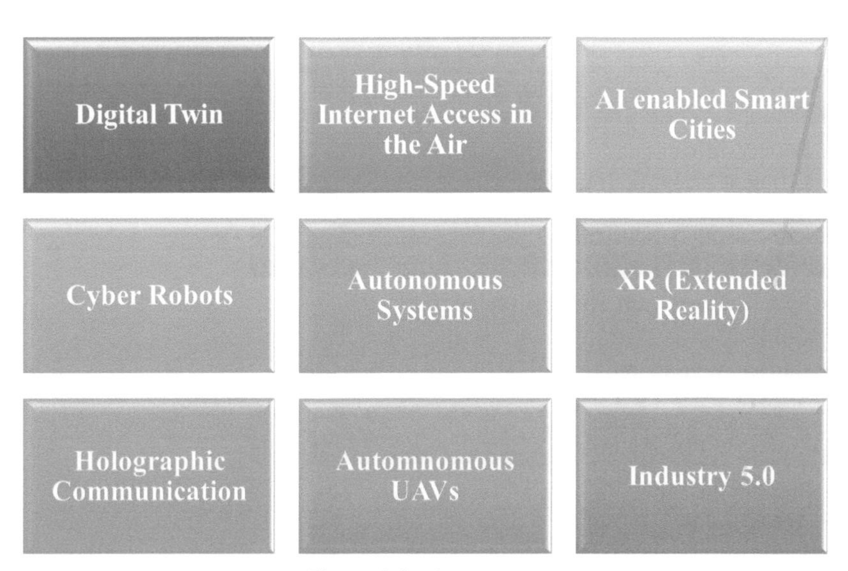

Figure 1.4 6G use cases

applications (AR/VR/XR) will rise to 85% of mobile broadband. 6G will cater to personalized technology to connect human emotions in blended spaces [3]. Data handling (like organizing, analyzing, storing, and using) from heterogeneous devices will be challenging. Standards and policies must comply with the user concerns and systems requirements. The considerations for 6G systems will encompass the components as depicted in Figure 1.5.

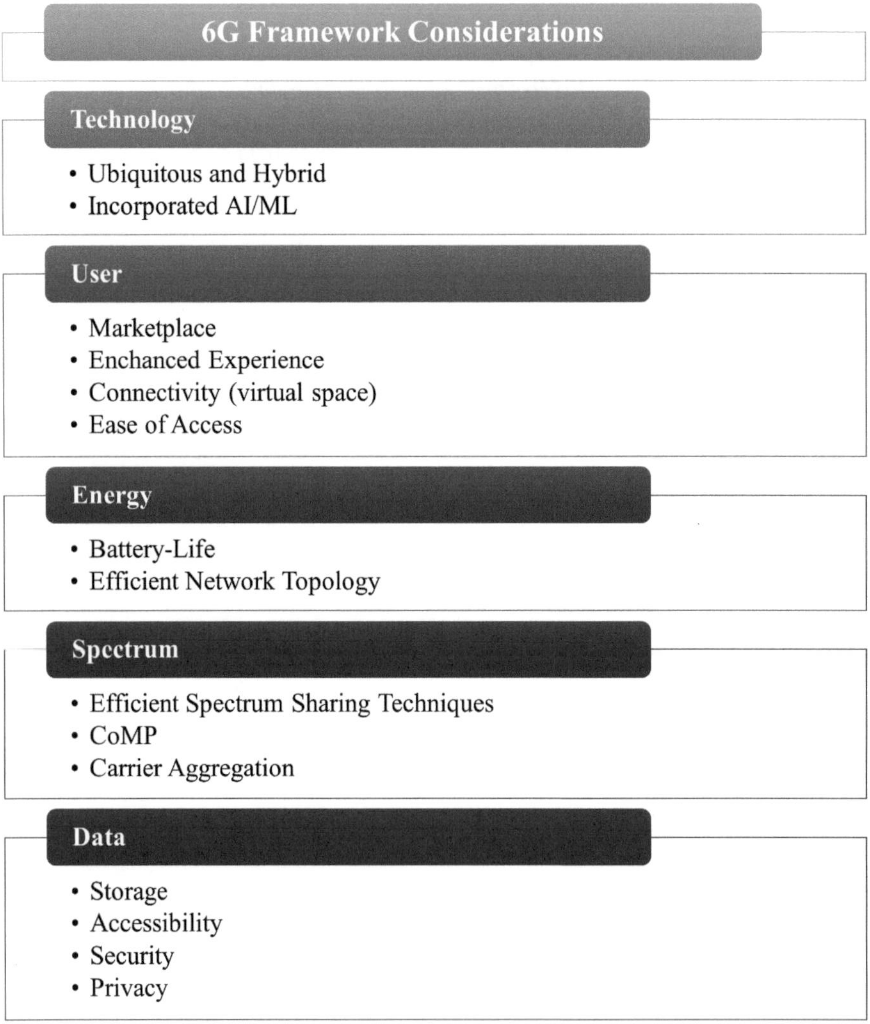

6G Framework Considerations

Technology

- Ubiquitous and Hybrid
- Incorporated AI/ML

User

- Marketplace
- Enchanced Experience
- Connectivity (virtual space)
- Ease of Access

Energy

- Battery-Life
- Efficient Network Topology

Spectrum

- Efficient Spectrum Sharing Techniques
- CoMP
- Carrier Aggregation

Data

- Storage
- Accessibility
- Security
- Privacy

Figure 1.5 6G framework considerations

1.3.2 Network Requirements

Figures 1.6 illustrates the mapping of 6G requirements to the extent of the network.

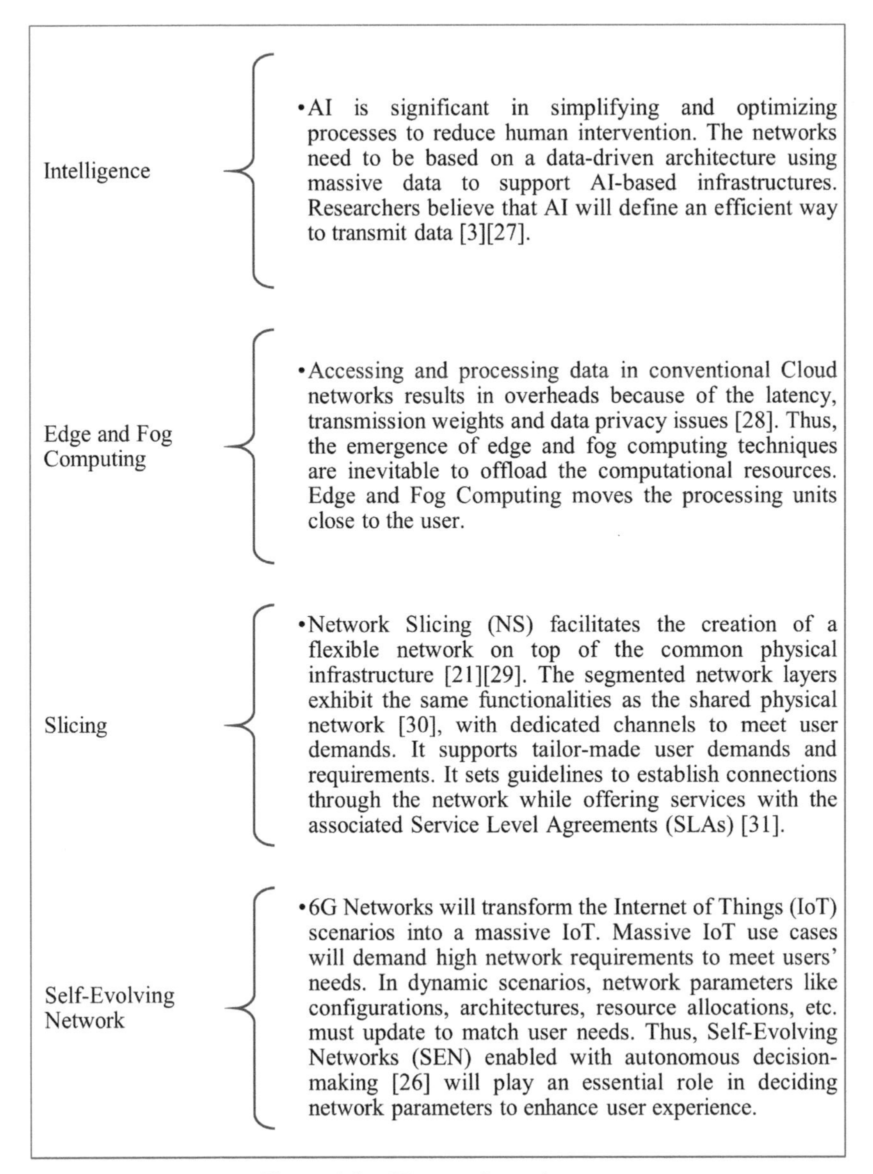

Figure 1.6 6G network requirements

1.4 6G System Architecture

6G networks are foreseen as massive IoT use cases and a magnificent amount of data. As discussed earlier, it is crucial to design the network components such that they autonomously work in parallel to provide ubiquitous connectivity and coverage. AI and ML are eminent in 6G system architecture, making it intelligent, self-evolving/configuring, self-organizing, and self-healing. The main identified pillars to building 6G infrastructure are as follows;

- Air Interface
- New Spectrum
- Application of AI and ML algorithms
- Interoperability
- Coexisting radio access technologies (RAT)

1.5 6G Standardization

Worldwide activities and alliances from industry and academia are taking initiatives towards 6G standards. The Next G Alliance is focused on 6G deployment, research and manufacturing standardization [32]. The European Commission radio spectrum policy group (RSPG) [33] has specified various requirements for the 6G initiatives in Europe. To add to the list, The 6G Flagship project (University of Oulu of Finland) [34], Vodafone Germany [35], NTT Docomo, Japan [36], Samsung-South Korea [37], NIIR [38] and many more are working on 6G standardization activities from various perspectives. 3GPP [22] is expected to release 6G specifications in its Release 18 by 2023 [39].

1.6 Challenges in 6G

The key challenges foreseen in 6G networks are illustrated in Figure 1.7. These challenges are covered gradually in the following chapters.

1.7 Book Overview

"*6G Enabling Technologies: New Dimensions to Wireless Communication*" intends to highlight the critical aspects of future wireless technologies. The motivation of this book is to bring a comprehensive view of research activities towards future mobile networks - 6G. In this book, we present contributions

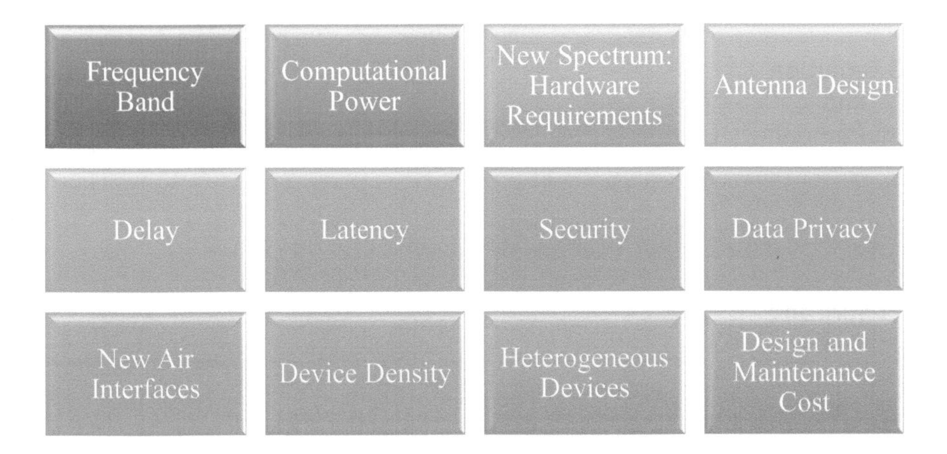

Figure 1.7 6G challenges

from pioneers across the globe to bring forward the research trends in 6G Wireless Networks. The book introduces critical technologies behind 6G wireless communication and mobile networking. It drafts the fundamental system requirements, enabling technologies, eminent drivers, evolved use cases, research requirements, challenges, and open issues expected to drive 6G research. It explains a general vision of 6G technology, including the motivation for conducting 6G research and recent progress in 6G research, followed by chapters on architectural evolution and enabling technologies, including the advancements in infrastructure. It targets students and young researchers involved in telecommunications and provides them with cutting-edge wireless networking technologies and market analysis. The motivation is to understand the research activities for future networks fully.

This book not only discusses the potential 6G use cases, requirements, metrics, and enabling technologies but also discusses the emerging technologies and topics such as 6G physical layer (PHY) technologies, reconfigurable intelligent surface (RIS), Millimetre-Wave (mmWave), and Terahertz (THz) communications, visible light communications (VLC), transport layer for Tbit/s communications, high-capacity backhaul connectivity, cloud-native approach, machine-type communications (MTC), edge intelligence, and pervasive artificial intelligence (AI), network security and blockchain, and the role of the open-source platform in 6G. The book is divided into sections, as illustrated in Figure 1.8.

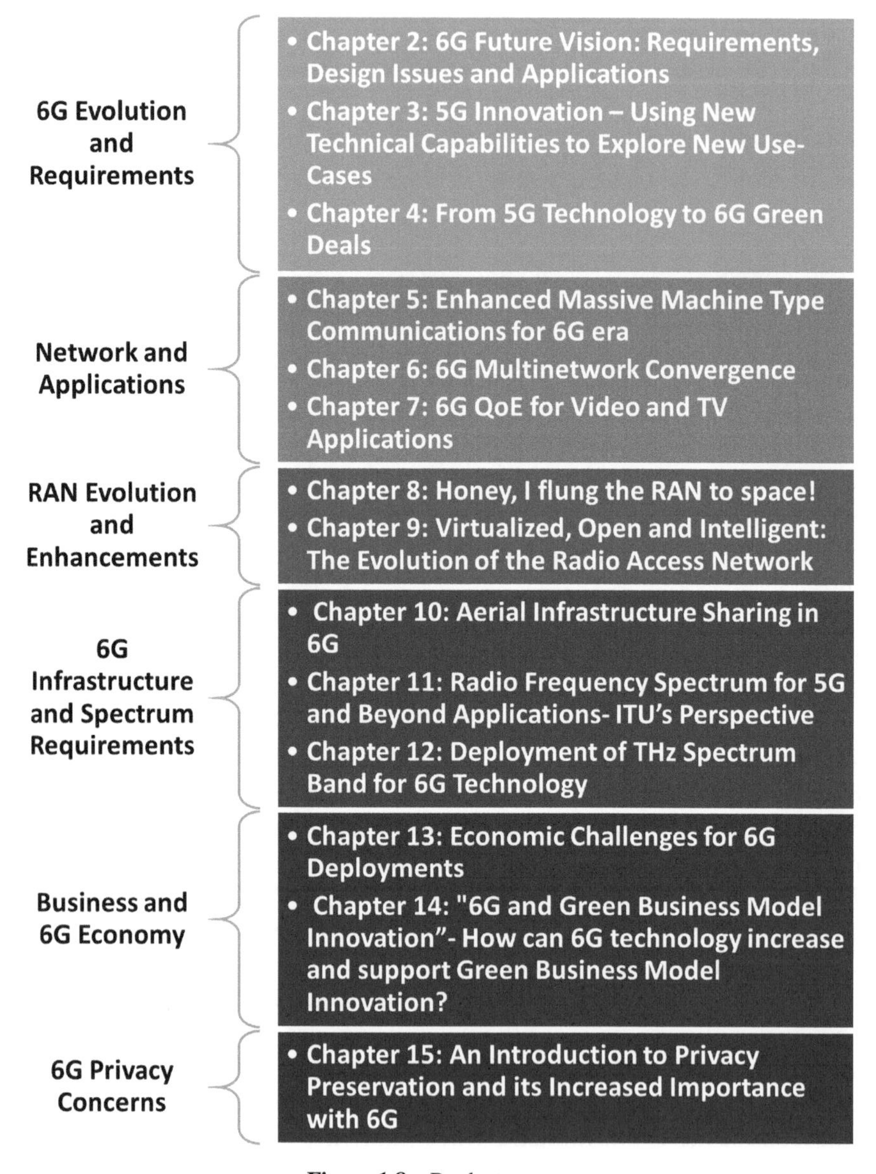

Figure 1.8 Book structure

- 6G Evolution and Requirements
- New Use-Cases
- Network and Applications

- RAN Evolution and Enhancements
- 6G Infrastructure and Spectrum Requirements
- Business and 6G Economy
- 6G Privacy Concerns

1.8 Conclusions

This book presented the international research effort towards 6G communication systems. In particular, the book focused on the various technology enablers of 6G.

The capabilities and design dimensions of future wireless networks (FWN) based on 6G are being driven by disruptive technologies such as AI, ML, deep analytics, software and, advanced computer technologies; advanced sensing, 3D imaging, AR/VR. This, in turn, drives the emergence of new applications that are manifested by stringent performance requirements.

The book comprises valuable discussions highlighting the need to efficiently and flexibly provide diversified services such as enhanced Mobile Broadband (eMBB) access, ultra-reliable low-latency communications (URLLC), and massive Machine-Type Communications (mMTC). The representative scenarios and services evolving from 5G (described in Chapters 2–4) are Further-enhanced Mobile Broadband (FeMBB), ultra-massive Machine-Type Communications (umMTC), extremely Ultra-Reliable and Low-Latency Communications (eURLLC), and Extremely Low Power Communications (ELPC). These scenarios demand peak data rates >1 Tbps, the latency of 10-100μs, and user experienced data rate >1 Gbps; such performance requirements are pretty beyond current 5G capabilities.

However, satisfying these requirements is essential for delivering the envisioned FWN applications that are emerging due to advances in technologies, such as sensing, imaging, displaying, and AI, with application scenarios that more and more will be based on AR/VR content. Related issues were discussed in Chapters 10–12. Enabling the use cases for FWN is an essential driver for the user demand and the value proposition, the basis of business model innovation, which is a topic discussed in Chapters 13, 14. Recently, many standardisation bodies have started establishing a 6G initiative focusing on 6G use cases and requirements.

With advances in AI, machines can transform data into reasoning and decisions that have the potential to enable the envisioned applications over 6G networks, as described in Chapters 5–7. At the same time, such advances introduce machines as a new type of user, in addition to humans, that make the demands on the performance requirements even more stringent. As today's domestic and industrial machines transform into swarms of multi-purpose robots and drones, new approaches based on human-machine haptic and thought interfaces to control them from anywhere should become an integral part of the future wireless networks. Network operators face considerable challenges in extending the network coverage and meeting the increased capacity demands while using a limited pool of capital and resources. Manual configuration of networks for management and orchestration (MANO) will make things even more complicated, time-consuming, susceptible to error, and expensive. AI can provide unprecedented opportunities to the MANO framework, unleash vast performance potentials of wireless networks, identify correlations and anomalies that cannot be observed by inspection, suggest novel ways to optimise network deployment and operation and facilitate making decisions with minimal human intervention. AI and data analytics (DA) will help solve traffic management issues and influence new antenna design, dynamic sharing of spectrum, and self-organised network architectures to improve the design and deployment of 6G. The deployment of AI and data analytics has already been proposed in various standards, including the 3rd generation partnership project (3GPP) network data analytics (NWDA) and the european telecommunications standards institute (ETSI) experimental network intelligence (ENI) NWDA utilises the approach of slice and traffic steering and splitting. At the same time, ENI uses a cognitive network management architecture and context aware-based approach.

Chapters 8, 9 presented research towards the opening of the radio access network (RAN) to enable interoperability of vendor solutions; for the efficient implementation of intelligence in the network and novel access solutions. The Open-RAN concept allows for the separation of the digital and radio components of the RAN infrastructure and, thus, enriches the resource management, control, and other functions within the RAN with AI.

The design of 6G networks needs to consider the QoE for the user into strong consideration. The general principle of QoE-awareness in network design should be implicitly supported in all layers of the functional architecture and take into account the subjective nature of the perceived experience. Therefore, it is crucial for QoE management to completely integrate precise QoE monitoring mechanisms focused on transparent real-time data collection and storage, effective big data and ML algorithms for data evaluation and functions that will ensure the adjustment and reshaping of collected data across the network components (Chapters 6, 7).

In conclusion, the 6G vision involves a human world of our senses, bodies, intelligence, and values; a digital world of information, communication and computing; and a physical world of objects and organisms. In addition, regulatory and technological trends are critical for the design and deployment of future networks. Therefore, one of the key concerns of the society coming with the development of 6G enabling technologies is that the related services must be accessible for all and everywhere they are needed. This digital inclusion requirement does not only mean "global coverage" (at an affordable cost in terms of deployment and for the customer as well) but also "easy-to-use technology".

References

[1] Jiang, W., Han, B., Habibi, M. A., & Schotten, H. D. (2021). The road towards 6G: A comprehensive survey. IEEE Open Journal of the Communications Society, 2, 334-366.

[2] Giordani, M., Polese, M., Mezzavilla, M., Rangan, S., & Zorzi, M. (2020). Toward 6G networks: Use cases and technologies. *IEEE Communications Magazine*, *58*(3), 55-61.

[3] Letaief, K. B., Chen, W., Shi, Y., Zhang, J., & Zhang, Y. J. A. (2019). The roadmap to 6G: AI empowered wireless networks. *IEEE communications magazine*, *57*(8), 84-90.

[4] J. A. del Peral-Rosado, R. Raulefs, J. A. López-Salcedo and G. Seco-Granados, "Survey of Cellular Mobile Radio Localization Methods: From 1G to 5G", in IEEE Communications Surveys & Tutorials, vol. 20, no. 2, pp. 1124-1148, Secondquarter 2018, doi: 10.1109/COMST.2017.2785181.

[5] T. Farley, "Mobile telephone history", *Telektronikk*, vol. 101, no. 3, pp. 22-34, 2005.

[6] F. Hillebrand, "The creation of standards for global mobile communication: GSM and UMTS standardization from 1982 to 2000", *IEEE Wireless Commun.*, vol. 20, no. 5, pp. 24-33, Oct. 2013.

[7] GSM Memorandum of Understanding, Copenhagen, Denmark, Sep. 1987.

[8] EIA/TIA, IS-95, "Cellular System Recommended Minimum Performance Standards for Full-Rate Speech Codes", May 1992.

[9] IMT-2000 radio interface specifications approved in ITU meeting in Helsinki, Geneva, Switzerland, Nov. 1999.

[10] "Requirements related to technical performance for IMT-Advanced radio interface(s)", 2008.

[11] Nidhi, A. Mihovska and R. Prasad, “Overview of 5G New Radio and Carrier Aggregation: 5G and Beyond Networks”, *2020 23rd International Symposium on Wireless Personal Multimedia Communications (WPMC)*, 2020, pp. 1-6, doi: 10.1109/WPMC50192.2020.9309496.

[12] Nidhi and A. Mihovska, “Small Cell Deployment Challenges in Ultradense Networks: Architecture and Resource Management”, *2020 12th International Symposium on Communication Systems, Networks and Digital Signal Processing (CSNDSP)*, 2020, pp. 1-6, doi: 10.1109/CSNDSP49049.2020.9249560.

[13] A. Ghosh, R. Ratasuk, B. Mondal, N. Mangalvedhe and T. Thomas, “LTE-advanced: Next-generation wireless broadband technology [invited paper]”, *IEEE Wireless Commun.*, vol. 17, no. 3, pp. 10-22, Jun. 2010.

[14] Network, E. U. T. R. A. (2011). S1 Application Protocol (S1AP)(Release 10). *Technical Specification, 36.*

[15] “LMU performance specification; network based positioning systems in E-UTRAN release 11 V11.4.0”, Oct. 2014.

[16] Johansson, T. (2013). 3GPP LTE Release 9 and 10 requirement analysis to physical layer UE testing.

[17] “New SI proposal: Positioning enhancements for E-UTRA”, Jun. 2013.

[18] “Requirements for support of radio resource management release 9 V9.22.0”, Dec. 2014.

[19] C.-X. Wang et al., “Cellular architecture and key technologies for 5G wireless communication networks”, *IEEE Commun. Mag.*, vol. 52, no. 2, pp. 122-130, Feb. 2014.

[20] Henrique, P. S. R., & Prasad, R. (2021). *6G The Road to the Future Wireless Technologies 2030* (pp. i-xxvi). River Publishers.

[21] B. Khan, Nidhi, A. Mihovska, R. Prasad and F. J. Velez, “Overview of Network Slicing: Business and Standards Perspective for Beyond 5G Networks”, *2021 IEEE Conference on Standards for Communications and Networking (CSCN)*, 2021, pp. 142-147, doi: 10.1109/CSCN53733.2021.9686125.

[22] K. Flynn, “A global partnership,” Mar 2020. [Online]. Available: https: //www.3gpp.org/release-16

[23] 5GPPP, “B5G/6G Research with StandarizationPotential Roadmap,” 5GPPP: 5G Infrastructure Public Private Partnership, November 2020.

[24] S. Elmeadawy and R. Shubair, "Enabling Technologies for 6G future wireless communications: Opportunities and Challenges," arXiv, 2020.

[25] Lv, Z., Lou, R., Li, J., Singh, A. K., & Song, H. (2021). Big data analytics for 6G-enabled massive internet of things. *IEEE Internet of Things Journal*, *8*(7), 5350-5359.

[26] Liu, B., Luo, J., & Su, X. (2021). The Framework of 6G Self-Evolving Networks and the Decision-Making Scheme for Massive IoT. *Applied Sciences*, *11*(19), 9353.

[27] Stoica, R. A., & de Abreu, G. T. F. (2019). 6G: the wireless communications network for collaborative and AI applications. *arXiv preprint arXiv:1904.03413*.

[28] Lovén, L., Leppänen, T., Peltonen, E., Partala, J., Harjula, E., Porambage, P., ... & Riekki, J. (2019). EdgeAI: A vision for distributed, edge-native artificial intelligence in future 6G networks. *The 1st 6G wireless summit*, 1-2.

[29] N. Alliance, *5G White Paper (Final Deliverable)*, 2015.

[30] H. Zhang, N. Liu, X. Chu, K. Long, A.-H. Aghvami and V. C. Leung, "Network slicing based 5G and future mobile networks: mobility resource management and challenges", *IEEE communications magazine*, vol. 55, no. 8, pp. 138-145, 2017.

[31] *Network slicing explained*, Nov 2020, [online] Available: https://www.nokia.com/about-us/newsroom/articles/network-slicing-explained/.

[32] T. 5. I. Association, "5GPPP: European Vision for the 6g Network Ecosystem," 7 June 2021. [Online]. Available: 5g-ppp.eu/european-vision-for-the-6g-network-ecosystem/. DOI: 10.5281/zenodo.5007671. [Accessed 29 July 2021].

[33] E. R. S. P. Group, "RSPG Additional spectrum needs and guidance on the fast rollout of future wireless broadband networks," March 2021. [Online]. Available: https://rspg-spectrum.eu/wp-content/uploads/2021/02/RSPG21-008final_Draft_RSPG_Opin. [Accessed July 2021].

[34] O. University, "6G Flagship," 12 July 2021. [Online]. Available: www.oulu.fi/6gflagship/. [Accessed 29 July 2021].

[35] M. Lennighan, "Vodafone Germany Launches a Network Offensive," June 2021. [Online]. Available: https://Telecoms.com/, 26 May 2021, telecoms.com/509917/vodafone-germany-launches-a-network-offensive/.

[36] T. Spanjaard, "Who Will Standardize 6G? Smartinsights," Smartinsights www.smartinsights.net/single-post/who-will-standardize-6g., 11 December 2020.

[37] A. Hernandez, "Techaeris: "South Korea Eyes 2028 to LAUNCH 6G, Samsung Leads the Charge."," 23 June 2021. [Online]. Available: techaeris.com/2021/06/23/south-korea-eyes-2028-to-launch-6g-samsung-leads-the-charge/. [Accessed 31 July 2021].

[38] Telecompaper, """Russian R&D Institute NIIR to Test 5g with Ericsson."," 2021. [Online]. Available: www.telecompaper.com/news/russian-randd-institute-niir-to-test-5g-with-ericsson--1353665. [Accessed 2021].

[39] Nidhi, Khan, B., Mihovska, A., Prasad, R., & Velez, F. J. (2021, December 20). Journal of ICT Standardization. Retrieved March 9, 2022, from https://doi.org/10.13052/jicts2245-800X.932

Biographies

Nidhi, PhD Student-CGC Research Lab, Department of Business Development and Technology, Aarhus University, Denmark. nidhi@btech.au.dk

Nidhi is an Early-Stage Researcher in the project TeamUp5G, a European Training Network in the frame of (MSCA ITN) of the European Commission's Horizon 2020 framework. Currently enrolled as a PhD student at Aarhus University in the Department of Business Development and Technology. She received her Bachelor's degree in Electronics and Telecommunication and master's degree in Electronics and Communication (Wireless) from India. Her research interests are small cells, spectrum management, carrier aggregation, etc.

Dr Albena Mihovska, Associate Professor CGC Research Lab, Department of Business Development and Technology, Aarhus University, Denmark. amihovska@btech.au.dk

Dr Albena Mihovska is a Senior Academic and Research Professional, currently an Associate Professor at the Dept of Business Development and Technology (BTECH) at Aarhus University, Herning, Denmark. She leads the 6G Knowledge Research Lab at the CTIF Global Capsule (CGC) research group at BTECH and is the Technical Manager of several EU-funded projects in Beyond 5G networks coordinated by Aarhus University. She is a member of IEEE INFORMS and has more than 150 publications.

Dr Anand R. Prasad, Partner at Deloitte Tohmatsu Cyber (DTCY), Tokyo, Japan & Board of Director at Digital Nasional Berhad, Malaysia.
anandraghawa.prasad@tohmatsu.co.jp

Dr Anand R. Prasad is a Deloitte Tohmatsu Cyber (DTCY) partner, where he leads connectivity security. He is also a board of directors at Digital Nasional Berhad, Malaysia. Before DTCY, Anand was Founder & CEO, wenovator LLC, which now forms part of Deloitte and Senior Security Advisor, NTT DOCOMO. He was CISO, Board Member of Rakuten Mobile, where he led all aspects of enterprise and mobile network security (4G, 5G, IoT, Cloud, device, IT, SOC, GRC, assurance etc.) from design, deployment to operations. Anand was Chairman of 3GPP SA3, where, among others, he led the standardisation of 5G security.

He is also an advisor to several organizations such as CTIF Global Capsule, GuardRails and German Entrepreneurship Asia. Anand is an innovator with 50+ patents, a recognized keynote speaker (RSA, MWC etc.) and a prolific writer with six books and 50+ publications. He is a Fellow of IET and IETE.

Dr Ramjee Prasad, Professor-CGC Research Lab, Department of Business Development and Technology, Aarhus University, Denmark. ramjee@btech.au.dk

He is the Founder and President of CTIF global capsule (CGC) and the Founding Chairman of the Global ICT Standardization Forum for India. Dr Prasad is also a Fellow of IEEE, USA; IETE India; IET, UK; a member of the Netherlands Electronics and Radio Society (NERG); and the Danish Engineering Society (IDA). He is a Professor of Future Technologies for Business Ecosystem Innovation (FT4BI) in the Department of Business

Development and Technology, Aarhus University, Herning, Denmark. He was honoured by the University of Rome "Tor Vergata", Italy as a Distinguished Professor of the Department of Clinical Sciences and Translational Medicine on March 15, 2016. He is an Honorary Professor at the University of Cape Town, South Africa, and KwaZulu-Natal, South Africa. He received the Ridderkorset of Dannebrogordenen (Knight of the Dannebrog) in 2010 from the Danish Queen for the internationalization of top-class telecommunication research and education. He has received several international awards, such as the IEEE Communications Society Wireless Communications Technical Committee Recognition Award in 2003 for contributing to the field of "Personal, Wireless and Mobile Systems and Networks", Telenor's Research Award in 2005 for outstanding merits, both academic and organizational within the field of wireless and personal communication, 2014 IEEE AESS Outstanding Organizational Leadership Award for: "Organizational Leadership in developing and globalizing the CTIF (Center for TeleInFrastruktur) Research Network", and so on. He has been the Project Coordinator of several EC projects, namely, MAGNET, MAGNET Beyond, and eWALL. He has published more than 50 books, 1000 plus journal and conference publications, more than 15 patents, and over 150 PhD. Graduates and a more significant number of Masters (over 250). Several of his students are today worldwide telecommunication leaders themselves.

Ramjee Prasad is a member of the Steering, Advisory, and Technical Program committees of many renowned annual international conferences, e.g., wireless personal multimedia communications symposium (WPMC); Wireless VITAE, etc.

2

6G Future Vision: Requirements, Design Issues and Applications

Nandkumar P. Kulkani[1], Dnyaneshwar S. Mantri[2], Neeli R. Prasad[3], Pranav M. Pawar[4], and Ramjee Prasad[5]

[1]Smt Kashibai Navale College of Engineering, Pune, India
[2]Sinhgad Institute of Technology, Lonavala, Pune, India
[3]TrustedMobi "VehicleAvatar Inc.", USA
[4]BITS Pillani Dubai
[5]Department of Business Development Technology, Aarhus University, Herning, Denmark
E-mail: npkulkarni.pune@gmail.com; dsmantri@gmail.com; neeli.prasad@smartavatar.nl; pranav@dubai.bits-pilani.ac.in; ramjee@btech.au.dk

Technological improvements and evolutions are required beyond fifth-generation (5G) networks for wireless communications, driven by increasing use cases in enormous future-radio-access-networks (FRAN). The future sixth-generation (6G) networks with trillions of connected devices would demand exceptionally high-performance interconnectivity, especially in dynamic circumstances such as varied mobility, extremely high density, and vibrant surroundings. Researchers must look at scalable, adaptable, and long-lasting wireless access mechanisms to handle a wide range of requirements and massive wireless connections to meet this need. Artificial intelligence (AI) techniques (e.g., machine learning (ML), deep neural networks (DNN), etc.), Big data, and Cloud Computing will most likely be a critical enabling factor for future 6G networks in terms of supporting performance, new functions, as well as new services and, will all play critical roles in attaining the highest efficiency and maximum advantages in the future 6G needs and features. They are already thought to be one of the essential missing elements in 5G networks. A rising market for these technologies and their use in various ICT applications can be observed in recent years. This chapter examined the essential requirements and enabling technologies of the 6G network and highlighted the critical design issues driving the 6G network applications. Finally, we proposed to enhance 6G network banking on the vision of AI, IoT forwards, and Cloud Computing to achieve the goal.

2.1 Introduction

Wireless networks have progressed from the first generation (1G) to the fifth generation (5G) networks, with different considerations such as data throughput, end-to-end latency, resilience, energy efficiency, coverage, and spectrum usage. The objective of these systems is to create a connected world in which everyone and everything is linked. Adhering to intelligent communication between things, People-to-People (P2P), Machine-to-Machine (M2M), Device-to-Device (D2D), Device-to-Machine (D2M), and People-to-Machine (P2M) over the internet, 5G networks can offer Mobile-Broadband-Connectivity, high reliability, and massive MIMO-type communication with low latency to improve Quality of Service (QoS) and network performance. Beyond 5G networks, industry and academia need to work hand in hand for various use cases suitable for 6G deployment to achieve data rate in Tbps, ultra-low-latency in micro-seconds, high connectivity even in case of high mobility may be in few thousand km per hour, and high bandwidth in THz [1–3]. It will be too complex and chaotic if the 6G networks are overlaid on current 4G/5G networks. Therefore, 6G mobile networks must be deployed in various markets to provide customers with various new services and vertical applications.

For that, AI, Big data, and Cloud will become obligatory and indispensable elements of the 6G network. In today's world, virtually every industry recognizes the importance of Big Data which is being used to boost many industries and technological advancements. Wireless big data is on the rise since there is a lot of data heterogeneity, communication, complexity, and privacy issues. Innovative outcomes of the primary research of Big Data, given the challenges regarding how to collect and analyze data and what data to store and delete, will be the bread and butter for many researchers and industries. Recently, AI's re-emergence, founded on big data, has generated new research possibilities in Wi-Fi and other wireless networks, particularly in the machine-, deep-, reinforcement- learning. One of the significant aspects that have led to the significant change in computing paradigms is providing QoS. In 6G, the focus has been shifted to decentralized paradigms like edge computing and fog computing. To overcome the constraints of cloud computing, the latter focuses on delivering various services at the network edge. This perspective is more than just a cloud extension; it acts as a blending platform for cloud and IoT, facilitating and ensuring successful connection.

This chapter aims to raise awareness of the most recent technological advancements and obstacles in enabling 6G networks and discusses the 6G technological wave, which includes the integration and convergence of new technologies, allowing for the provision of 6G services anytime, anyplace, and for anything. Finally, the chapter offers a 6G network enhancement plan to keep pace with the wireless domain growth. We provided a detailed overview of 6G technology and discussed several unresolved concerns related to the transition to 6G from IoT, 5G, and cloud. The chapter's main intention is to outline 6G enabling technologies and use cases to show the necessity for integrating different primary approaches such as IoT+5G+Cloud+AI/ML. The suggested design and technology road map demonstrate the fundamentals of a 6G network. The article examines possible situations based on the most common use cases [4, 5]. According to the authors' research, fusion, convergence, and integration of communication technologies, network designs, and deployment patterns are required to satisfy the rising needs of users and 6G networks.

The chapter is divided into several sections. The second section gives an overview of current work in 6G networks. The third section contains information on the 6G network's fundamental requirements. The fourth section focuses on 6G network design issues and applications and enabling technologies. The fifth section compares and contrasts first-generation to sixth-generation networks. The sixth section presents the 6G network improvement idea and underlines the article's contribution. Section seven ends with concluding remarks.

2.2 Related Work

The section gives different researchers' findings and contributions to developing 6G and future network requirements for 6G. It also offers ideas for integrating various technologies to meet the objectives of 6G and the future direction of the new proposal.

The centralized service has many problems and is unsuitable for high-speed and intelligent networks like 6G. The work discussed in [6] gives an overview of the requirements of decentralized architectures for the 6G network. It also provides the design of such architectures for the seamless working of 6G as an intelligent network.

The 6G network has capabilities to provide a multi-terabyte per second. However, to achieve it requires an intelligent and autonomous network. The research considered in [7] proposes the multi-dimensional autonomous network to provide efficient communication across air, underwater, space, and ground. The paper also provides a brief discussion about different technologies like 6G as THz communications, Supermassive Multiple –input and Multiple-output system (SM-MIMO), holographic beam forming (HBF), and Large Intelligent Surfaces (LISs) to form a robust 6G ecosystem.

To increase the horizon of the network, it is necessary to provide good network connectivity in rural and underdeveloped areas. The research in [8] talks about the 6G network issues for connectivity in rural and underdeveloped regions. The paper provides the possibility of applying different fronthaul and backhaul networks for 6G rural connectivity. Furthermore, the research gives a brief overview of the requirements to maintain energy and a cost-efficient 6G network. The vital initiative, key challenges, and significant case studies for 6G network applications are also highlighted in the paper.

[9] proposes an IoT-based ubiquitous learning system, which needs a 6 G-based autonomous learning system. The author implemented four different versions of the proposed architecture using a standard IoT edge unit. Each of the versions uses an advanced message queuing protocol (AMQP), the constrained application protocol (CoAP), message queuing telemetry transport (MQTT), and an extensible messaging and presence protocol (XMPP), respectively.

Software-defined networks (SDN) and virtualization are considered promising technologies in 6G. The research in [10] proposes an innovative, collaborative tracking system by taking motivation from SDN and function virtualization capabilities of edge-cloud interplays. The proposed work shows stability and better tracking accuracy than the existing candidate stated in the paper.

[11] discusses 5G New Radio (5G NR), which acts as the migrating architecture from 5G to 6G. The paper also discusses the AI-based 6G network, combining other technologies such as cloud computing, quantum computing, and tactile services. Finally, the work provided the proposed virtualized slicing architecture for future 6G networks.

The 6G network is mainly considered the convergence of many emerging technologies. The authors in [12] discuss the applications case study and business model by considering the convergence of VANET and the cloud. The work also provides the communication and storage algorithm for such promising converged networks.

Smart healthcare is one of the promising applications of the 6G network. The research in [13] gives a brief overview of different case studies and related work in smart healthcare. The work also proposes a robust health care monitoring system using convergence IoT and cloud.

2.3 Essential Requirements of 6G Networks

6G will be the dominant technology in the future, not only in economic revolutions but also in social means, as human intelligence (HI) enables personalization. A wide range of smart applications for environmental, personal, and societal survival are included. It's challenging to organize, analyze, store, and use data for future reference when billions of heterogeneous devices are connected in a single network and communicating with one another. For the policies to be implemented, the entire system requires organizational support. Figure 2.1 depicts the necessary needs as well as management policies. The 6G requirement framework is divided into four major components: technology, Human, energy, and Data.

Technology Framework: The development of a 6G network depends on a technological infrastructure that allows humans to engage with it and support new ideas to be implemented. Digital, Analogue, Intelligent, Ubiquitous, Wired, Wireless, and Hybrid technologies are all included. These

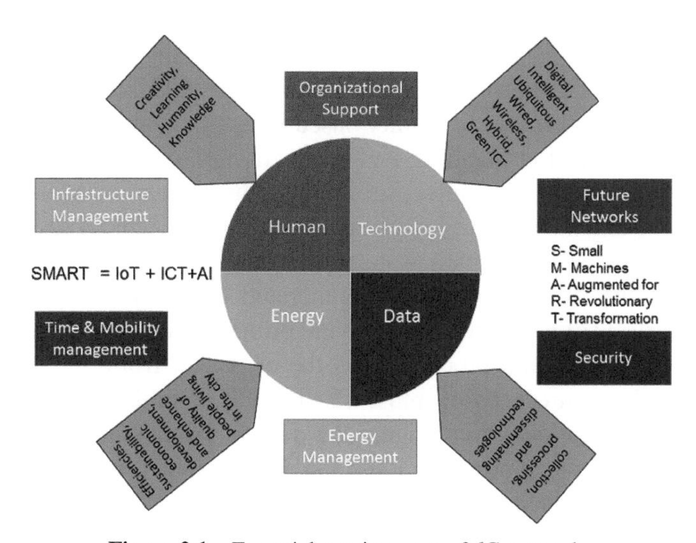

Figure 2.1 Essential requirements of 6G network

technologies enable public services to be accessed via any connected device in the IoT network for monitoring and control. In addition, artificial intelligence and machine learning are intelligent technologies that may be used to analyze data and connect people and devices in a network.

Human Framework: In terms of economy, knowledge, learning, and creativity, technological advancements have a positive impact on citizens' lives. To create long-term learning and creativity networks, it is necessary to have an education system with adequate resources, a skilled workforce, and skilled management.

Energy Framework: A single network of interconnected devices may handle a large number of heterogeneous, diversified devices, such as home appliances, security cameras, monitoring sensors, actuators, displays, automobiles, D2D, M2M, P2M, and M2P communications, and so on, in 6G networks. Each consumes energy, and extra attention must be paid to maintain the balance so that the network can operate at any moment.

Data Framework: The data generated by a wide range of applications using connected sensors must be redundancy-free. The 6G network combines interoperated networks with data collecting, processing, and dissemination technologies. The data collecting and computation processes aid in the decision-making process.

The requirement frame is supported by 1. Organizational Support, 2. Future Network evaluations 3. Security Aspects, 4. Energy Management, 5. Time and Mobility management, and 6. Infrastructure management. To make the network smart, both must support each other. The organization's support requires network architecture flexibility based on applications. Energy management encompasses the planning and production of required units while preserving resources and reducing costs, among other things. GSM networks are used by time and mobility management to track the devices. It also makes use of an intelligent transportation system (ITS).

The safety and privacy of the data generated is a major priority. The security and privacy of generated data are vital priorities, and network designs follow these principles and processes. As the evolution from 4G to 5G to 6G continues, the future network must transition from traditional networks to IP-based and sensor networks. Figure 2.2 depicts the essential factors the network must address to combat this issue and develop the 6G network. These parameters are vital when designing and analyzing data: minimal latency, quick access to all services, good service providers, and applications that use less energy.

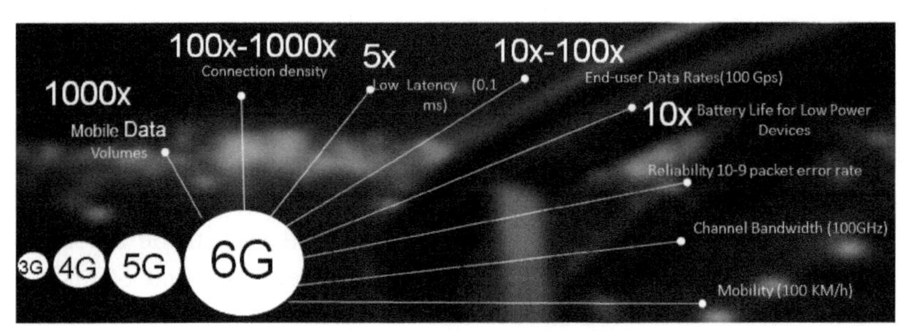

Figure 2.2 Parameters of 6G network

2.4 Applications, Enabling Technologies, and 6G Network Design Issues

The expected 6G network has an infinite number of applications in each field. The details of applications and enabling technologies are given in Figure 2.3. The applications are related to the IoT and AI. Internet of Things and ICT support for the theme of 5G and with added computational intelligence using AI/ ML algorithms meets the requirement of 6G. The 6G applications are unlimited and have a broad scope in several fields; including

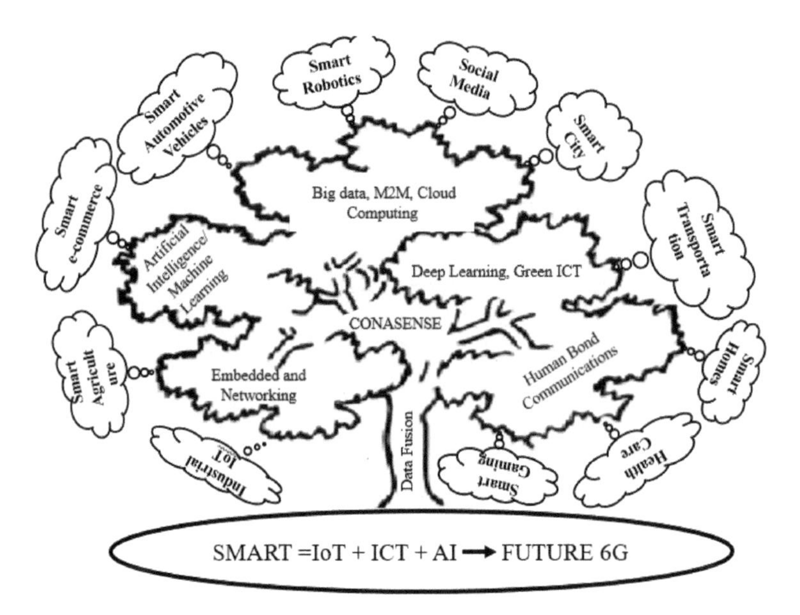

Figure 2.3 Applications and enabling technologies of 6G

Telemedicine and health care, Smart city, Smart Homes, Smart Transportation, Industrial IoT changed to PIoT (Industry 5.0), Smart Agriculture, Smart e-commerce, Finance, Banking, Smart Automotive (V2X communications), Smart Robotics and automation, Social Media, Data Security, Entertainment, Gaming, Smart education, and Astronomy, etc. In the current pandemic of COVID 19, AI and ML algorithms are playing a vital role in the detection of diseases and analysis of data with security. As we know that 6G is the integration of various technologies such as

Embedded and networked: used for V2X communications, sensor networks for collections of data from on-field,

AI/ML: works at the top of IoT for analysis and precise computation of data used in further references.

Green ICT: used for reduction of carbon footprints generated due to e-garbage.

M2M, D2D Communication: in the industrial revolution, 5.0 role of M2M and D2D communication is essential. The advanced features of IoT and AI are embedded in all kinds of machines and devices used for network formation and data communication.

Cloud Computing: It supports software and hardware platforms and provides virtual data storage services.

Human Bond Communication (HBC): upcoming technology with five human senses as Test, Smell, Touch, Sound, and Sight is used for the detection and communication of information. It is primarily helpful in health care applications.

CONASENSE: Uses the Sensor Network for all types of communication, location finding, Navigation, and Sensing Services used by the network and could be the main base for 6G.

2.5 Design Issues in 6G

The difficulties are considerably greater than 6G services delivery. It is predicted that most of the challenges with 5G would be overcome by 6G. However, the demands are increasing at an exponential rate. As a result, the problems of 6G are far greater than those of 5G. Figure 2.4 illustrates the 6G network's design issues.

Figure 2.4 Network design issue

2.5.1 Frequency Band

The Terahertz (THz) frequency from the ultra/super-high frequency range will be used for communication in 6G, where the wavelength is 300 μm. However, THz signal generation is more complex and expensive due to the more stringent standards. Moreover, for short-range transmission, attenuation is relatively high. Therefore, new communication options are also being investigated. VLC (Visible Light Communication) is a superior choice that employs low-cost light-emitting diodes (LED) to attain higher frequency bands than THz signals. However, it suffers from noise interference from other light sources.

2.5.2 Heavy Computation

6G wireless networks are massive, complicated, and multifaceted. Here, communication and computing will be combined. Many computations and communication are required for automatic network configuration, traffic management, mobility management in case of a node failure, or disaster management. New sophisticated technologies such as edge/ fog computing,

AI, and others will be used. The merger of these diverse mechanisms and massive computations and communication is required for interoperability.

2.5.3 Design of Transmitter and Antenna

The 6G standard necessitates the use of highly efficient transmitter and receiver antennas. The data rate is also influenced by the materials used to create the antennas. Nanomaterials, bio-based materials, foams, room-temperature manufactured materials, and ultra-low permittivity are among the technologies being developed.

2.5.4 Delay and Reliability

The 6G data needs are 1 Tbps, and it will allow a variety of applications, including virtual/augmented reality, smart healthcare, unmanned aerial vehicles, smart electric vehicles, and smart cities, to mention a few. These applications need a high data throughput with negligible latency. Therefore, smaller data packets, flat network architecture, and forward error correction (FEC) mechanisms must be utilized to obtain a shorter end-to-end latency. This issue must be addressed at the physical, Datalink and network layers.

2.5.5 Underwater Communication

The goal of 6G is to offer communication beneath the water. The underwater world is complicated and unexpected, with significant signal attenuation and equipment damage. In salty water, radio transmissions are significantly weakened. As Underwater-sensor-node (USN) deployment is costly and complex, node density must be modest. The flow and density variations in the water will substantially affect the node mobility. The USNs require complicated transceivers, more memory, and a vast power source (as solar power cannot be used). They are costly. Optical fibre is the most excellent option, but it is also the most expensive. Quantum communication is another alternative, although it is still in its early stages of development.

2.5.6 Capacity

6G will link billions of smart gadgets and smart wearables as a part of IoT. These IoT devices will generate enormous traffic, and maintaining the latency of < 1 ms is a big challenge in 6G. As a result, the 6G network's traffic carrying capacity should be very high. The capacity of a 6G network may be improved by increasing spectral bandwidth and its efficiency. Also, by utilizing novel channel coding and modulation mechanism and by reuse of spectrum, the traffic carrying capacity of 6G can be enhanced. Adding mobility to the nodes in the hotspot region will reduce the burden on the node. Nevertheless, AI can deal with the problems of repositioning the node and handing- and taking- over control.

2.5.7 Density and Cost

Density refers to the number of nodes placed per square kilometre. On the other hand, 6G will have a greater density for a smooth and high QoE. In congestion control, scheduling, synchronization, and failure detection, communication costs will rise. The objective of 6G is to make services more affordable. The economic cost of 6G will be substantial since it comprises both non-terrestrial and terrestrial nodes. Non-terrestrial nodes are incredibly costly as compared to terrestrial nodes. At the same time, Satellites, drones, and other mobile nodes are highly costly non-terrestrial nodes. Launching satellites into orbit is expensive, and their repair and maintenance are another concern. UWS needs costly infrastructure. Furthermore, costly high-quality equipment is required to maintain a high level of QoS/QoE.

2.5.8 Coverage

6G will rely on a low-earth orbit (LEO) satellite to reduce route loss and transmission delay worldwide. In LEO transmission, Doppler -fluctuation, -shift, transmission delays, and route loss issues must be considered while deploying 6G.

2.5.9 Energy

The individual energy requirements of 6G devices are pretty high due to the large amounts of data processing and implementation of evolving technologies like Edge and AI. Thus, the 6G-enabled devices should use cutting-edge signal- and data-processing schemes to reduce energy consumption. For example, low peak average power ratios (PAPR) are required to generate novel waveforms and modulation. In addition, 6G must handle issues such as energy harvesting, charging, and conservation.

2.5.10 Heterogeneity

The 6G network must link a wide range of smart devices, including smartphones, smart TVs, smart watches, sensors, and self-driving cars.

Furthermore, the communication network will change depending on the coverage. It is difficult to link the entire planet with single worldwide coverage. Therefore, heterogeneous sub-networks will be utilized for communication. Furthermore, 6G will use diverse non-terrestrial and terrestrial communication concerning devices and communication mechanisms. Accordingly, 6G communication layered architecture needs diverse protocols to combine multiple heterogeneous components.

2.5.11 CIA TRIAD

The essential aspects of 6G are confidentiality, integrity, and availability (CIA). Because of THz communication, 6G is intended to provide the highest degree of security. In addition, quantum cryptography and artificial intelligence can be used to offer encryption. This might improve security in the healthcare and financial industries. Also, 6G will be eavesdropping and jamming proof.

2.6 Waves of Technological Developments

Wireless communication, like wired communication (optical fibre), aims to deliver high-quality, dependable communication, and each new generation of services implies a significant advance (or jump) in that direction. The evolution journey in technological developments added new features according to demands from end-user and network operators, such as mobile data volumes, number of connected devices, low latency, end-user data rates, battery life for low power devices, reliability, and channel bandwidth, and mobility of devices. All these parameters are essential to meet the requirement of 6G networks as a fully Digital and Connected World providing data from flexible service platforms. The wave of technological developments to meet the need for 6G technology is shown in Figure 2-5.

The distinctions between 2G, 3G, 4G, 5G, and 6G are readily visible in Table 2.1. Comparing 2G, 3G, 4G, 5G, and 6G also reveals that 6G will be one of the most ambitious technological advances.

2.7 Proposal for 6G Enhancements

The Internet of Things (IoT) is envisioned as a combination of numerous approaches exploited in complimentary advancements, with no single technology serving all purposes. It connects the worlds of the physical, virtual,

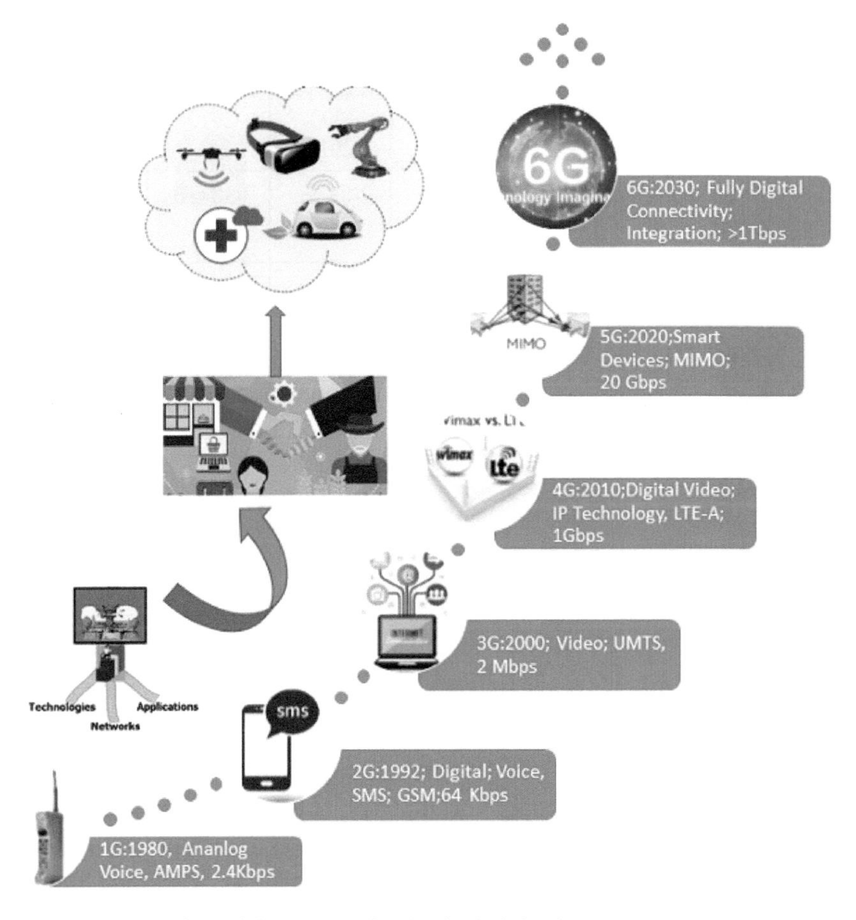

Figure 2.5 Waves of technological developments

and imaginative (HBC). The integration and convergence of numerous technologies in 6G technology lead to a completely digitized and linked world. The capabilities of the ubiquitous networks used in 6G include and are not only limited to,

- Cooperative and communication technology
- Network addressing and identification
- Sensing and actuation
- Processing of information at hardware and software level
- Network localization
- User interfaces and auto-updates

Table 2.1 Comparison of the evolution of technologies [1-4]

Parameter	1G	2G	3G	4G	5G	6G
Evolution	1970-1984	1980-1999	1990-2002	2000-2010	2010-2015	2020-2030
Frequency	30 KHz	1.8 GHz	2 GHz	2-8 GHz	3-30 GHz	1-3 THz
Bandwidth	2 kbps	64 kbps	2 Mbps	1 Gbps	10 Gbps	3 Tbps
Technology	AMPS	GSM	WCDMA	LTE, WiMAX	MIMO, mmWave	THz Comm.
Access System	FDMA	TDMA/ CDMA	CDMA	CDMA	OFDM	Convergen ce
Core Networks	PSTN	PSTN	Packet	Internet	Internet	IoT, AI/ML
Application	Voice	Voice Text	Voice Text Website	Voice Text Website Video	Smart devices	Digitally connected world

Figures 2-6 depict the road map for enhancing the user industry interface revolution, extending functionality to 6G technology. The end-user expects an entirely digitized and connected world of things in today's digital environment. Conversions and integrations of smart technologies such as the Internet of Things, Information Communication Technology, and Cloud constitute the foundation for 5G, or fifth generation and become SMART, expressed in Equation (1)

$$SMART \approx 5\,G \approx IoT + ICT + Cloud \tag{2.1}$$

5G can provide data at high rates (Gbps) and does not have communication bandwidth limitations; additionally, the infrastructure is exceptionally flexible, allowing for a wide range of applications such as e-Health, smart cities, and public safety in D2D and M2M interfaces with LTE-A. Furthermore, increased spectral efficiency and the capacity to store data virtually help lessen the end-to-end delay.

Data+ Technology+ Communication+ Intelligence + Storage

SMART ≈IoT + ICT + Cloud + AI ≈ future 6G

INTERNET OF THINGS

Figure 2.6 Enhancement of 6G

With the addition of data analysis in 5G and a more comprehensive range of applications, as shown in Figure 2-6, the 6G might be a mix of many technologies converged together and expressed as,

$$6G \approx Virtual + Imagination + Cognitive \approx IoT + 5G + Cloud + AI/ \quad (2.2)$$

Following the global deployment of 5G networks, the next-generation network, 6G, with extremely low latency and AI capabilities, will be recommended for future evolutions. The growth of intelligence communication will be influenced by 6G, which includes several connectivity elements such as intelligent, deep, holographic, and ubiquitous.

2.8 Conclusions

This article explores some of the significant technologies that will make future 6G more efficient. 6G networks seek to deliver a very high data rate and capacity and reduce latency regardless of the number of connected devices. There will be no decrease in QoS or QoE from the user's perspective. In connection to creating 6G standards, the article also links numerous technologies like AI, Cloud, Big Data, and IoT. The importance of the VHF/SHF (THz) frequency range in a 6G network is explored. Contrasting to previous generations and the 6th generation communication system, new use-cases and their requirements concerning upcoming technologies are significant contributions. The article also discusses some of the primary design challenges/requirements for a 6G network, such as computing, energy, storage, security, etc.

References

[1] Mohammed H. Alsharif, Anabi Hilary Kelechi, Mahmoud A. Albreem, "Sixth Generation (6G) Wireless Networks: Vision, Research Activities, Challenges and Potential Solutions", Symmetry 2020, 12, 676; doi:10.3390/sym12040676.

[2] Ian F. Akyildiz, (Fellow, IEEE), Ahan Kak, and Shuai Nie, "6G and Beyond: The Future of Wireless Communications Systems", IEEE Access, Vol 8, 2020. PP. 133995-134030.

[3] Yang Lua,, Xianrong Zheng, "6G: A survey on technologies, scenarios, challenges, and the related issues", Elsevier Journal of Industrial Information Integration 19 (2020) 100158.

[4] Fengxiao Tang, Yuichi Kawamoto, Nei Kato, and Jiajia Liu, "Future Intelligent and Secure Vehicular Network Toward 6G: Machine-Learning Approaches", Proceedings of the IEEE, Vol.108, No. 2, pp 292-307, February 2020 doi: 10.1109/JPROC.2019.2954595.

[5] Marco Giordani,, Michele Polese, Marco Mezzavilla, Sundeep Rangan, Michele Zorzi, "Towards 6G Networks: Use Cases and Technologies", arXiv:1903.12216v2 [cs.NI] 4 Feb 2020

[6] Xiuquan Qiao , Schahram Dustdar, Yakun Huang, Junliang Chen, "6G Vision: An AI-Driven Decentralized Network and Service Architecture", Department: Internet of Things, People, and Processes, IEEE Computer Society, pp 33-40, Sept 2020, Doi: 10.1109/MIC.20 20.2987738.

[7] Z. Zhang, Y.Xiao, Z.Ma, Ming Xiao, Z.Ding, X Lei, G.K. Karagiannidis and P Fan "6G WIRELESS NETWORKS, Vision, Requirements, Architecture, and Key Technologies", IEEE Vehicular, Technology Magazin, pp 28-41, SEPT 2019, doi: 10.1109/MVT.2019.2921208.

[8] Elias Yaacoub, and Mohamad Salim Alouini, "A Key 6G Challenge and Opportunity—Connecting the Base of the Pyramid: A Survey on Rural Connectivity", Proceedings of IEEE, Vol. 108, No. 4, April 2020 |pp. 533-582 doi: 10.1109/JPROC.2020.2976703.

[9] Salsabeel Y. Shapsough and Imran A. Zualkernan, Member, IEEE "A Generic IoT Architecture for Ubiquitous Context-Aware Learning", IEEE TRANS. On Learning Methodologies, Vol. 13, N0. 3, pp. 449-464, JULY-SEPT 2020. Doi; 10.1109/TLT.2020.3007708

[10] Fei Song , Mingqiang Zhu, Yutong Zhou, Ilsun You, and Hongke Zhang, "Smart Collaborative Tracking for Ubiquitous Power IoT in Edge-Cloud Interplay Domain", IEEE, Internet of Things Journal, Vol. 7, No. 7, pp 6044-6055, JULY 2020, doi: 10.1109/JIOT.2019.2958097.

[11] 17Anutusha Dogra, Rakesh Kumar Jha, and Shubha Jain, "A Survey on beyond 5G network with the advent of 6G: Architecture and Emerging Technologies", IEEE Access, pp. 1-37, doi: 10.1109/ACCESS.2020.30 31234.

[12] Nandkumar Kulkarni, Neeli Rashmi Prasad, Tao Lin, Mahbubul Alam, Ramjee Prasad, "Convergence of Secure Vehicular Ad-Hoc Network and Cloud in Internet of Things", River Publications, Role of ICT for Multi-Disciplinary Applications in 2030, Book Chapter, ISBN: 9788793379480, 2016.

[13] S. Mohapatra, S. Mohanty, and S. Mohanty, "Smart healthcare: An approach for ubiquitous healthcare management using IoT", in Big Data Analytics for Intelligent Healthcare Management. Amsterdam, The Netherlands: Elsevier, 2019, pp. 175-196.

Biographies

Dr N. P. Kulkarni received his PhD from Aarhus University, Denmark (2019), a Master's degree from COEP, Pune, India (2007), and a Bachelor's degree from Walchand College of Engineering, Maharashtra-India (1996). In addition, he received Diploma in Advanced Computing (C-DAC) from MET's IIT, Mumbai (2000), and received recognition as a Microsoft Certified Solution Developer (2002).

He has 23 years of experience both in industry and academia. He was associated with Electronica, Pune, from 1996-2000 (retrofits, CNC machines, and PLC programming). He also worked as a software developer (CITIL, Pune) and system analyst (INTREX, Mumbai). He worked as a faculty in Savitribai Phule Pune University, Pune (2002-2007). Since 2007, he has been working with SKNCOE, Pune, as a faculty in IT Department. His area of research is in WSN, VANET, and Cloud Computing. He has published papers in 18 International Journals, 15 International IEEE conferences, and 3 National Conferences.

Dr Dnyaneshwar S. Mantri graduated in Electronics Engineering from Walchand Institute of Technology, Solapur (MS) India, in 1992 and received a master's from Shivaji University in 2006. He has awarded PhD. in Wireless Communication at the Center for TeleIn-Frastruktur CTIF), Aalborg University, Denmark, in March 2017. He has teaching experience of 25years.

From 1993 to 2006, he worked as a lecturer in different institutes [MCE Nilanga, MGM Nanded, and STB College of Engg. Tuljapur (MS) India]. Since 2006 he has been associated with Sinhgad Institute of Technology, Lonavala, Pune, and presently working as a Professor in the Department of Electronics and Telecommunication Engineering. He is a member of IEEE, a Life Member of ISTE and IETE. He has written three books and published 15 Journal papers in indexed and reputed Journals (Springer, Elsevier, IEEE etc.) and 23 papers in IEEE conferences. He is a reviewer of international journals (Wireless Personal Communication, Springer, Elsevier, IEEE Transaction, Communication society, MDPI etc.) and conferences organized by IEEE. He worked as a TPC member for various

IEEE conferences and organised the IEEE conferences GCWCN2014 and GCWCN2018. He worked on various committees at University and College. His research interests are in Adhoc Networks, Wireless Sensor Networks, Wireless Communications, VANET, and Embedded Security, focusing on energy and bandwidth.

Dr Ir. Neeli R. Prasad, CTO of SmartAvatar B.V. Netherlands and VehicleAvatar Inc. USA, IEEE VTS Board of Governor Elected Member & VP Membership. She is also a full Professor at the Department of Business Development and Technology (BTech), Aarhus University. Neeli is a cybersecurity, networking and IoT strategist. She has, throughout her career, been driving business and technology innovation, from incubation to prototyping to validation and is currently an entrepreneur and consultant in Silicon Valley.

She has made her way up the "waves of secure communication technology by contributing to the most groundbreaking and commercial inventions. She has general management, leadership and technology skills, having worked for service providers and technology companies in various key leadership roles. She is the advisory board member for the European Commission H2020 projects. She is also a vice-chair and patronage chair of IEEE Communication Society Globecom/ICC Management & Strategy Committee (COMSOC GIMS) and Chair of the Marketing, Strategy and IEEE Staff Liaison Group.

Dr Neeli Prasad has led global teams of researchers across multiple technical areas and projects in Japan, India, and throughout Europe and USA. She has been involved in numerous research and development projects. She also led multiple EU projects, such as CRUISE, LIFE 2.0, ASPIRE, etc., as project coordinator and PI. She has played key roles from concept to implementation to standardization. Her strong commitment to operational excellence, innovative approach to business and technological problems and aptitude for partnering cross-functionally across the industry have reshaped and elevated her role as project coordinator, making her a preferred partner in multinational and European Commission project consortiums.

She has four books on IoT and Wi-Fi, many book chapters, peer-reviewed international journal papers and over 200 international conference papers. Dr Prasad received her Master's degree in electrical and electronics engineering from the Netherland's renowned Delft University of Technology, focusing on personal mobile and radar communications. She was awarded her PhD degree from Universita' di Roma "Tor Vergata", Italy, on Adaptive Security for Wireless Heterogeneous Networks.

Dr Pranav M. Pawar graduated in Computer Engineering from Dr Babasaheb Ambedkar Technological University, Maharashtra, India, in 2005, received a Master in Computer Engineering from Pune University in 2007 and received a PhD in Wireless Communication from Aalborg University, Denmark, in 2016. His PhD thesis received a nomination for the Best Thesis Award from Aalborg University, Denmark. Currently, he is working as an Assistant Professor in the Dept of Computer Science, Birla Institute of Technology and Science, Dubai; before to BITS, he was a postdoctoral fellow at Bar-Ilan University, Israel, from March 2019 to October 2020 in the area of Wireless Communication and Deep Leaning.

He is the recipient of an outstanding postdoctoral fellowship from the Israel Planning and Budgeting Committee. He worked as an Associate Professor at MIT ADT University, Pune, from 2018-2019 and as an Associate Professor in the Department of Information Technology, STES's Smt. Kashibai Navale College of Engineering, Pune from 2008-2018. From 2006 to 2007, was working as System Executive in POS-IPC, Pune, India. He received Recognition from Infosys Technologies Ltd. for his contribution to the Campus Connect Program and received different funding for research and attending conferences at the international level. He published more than 40 papers at the national and international levels. He is IBM DB2, and IBM RAD certified professional and completed NPTEL certification in different subjects. His research interests are energy efficient MAC for WSN, QoS in WSN, wireless security, green technology, computer architecture, database management system and bioinformatics.

Dr Ramjee Prasad is the Founder and President of CTIF Global Capsule (CGC) and the Founding Chairman of the Global ICT Standardization Forum for India. Dr Prasad is also a Fellow of IEEE, USA; IETE India; IET, UK; a member of the Netherlands Electronics and Radio Society (NERG); and the Danish Engineering Society (IDA). He is a Professor of Future Technologies for Business Ecosystem Innovation (FT4BI) in the Department of Business

Development and Technology, Aarhus University, Herning, Denmark. He was honoured by the University of Rome "Tor Vergata", Italy as a Distinguished Professor of the Department of Clinical Sciences and Translational Medicine on March 15, 2016. He is an Honorary Professor at the University of Cape Town, South Africa, and KwaZulu-Natal, South Africa. He received the Ridderkorset of Dannebrogordenen (Knight of the Dannebrog) in 2010 from the Danish Queen for the internationalization of top-class telecommunication research and education. He has received several international awards, such as the IEEE Communications Society Wireless Communications Technical Committee Recognition Award in 2003 for contributing to the field of "Personal, Wireless and Mobile Systems and Networks", Telenor's Research Award in 2005 for outstanding merits, both academic and organizational within the field of wireless and personal communication, 2014 IEEE AESS Outstanding Organizational Leadership Award for: "Organizational Leadership in developing and globalizing the CTIF (Center for TeleInFrastruktur) Research Network", and so on. He has been the Project Coordinator of several EC projects, namely, MAGNET, MAGNET Beyond, and eWALL. He has published more than 50 books, 1000 plus journal and conference publications, more than 15 patents, and over 150 PhD. Graduates and a more significant number of Masters (over 250). Several of his students are today worldwide telecommunication leaders themselves.

Ramjee Prasad is a member of the Steering, Advisory, and Technical Program committees of many renowned annual international conferences, e.g., Wireless personal multimedia communications symposium (WPMC); Wireless VITAE, etc.

3

5G Innovation – Using New Technical Capabilities to Explore New Use-Cases

Christian Kloch[1] and Thomas Uraz[2]

[1]5G Innovation Architect, TDC Net,
[2]5G Core Partner Lead, Ericsson Teglholmsgade 1, 2450 København C, Denmark
E-mail: chkl@tdcnet.dk; thomas.uraz@ericsson.com

Typically, when new communication technology is introduced, it is often referred to as a potential game-changer for the Telecommunication Service Provider (hereafter referred to as "TSP"). This is also the situation with the introduction of 5G. However, most of the current revenue for the TSP comes from the transportation of bits and voice calls which limits the possibility for additional revenue with increasing traffic. The revenue per bit has decreased over time, while the number of bits per user has increased [1], and the expectation of the minimum bit rate has increased [2]. This means that the TSP needs to increase the capacity in the mobile infrastructure continues to meet the customer demands, but it may not necessarily increase revenue [3].

3.1 5G – A Significant Market to Reap

With the introduction of 5G, a new era may come to the TSPs [4]. 5G may be one of the key components, together with the cloudification and the introduction of mobile edge computation, that allows the TSP to get a central role in the digitization of the industry, e.g., via Industry 4.0, by providing the main platform for future services to be built upon.

In 2026, Arthur D. Little predicts that the total addressable market based on 5G will be 582 USDbn. This market potential can be realized through the support of the ongoing digitalization of society and the introduction of new technical capabilities that may be provided through the introduction of 5G. When referring to 5G, it is essential to have in mind that 5G is not a single release, but the term "5G" shall be considered as a denominator for the technical capabilities that will be made available in the cellular network over ten years until 6G will be introduced. The first commercial launches of 5G networks were in 2019 [5].

Figure 3.1 shows that the global addressable market based on 5G varies vertically. Furthermore, the figure shows how the revenue from the 5G related market may be distributed when considering digitalization maturity within each industry and how well 5G, as cellular technology, may support the future automation within each vertical. The crucial part here is to look at the various working processes within the industry and understand if 5G will make a real difference in each working procedure.

Globally, it is seen from the numbers in Figure 3.1 that the most significant market potentials are within Manufacturing and Energy & Utilities. However, this may not be the case when focusing on the individual local markets. It depends on the regional distribution of the industries, the general level of digitalization, and if better alternative solutions already exist in the individual local markets.

At the same time, it is also dependent on which roles the different players in the individual industry vertical decide to take.

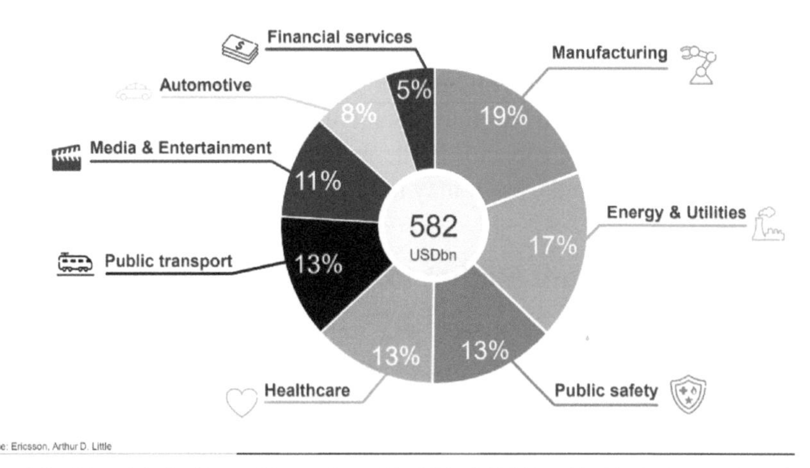

Figure 3.1 The global addressable market with 5G in 2026 identified by Ericsson & Arthur D. Little [6]

The TSPs can decide to become network providers, service enablers, or service creators. The size of the market potential increases the closer to the service creator role the operator decides to take.

To realize the market potential of 5G, the TSP plays an active role. The TSP enables the fundamental building blocks such as providing the network infrastructure, ensuring the cellular coverage, and applying the licensed spectrum.

The licensed spectrum can also be rented out to the industries themselves, challenging the role of the TSP. However, renting the spectrum, and hence determining how to use the licensed spectrum, also comes with a set of obligations as the authorities expect the licensed spectrum to be actively used [7].

Therefore, an industry that gets a licensed spectrum, compared to getting an unlicensed spectrum, is also obliged to use the licensed spectrum. This is putting an additional burden on the individual industries. They need to acquire the competencies required to deploy, operate, and continuously optimize a 5G network with the network capabilities required to support the digital processes in the individual industry.

Several different constellations can be considered when it comes to the rollout of 5G coverage and infrastructure for a given purpose, e.g., operation of machines on the factory floor, deploying service robots for monitoring [8], or providing general coverage for more generic usage of cellular technology such as smartphones, drones for logistic purpose, or infrastructure for intelligent tractors optimizing the usage of the farm resources [9].

Two examples of these constellations are:

- A manufacturing company decides to establish a private network in the factory using its spectrum.
- A service provider decides to use the macro network provided by the TSP.

In the first constellation, a manufacturing company establishes a private network. In this case, the private network is a cellular network that enables, e.g., all production machinery to be connected wirelessly instead of, as today, being connected via wires. Wires guarantee high reliability and low latency, but they challenge every reconfiguration of the production process since cables/physical connections need to be moved. Moreover, reconfiguring cables on the production floor may pause the production for up to 6 months.

However, it will be challenging since the company needs to rent the spectrum and acquire additional competencies to design, build, and operate the network infrastructure.

The service provider uses the macro network for the industries' services in the second constellation. This could be for services like large area drone logistics or offering an infrastructure for inspection robots. These new services will impact the macro network with new demands to design, capacity and capabilities, but it is the responsibility of the TSP to facilitate that the macro network can fulfil these demands in case a service level agreement (SLA) is signed.

A drone for logistic purposes is an excellent way to exemplify how these new services will impact the macro network. To serve the drone efficiently, it must be connected to a control centre, be continuously followed, and offload its data for real-time postprocessing in the cloud. These demands come from the fact that it shall be possible to (re)direct the individual drone dependent on other airborne entities' positions and get real-time access to the data.

The demands to the network will evolve due to the continuous increase in expectations for what a good service looks like, but also due to the new technical capabilities offered by the network, e.g., better positioning accuracy, shorter response times, and faster bitrate.

For the 5G digitalization market to mature and release its potential, the TSPs must understand when there is a market pull and where the 5G technology is ready to support the different use-cases identified either by collaboration with the ecosystem. The local market pulls and the maturity of the 5G technology may differ from country to country, e.g., due to the local spectrum auction, the actual 5G deployments, or the availability of alternative technical solutions and industry readiness.

Furthermore, the TSPs need to create a product portfolio that allows them to get positive business cases. This implies that the cost of establishing the new capabilities shall remain attractive compared to the potential revenue of enabling new business opportunities based on the newly enabled capabilities.

3.2 5G Use-Cases Mature Towards 6G

With the introduction of 6G, new use-cases such as high-fidelity holograms, multi-sensory communication, and pervasive artificial intelligence (AI) are often mentioned. But the 6G technology is not expected before 2030 [10], and history has shown that there must be a market pull for these new use-cases to mature. Moreover, the envisioned use-cases for 6G are far from the network capabilities provided in today's realized use-cases such as Mobile Broadband, Massive/Broadband Internet of Things (IoT), and voice services that are widely deployed using 4G.

	Current	On the road to 5G	5G experience
Enhanced Mobile Broadband	Browsing, social media Music, video	Fixed Wireless Access Interactive live concerts and sport events	4K/8K videos, Mobile AR/VR gaming and immersive media
Automotive	WiFi Hotspot, On demand GPS map data Over-the-air SW updates	Predictive vehicle maintenance Capturing sensor data for real-time traffic, weather, parking, and mapping services.	Autonomous vehicle control Cooperative collision avoidance Vulnerable road user discovery
Manufacturing	Connected goods Intra-/inter enterprise communication	Collaborative robots Distributed control system Remote quality inspection	Remote control of robots AR in training, maintenance, construction and repair
Energy & Utilities	Smart metering Dynamic and bidirectional grid	Distributed energy resource management Distribution automation	Control of edge-of-grid generation Virtual power plant Real time load balancing
Healthcare	Remote patient monitoring Connected ambulance Electronic health records	Telesurgery Augmented reality aiding medical treatment	Precision medicine Remote robotic surgery Ambulance drones

Figure 3.2 Development of use-cases within different verticals [6].

But there is also a timeframe of 10 years until 6G starts to materialize. These ten years shall be used to materialize the visions and ideas with 5G, visions and ideas that are much closer to the services expected with 6G than the services that have been materialized throughout the development of 4G.

Examples of how use-cases are going mature within the different industry verticals are shown in Figure 3.2. An example of how services can become more intelligent is the development of new services related to Automotive. For example, some current use-cases cover Over-the-air SW updates and enable in-car Wi-Fi hotspots.

On the road to 5G, cellular technology is expected to enable services like predictive vehicle maintenance due to real-time access to cloud data resources. Furthermore, when 5G is fully matured, it is expected that the capabilities of 5G can provide a platform for autonomous vehicle control.

Another vertical often mentioned in 5G is healthcare, where the use-case develops from remote patient monitoring via telesurgery to remote robotic surgery. Again, this development shows a significant shift from currently thought services to more futuristic, but not impossible, benefits due to the improvement of the 5G technology.

To unleash the potential of 5G, it is essential to demonstrate the different aspects of 5G in different use-cases. The demonstration phase is a fair assumption that the technology can provide the necessary reliability and support an increasing number of devices. During the time it takes for the use-case to mature from demonstration mock-up to actual commercial products, the technology has evolved, and hence the products to be launched are supported concerning network capabilities and service experience.

To determine whether a use case needs 5G or if 4G will be enough, it is paramount to consider the requirements concerning latency and throughput.

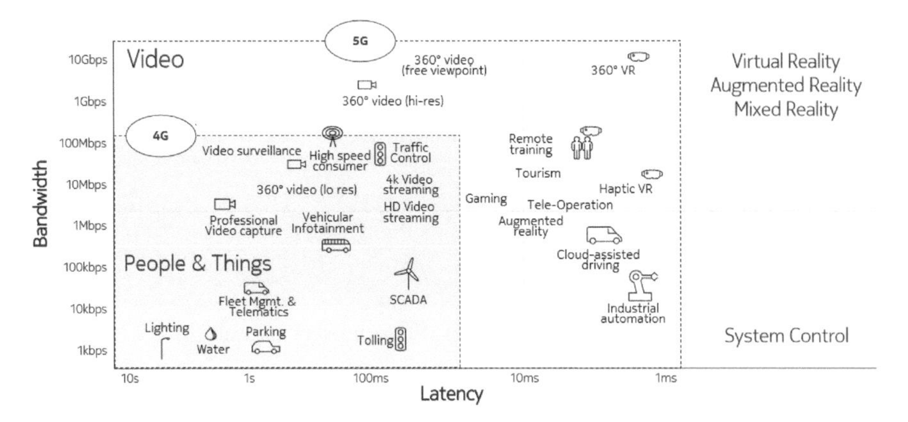

Figure 3.3 Bandwidth and latency demands from various use-cases and whether they can be implemented via 4G or 5G.

Source: Nokia

However, latency and throughput are not the only dimensions to consider when evaluating whether to use 4G or use 5G to realise the use case.

Several use-cases have been mapped in Figure 3.3 concerning latency and bitrate requirements against the capabilities in 4G and 5G to show which type of use-case that can be realized in the current 4G network and which use-cases require the new stuff that can be provided via the 5G technology.

Use-cases like smart meters, tolling, traffic control, and 4K video streaming are all possible today with the existing capabilities of 4G, while 5G is needed to meet the capacity demands of the mobile network as the usage of these services is going to increase. Therefore, the maximum user bitrate on 4G is indicated to be 150 Mbps. The 150 Mbps reflects the peak rate available on a single 20 MHz frequency carrier. Of course, 4G can offer much higher bitrates through higher modulation and carrier aggregation, but it is essential to remember that the 150 Mbps is the user bitrate and not the peak rate of 4G.

More bandwidth-demanding services like 360 degrees high-resolution video service will require a bitrate higher than 150 Mbps, hence 5G. At the same time, enabling latency lower than currently is achievable in 4G allows for new use-cases within industrial automation, e.g., cutting the cables between the machines on the factory floor, allowing for more agile production processes.

Some new services put requirements on both shorter latency and higher bitrate. For example, new services could be teleoperation based on haptic VR or 360 degrees VR. Sometimes new use-cases are being developed further, e.g., by combining haptic VR and Augmented reality, offering user experiences with dragons flying around the stadium or the possibility to select which player you would like to follow in the field.

Figure 3.4 adds other dimensions to the use-cases beyond the latency and bandwidth dimension shown in Figure 3.3. In preparing the network for the new use-cases, it is also essential to consider the number of devices and the demand for stability.

How widely the different services can be made available depends on the rollout of the 5G network infrastructure. The deployment of a 5G network varies from country to country and city to city. Therefore, the local rollout impacts the user experience as it may vary due to the different density of radio sites and allocation of the spectrum on the individual sites. This is important to have in mind, as the capability of the network to support the new use-cases foreseen with 5G does not only depends on the rollout of the 5G infrastructure but also on how well the network is dimensioned to meet both the traffic load from existing services, their growth rate as well as the demands coming from the introduction of the new services.

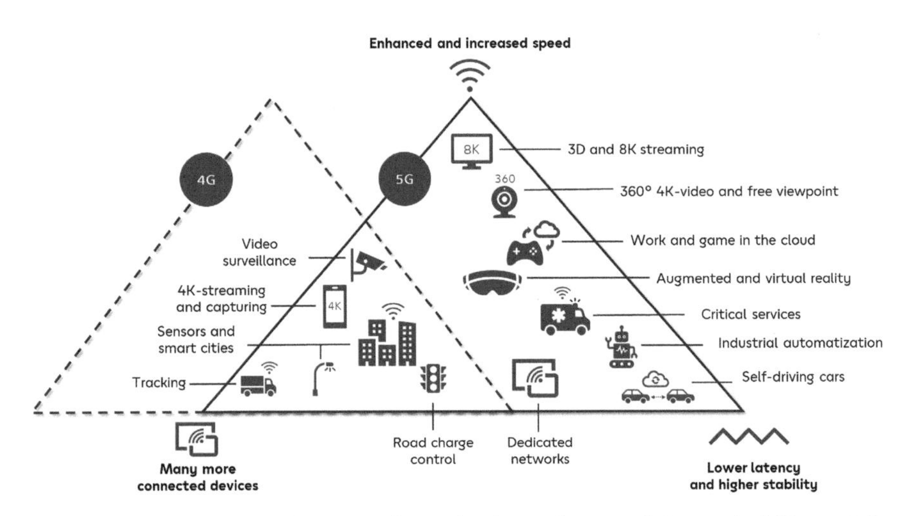

Figure 3.4 Use-case mapping according to its demand to speed, latency/stability, and the number of devices.

3.3 The Major Performance Differences between 4G and 5G

There is a vast diversity in the use-cases being enabled by introducing the 5G technology due to the multi-dimensioned development of the cellular network capabilities in the transition from 4G to 5G, as indicated in Figure 3.4.

Figure 3.5 shows the significant differences between the expectations for 4G and 5G. The inner-circle represents the expectations for IMT-Advanced (or 4G). It may be noticed that some of the expectations for 4G have been exceeded, such as downlink bitrates above 1 Gbps (e.g., in TDC Net's 4G infrastructure [11]) or the accelerated support for massive IoT with the launch of NB-IoT and LTE-M in 3GPP Rel 13 [12].

The expectations for the performance levels in 5G in 2030 are shown in the outer circle, here shown as IMT-2020. The numbers indicate which performance levels could be expected as 5G matures, but the possible values depend on the progress in the standardization bodies, like 3GPP, and the market demands.

When considering the different parameters, they are here considered as individuals. Multiple of them will be used in combination in most use-cases, but they are mentioned as individual parameters here.

- Connection density (devices/km^2) represents that many devices will be connected to the network. The increase in the number of devices appears with the introduction of more intelligent services, beyond the smartphone and basic tagging IoT services, like an increasing amount of monitoring devices (on the body, in the building, etc.) and the use of robots for various purposes, e.g., delivery robots, service robots, drones, etc.
- Latency is expected to go down towards 1 ms to support a fast reaction time. But the demand for latency is not only about short latency but also about controllable latency. 5G enables bounded latency. This ability provides much value to use-cases since a bounded latency helps the producer of, e.g., the robots, compensate for the variations in latency in the application development.
- Mobility is expected to increase from 350 km/h to 500 km/h, supporting high-speed trains, fast drones, etc.
- User experience data rate is expected to increase ten times from 10 Mbps to 100 Mbps. 15 Mbps is enough for 4K streaming on most smartphones, but the demands increase due to broader viewing areas and better resolution on the screen, e.g., larger displays on smartphones or 8K 360 degrees views in special glasses.

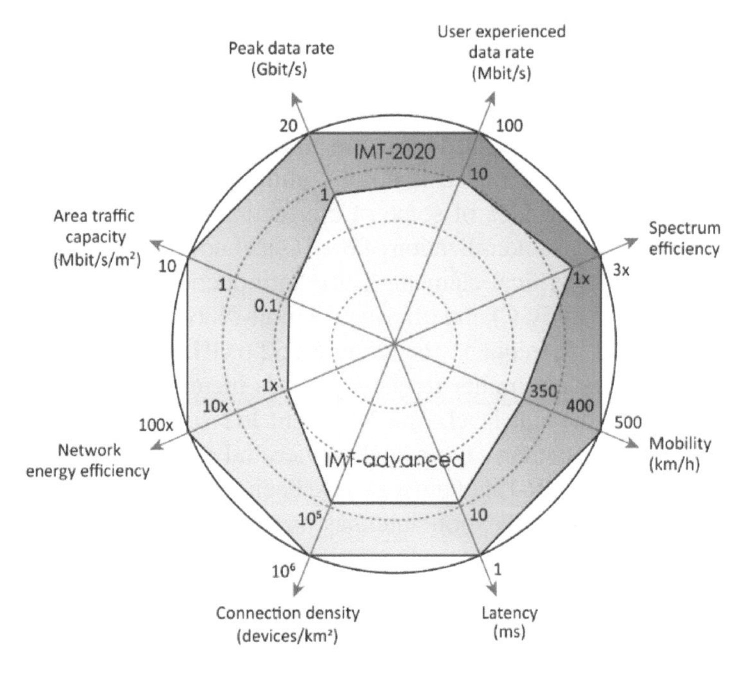

Figure 3.5 The ambition for the technical parameters within IMT advanced (4G) and IMT-2020 (5G)

- The increase in peak data rate to 20 Gbps will support the increasing demand for capacity.

For the network to provide these excessive performance capabilities, the network technology needs to be improved compared to the implementation of today.

- The spectrum efficiency is expected to increase by 3x due to better coding. Furthermore, with 5G is possible to access the frequency spectrum up to 100 GHz.
- The Network energy efficiency will be increased to match the more sustainable handling of the increasing traffic. The entire telecommunication industry is responsible for 1.5 % of the total CO2 emission [13]. With the introduction of 5G, studies have shown that the traffic in the mobile infrastructure can increase, and still, the CO2 emission will be lower than today's emission from 4G [14].
- Area traffic capacity is expected to increase 100x times, ensuring that the total power (from new features, more sites, more spectrum) can meet the

foreseen increase in data traffic (with a 40 % increase per year, the total traffic will increase by a factor of ~30x over ten years).

Figure 3.6 illustrates some of the main capabilities expected to be part of the 5G standardization today, in the short term, the mid-term, and 3GPP releases, including the first flavours of 6G. It is not a committed roadmap for what will happen, but it gives an idea of some of the focus areas expected to be considered in the ongoing standardization of the 5G technology. For example, in release 16, one of the critical elements is the introduction of a 5G Stand Alone Core (referred to as 5GC) and offering accurate Network Slicing.

In 3GPP Rel 17 and 18, frequency bands above 52.6 GHz and NR light are being introduced. Accessing higher frequency bands increases the potential capacity and bitrate at the cell level. The NR-Light IoT standard offers IoT devices with less capability than a typical NR device and reduced energy consumption. In addition, the NR-Light devices have been simplified by reducing the number of transmission/reception antennas on the device side, reducing the minimum required bandwidth, and supporting devices only capable of half-duplex operation in the paired spectrum. The idea with NR-Light is to address new types of use-cases in Industrial IoT, such as:

- Industrial wireless sensors with low latency (5-10ms) and medium data rate (<2 Mbps)
- Medium to high data rate (2-25 Mbps) video transmission
- High data-rate wearables (5-50 Mbps) with long battery life (1-2 weeks)

In 3GPP Rel 19 and Rel 20, some of the new features are the highly accurate positioning (< 1m) and the integration to non-cellular technologies, while the introduction of 6G is expected to provide even higher speed, better accuracy, lower latency opening for even more advanced use-cases.

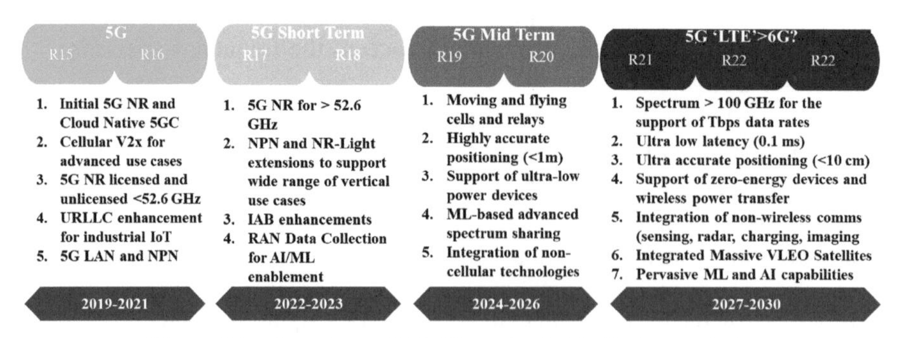

Figure 3.6 Overview of 3GPP Rel 15 to 3GPP R22 towards 6G

With the improvements in 5G technology compared to the capabilities in 4G, the 5G technology provides new capabilities that can be used for new use-cases. These new use-cases can be the digitalization of current manual processes, a wireless representation of currently wired use-cases, or simply slight improvements of the well-known use-cases such as mobile voice, mobile data, and mobile IoT meeting the increasing demands from the end-users.

3.4 Innovation Hubs: Key to the Success of 5G

The introduction of 5G provides several new technical opportunities that are very different from the typical cellular network use. Furthermore, the announcement of 5G being nationally available in Denmark [15] has re-initiated the digitalization process by creating a need for cellular-based services that in principle could have been realized in 4G.

For 5G to materialize beyond mobile broadband and voice, there must be a decisive driving factor for the extended capabilities, either in the form of a market pull from a large service provider like Apple's introduction of the iPhone or a market push from the TSPs capturing a more significant part of the value chain and not only remain the bit-pipe provider. If the TSPs remain a bit pipe provider, there is a risk that the development of the telecommunication infrastructure will happen based on a minimum set of requirements that only ensure a good user experience for services like the ones that are possible today.

As part of the investigation on how the TSPs can get a more active role in the 5G eco-system, several operators have created "Innovation Hub" concepts, e.g., Vodafone (Digital Innovation Hub [16]), TDC (5G Innovation Hub [17]) or Telia (Division X [18]).

Their purpose is to make 5G tangible by implementing new use cases and driving the surrounding ecosystems into new business development. This is done by:

- Developing new use-cases for new 5G business opportunities related to current and future network capabilities.
- Leveraging the home-market ecosystem and engaging with enterprises, academia, 5G communities, and experts to identify where the TSP can take an active role and formulate a commercial product.
- Guiding the 5G Innovation roadmap in a strategic partnership with the network technology vendors to materialize use-cases before the commercial launch of the new capabilities.

- Supporting communication to impact 5G perception internally and externally through knowledge sharing and inspirational talks.

An example of how an Innovation Hub has been developing new 5G business opportunities in the dialogue with the industry verticals about the requirements for shorter latency. The dialogue has shown significant differences between latency requirements for applications on the factory floor compared to the latencies required for 5G controllable robots performing inspection, even though their speed may be similar.

For the factory floor, the latency range must be below 10 ms, while for the 5G controllable robot, the latency must be below, e.g., 300 ms. The latter has been demonstrated at H.C. Andersen Airport in Denmark, where the four-legged robot dog SPOT from Boston Dynamics has been controlled via 5G performing inspection along the airport's fence [8].

Even though the trial with SPOT was about the inspection of the fence around H.C Andersen's airport, the fact that the robot can be remotely controlled using 5G has led to several equivalent use-cases, like robots in the construction sector and cleaning robots in cities, that can all be handled by the 5G macro network.

In other words, in the case of the SPOT robot and similar use-cases, the 5G macro network has the true power to remove the short leash implied by Wi-Fi's requirement of having a robot's controller next to it.

The 5G macro network can enable robotics to become truly autonomous.

Through genuine engagement with the ecosystem, the Innovation Hubs can identify the use-cases that can be realized in a reasonable timeframe, which varies from industry vertical to industry vertical depending on the technical demands and the market potentials.

3.5 3GPP Rel 16 and Rel 17, The First True 5G Releases

5G was introduced with 3GPP Rel 15. The 5G-related improvements were mainly at the radio interface with a new radio (NR) Interface built upon advanced channel coding, sub-6 GHz with massive MIMO, mobile mmWave, and scalable OFDM-based air interface. These improvements provide more capacity and higher speed, so the 3GPP Rel 15 mainly focuses on enhanced mobile broadband (eMBB) services such as higher data rates for smartphones or even the introduction of fixed wireless access (FWA).

In addition to the improvements on the radio interface, 3GPP Rel 15 introduced the ability to combine the 5G NR with the current 4G Core by using 4G as the control channel into the 4G EPC with 5G NR for the

User Plane. Thus, at this moment, the terminals can combine the usage of frequency bands dedicated for either 4G or 5G, even though the functionality is limited to the 4G Core capabilities.

On the core side, the evolution with 3GPP Rel 15 was mainly limited to the ongoing "Softwarization", replacing the dedicated physical hardware with similar functions in software that can now be executed on general-purpose hardware.

Even though the introduction of 3GPP Rel 15 is mainly about reducing the latency on the radio interface and adding additional frequency bands that improve both the downlink bitrate and the uplink bitrate, the commercial launch of 5G, like for TDC Net, provides a platform for discussion of the possibilities for innovation due to its nationwide presence. Furthermore, this dialogue with the ecosystem can lead to a better understanding of how to transform the capabilities of 5G into new commercial products, such as better internet connection in trains from Swisscom [19] or applications for smart cities [20].

This means that the innovation aspect is not limited to what is available with currently implemented versions of 5G, but the new use-cases are also forward-looking towards what will be available with 3GPP Rel 16 and 3GPP Rel 17. Of course, in parallel to the addition of new capabilities, the new releases will include improvements to existing capabilities, but this is not the focus here in the chapter.

With the introduction of 3GPP Rel 16, as the first actual 5G release, the network topologic changes from a non-standalone core to a standalone core. As a result of this, Network Slicing becomes genuinely available, and allows the TSPs to more efficiently assign resources for the various service, e.g., EDGE for short latency, high amount of spectrum bandwidth for extreme bitrates or assign a more reliable transmission path to provide a communication platform with high reliability.

On top of this, the 3GPP Rel 16 opens for:

- 5G NR Industrial IoT with eURLLC reduces latency and improves the reliability of critical IoT services.
- 5G NR Cellular V2X for interaction between vehicles or between the vehicle and its surrounding infrastructure. With 5G, it is expected that the degree of interconnectivity will increase and enable more advanced use-cases.
- 5G NR in the unlicensed spectrum allows the unlicensed band to be used for additional capacity.

- 5G broadcast, and potentially better coordination between broadcasters and TSPs, allowing broadcasters to access the multitude of streaming platforms fed through TSPs' cellular network(s) and for the TSPs to use broadcasters' spectrum for broadcasted content.
- 5G massive IoT supports the expected number of IoT devices significantly higher than feasible with NB-IoT and LTE-M.
- Positioning across use-cases, and hence positioning accuracy beyond cell level.
- Integrated access backhaul (IAB) enables that part of the spectrum used in the cell can be assigned as transmission resources.

It is difficult to predict which new capabilities are the most exciting additions to 3GPP Rel 16 from a use-case perspective and business potential point of view. However, it may be the improved positioning accuracy, cellular V2X, and eURLLC. These seem to be the most interesting ones as they are essential for driving the development of new use-cases within service, monitoring, and logistic robots that can be applied in most industry verticals at ground level and above.

In the study items listed for 3GPP Rel 17, some of the additions mentioned are

- P2V (person to vehicle) and IoT relay. These features may enable better user interaction with the vehicle and improve the coverage range.
- New spectrum above 52.8 GHz to address spectrum bands up to 100 GHz.
- NR-lights, e.g., wearables, and industrial sensors, close the gap between LTE-M and smartphones.
- Centimetre accuracy in very focused areas where mmWave can be used to connect to, e.g., Industrial IoT (IIoT) devices.

Predicting which new features in 3GPP Rel 17 are most interesting in new use-cases is too early as it will depend on the market pull and the willingness to pay for the services being developed as part of 3GPP Rel 16 implementation.

3.6 Service-Based Architecture: An Enabler for New Use-Cases

Until recently, the mobile core network was considered one monolithic network established on purpose-built hardware for the individual parts of the mobile core.

The introduction of cloud/NFV/SDN functionality and later orchestration and programmable APIs enabled the virtualization of the core and RAN functions. It resulted in a general-purpose hardware platform, while the specific features and functionality in the mobile infrastructure were implemented in a software platform (see Figure 3.7).

The functionality of very different core components can run on the same hardware platform, thereby providing a comprehensive platform for a more flexible design of the cellular infrastructure. As a result, the cloud concept is introduced in telecommunication networks giving the ability to move the network functionality within the cloud network and transport.

Furthermore, data centres and edge computation capabilities are introduced in the network allowing the core functionality to be placed closer to the network's edge. The distance between the end-user and the mobile core functionality is significantly shortened.

The introduction of data centres and edge computing opens for new short-latency services using the computational power in the cloud, closer to end-users than ever before. New short-latency benefits can be realized due to the shorter distance obtainable inherited by distributing the core functionality. The usage of the edge cloud enables higher bandwidth and better response times. The application/customer feels like the data is stored/processed locally at the device even though the data is processed in the cloud.

The virtualization of the mobile core functionalities gave the foundation for the Service Based Architecture as part of 3GPP Rel 15/16 with the introduction of 5G Core. In the 3GPP Rel 15/16, the telecommunication network functions can be exposed to the end-users allowing for unprecedented flexibility around new service creation. In principle, the service based architecture (SBA) is built like a bus architecture enabling all the different functionalities to get the most accessible access to all other functionalities in the core. This opens for a more flexible design of new services that can be offered to the end-users, and these new services can easier be upgraded with the latest available features. Due to the Softwarization of the different functionalities and the introduction of Cloud Native concepts, the upgrade processes have changed. Individual software components can now be upgraded more often than before where the special purpose-built hardware needs to be replaced as part of the upgrade.

With Service Based Architecture, the functionality in the core is split into a User Plane, a Control Plane, and a Data Layer, see Figure 3.9. Therefore, service-Based Architecture is defined in Control Plane and Data Layers. The

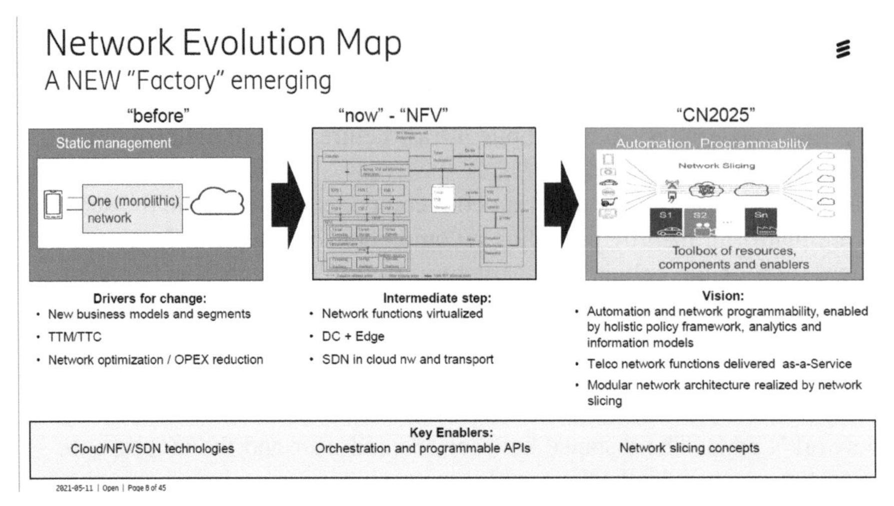

Figure 3.7 Evolution of the mobile core network from monolithic hardware, enabled by Cloud/NFV/SDN technologies, to NFV based platform built on orchestration and programmable APIs.

following will focus on the 5G specific parts, but naturally, 4G parts, SGI-LAN functions, and Security functions are crucial parts of a well-functioning cellular network infrastructure.

3GPP defines Network Functions and the interaction necessary between them. Implementation of functionality into products/software is left to the vendor community. It is common for vendors to combine 3GPP-defined Network Functions into products enabling more flexibility. The example shown in Figure 3.8 is one such approach by Ericsson.

The User Plane consists of packet core gateways (PCG) that forward user data (payload) between the RAN and the internet. For example, Figure 3.8 shows that the specific 5G part is the UPF (User Plane Function).

In the example, the Control Plane consists of (Pure 4G Nodes are not explained):

- A Cloud Core Resource Controller was built up by the network slicing selection function (NSSF) and the network repository function (NRF).
- A Cloud Core Subscription Manager built up by the user data management (UDM), the authentication server function (AUSF), and the 5G equipment identity register (5G-EIR).
- A Cloud Core Exposure Server consists of a network exposure function (NEF).

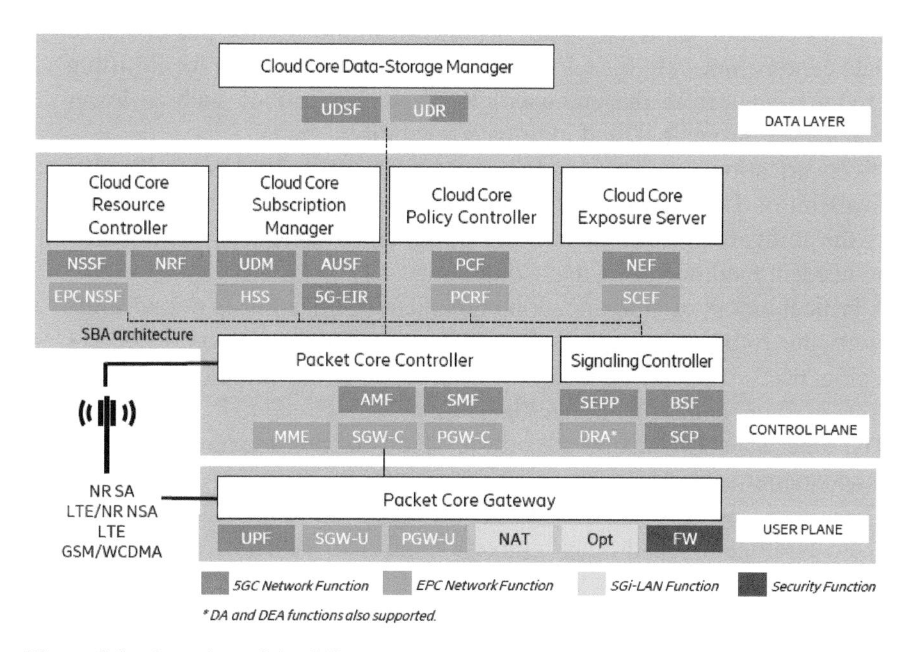

Figure 3.8 Overview of the different components in a service-based architecture (an example from Ericsson based on 5G dual-mode core – EPC and 5GC)

- A Packet Core Controller consists of an access and mobility management function (AMF) and an session management function (SMF).
- A Signaling Controller consists of a security edge protection proxy (SEPP), a binding support function (BSF), and an service communication proxy (SCP).

The third layer in the Service Based Architecture is the Data layer. The Data Layer includes the Cloud Core Data and the Storage Manager established with the UDSF (User Data Service Function) and the UDR (User Data Repository).

The Softwarization of the mobile core and the mobile RAN functionalities enables a more flexible design.

With the introduction of a Service-Based Architecture and the network's cloudification, the data centres' deployment becomes more flexible, as shown in Figure 3.9. Instead of having a limited number of core locations, the mobile core node functionality will be deployed across the network spanning the Cloud Data Centers and the access network. Furthermore, the cloudification of the network does also provide a platform for introducing Cloud RAN.

The telecommunication network benefits from the flexible placement of the data centres, not only for telco Network Functions but also for fulfilling new requirements from the end-user/enterprise applications such as lower latency, higher throughput, and improved reliability.

"Softwarization" of Network Functions will drive cloud enhancements and availability. This will directly benefit future end-user/enterprise use cases due to the ability to place computational power and mobile core functionality where needed, facilitated by Edge Computing.

A critical aspect of cloudification is the better utilization of the different resources due to better coordination on how and when the network resources are being used. This is realized by better operational methods, e.g., via machine learning and artificial intelligence.

The other aspect, which is the primary focus of this section, is offering new and smarter services.

These new services could relate to the smart building, the smart factory, and intelligent self-driving/remotely controlled vehicles, as shown in Figure 3.9.

One of the fascinating aspects of Service Based Architecture is exposing network capabilities to the end-users. Drone inspection could be a use-case that may benefit from this since the requirements to the network from the drone depend on where it is in its mission:

- For take-off and flying to the area of inspection, accurate information of the actual position is essential to manoeuvre the drone relative to the other flying objects. At the same time, the demand for the uplink bitrate

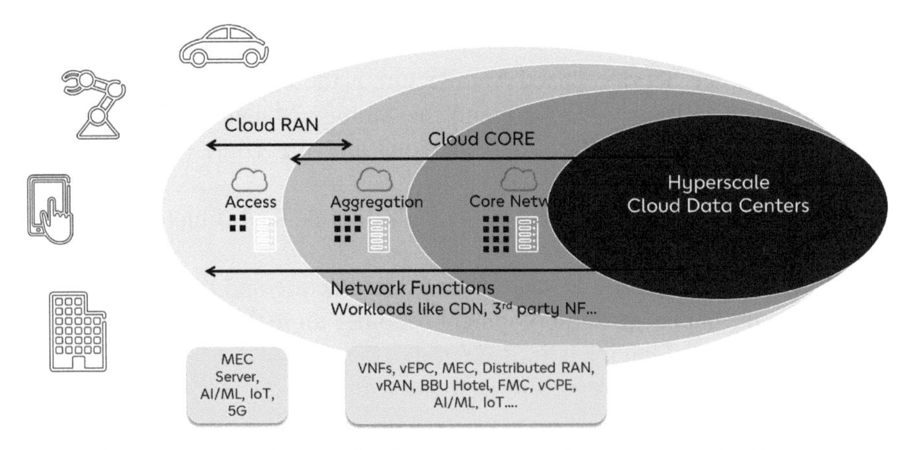

Figure 3.9 Location of regional data centers provides more than short latency

is limited since the purpose of the pictures being sent is only to identify if there is an obstacle on the flying path.

- When arriving at the inspection area, the uplink bitrate demand increases since high-quality pictures are essential to identify if maintenance activities are required. However, the need for accurate positioning may be lower since the drone is expected to be manually controlled when close to the inspection target.
- On the return to the landing zone, and while the drone performs an automatic flight, the demands for the positioning accuracy increase to manoeuvre the drone relative to other flying objects, while the picture resolution can be decreased.

In this case, exposing the capabilities is essential as it allows the drone operator to access the capabilities required for the individual parts of the mission. At the same time, it provides an advantage to the TSP as it minimizes the need for network capacity resources.

3.7 Network Slicing: Efficiently offer QoS and Customization

In Figure 3.7, NFV was introduced as the first step after submitting one monolithic network. The subsequent step to the NFV is Network Slicing, as shown in Figure 3.10. 5G Network Slicing was introduced in 3GPP Rel 15. 3GPP mandates that a 5G network must at least have one Network Slice. Network Slicing is expected to be widely deployed as a part of 3GPP Rel. 16 enhancements. The Network Slicing concept enables optimized resource orchestration provided by automation and network programmability, enabled by holistic policy framework, analytics, and information models.

Furthermore, with Network Slicing and the introduction of logical, isolated end-2-end networks running on shared infrastructure, varying network architectures are possible. Flexible architectures, yet all running on the same underlying infrastructure, allow meeting diverse, often conflicting requirements, offering additional security isolation if required while maintaining efficiencies.

Network Slicing is often referred to as a feature of the 5G Core introduced in 3GPP Rel 15. In reality, Network Slicing has been possible earlier. In previous versions of the cellular technology, one of the challenges has been that it has been necessary to build up a physical core for the mobile broadband traffic and another physical core for the IoT traffic. IoT in the past, has

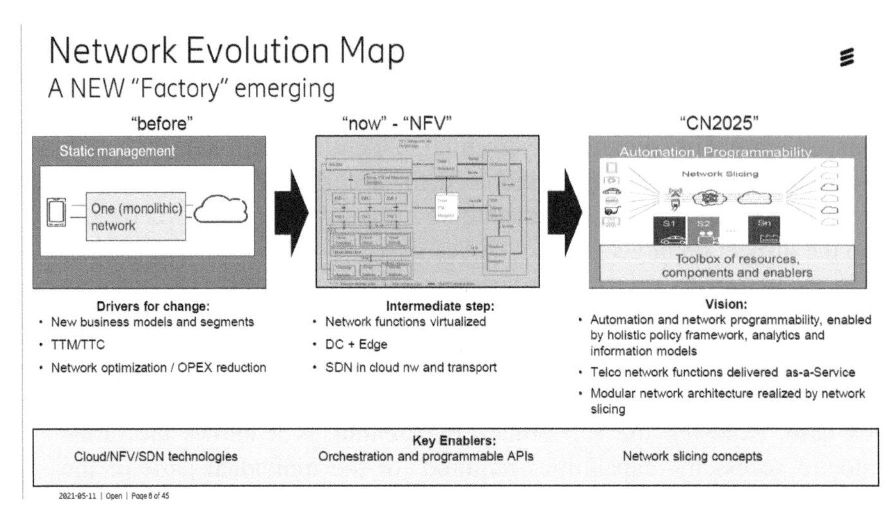

Figure 3.10 Development of the mobile core network from the intermediate step with NFV to the target network based on service based architecture and network slicing to offer telecommunication network functions delivered as a service

been mainly IoT applications with minimal data demands, such as smart meters (measuring temperature, water consumption, etc.) or parking solutions registering whether there was a car parked at a given parking lot.

As part of the roll-out of 5G, it is expected that the IoT applications will become more intelligent. Hence, the applications imply remote control of robots that needs a bounded latency (and therefore no allowance for latency outliers) or drones flying beyond visual line of sight where the connection to the drone is mandatory for the drone to remain airborne.

Other more advanced IoT applications could be service robots, self-driving trains, self-controlled mining trucks, or even applications based on AR/VR that put requirements both on reliability, data bandwidth, and latency.

An overview is shown in Figure 3.11, where several different use-cases and their mapping into the categories Massive IoT, Broadband IoT, Critical IoT, and Industrial Automation IoT. All these use-case categories put very other requirements on the infrastructure.

For the TSPs to offer these services without over-investing in the telecommunication infrastructure, Network Slicing is a way forward as it allows the different types of traffic profiles to be disaggregated. The alternative would be to build excessive capacity, so there is enough capacity in rare cases where the service requirements are stringent for the IoT devices to work correctly.

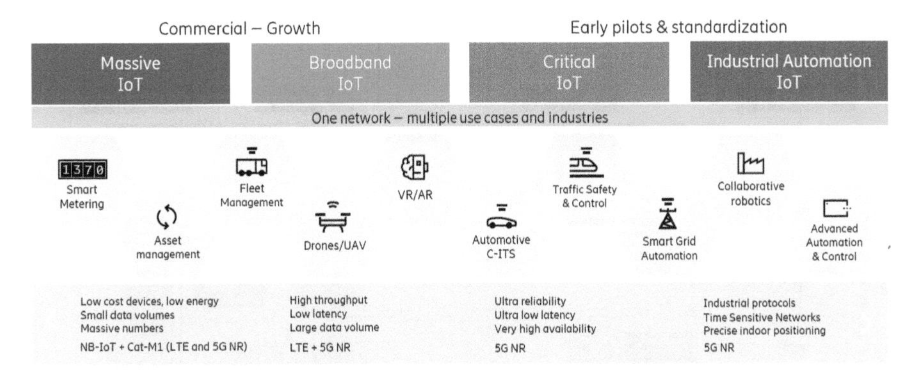

Figure 3.11 One network – multiple use-cases and industries (Source: Ericsson)

The consequence of not adding capacity to the network is that there will not be the capacity to simultaneously fulfil the demands of bitrate, latency, and reliability.

Figure 3.12 illustrates different services that set additional throughput and latency requirements. Latency requirements above 20 ms are considered typical latency, while latency requirements between 5 and 20 ms are considered low. When a latency requirement is 1 ms to 5 ms, it is regarded as ultra-low latency.

The requirements can be fulfilled by setting up logical end-to-end (E2E) networks with shorter or longer delays depending on how the signal is

	Edge computing	Network slicing	Throughput	Latency (RTT)
Fixed wireless access (residential, SoHo)		●	>100 Mbps	Internet service
Enterprise VPN (secure VPN service)		●	>10 Mbps	<100 ms
Public safety (security, body cameras)	●	●	>10 Mbps	<50 ms
Connected car (telematics, infotainment)		●	>5 Mbps	<100 ms
Low latency service (mobile gaming)	●	●	>10 Mbps	<20 ms
Broadband IoT (AR, surveillance)	●	●	>10 Mbps	<20 ms

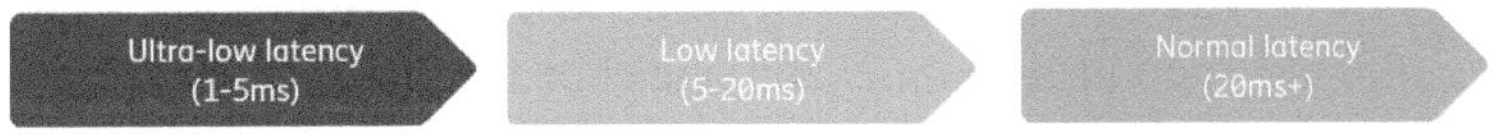

Ultra-low latency (1-5ms) | Low latency (5-20ms) | Normal latency (20ms+)

Figure 3.12 Examples of different services with different requirements concerning throughput and latency (RTT)

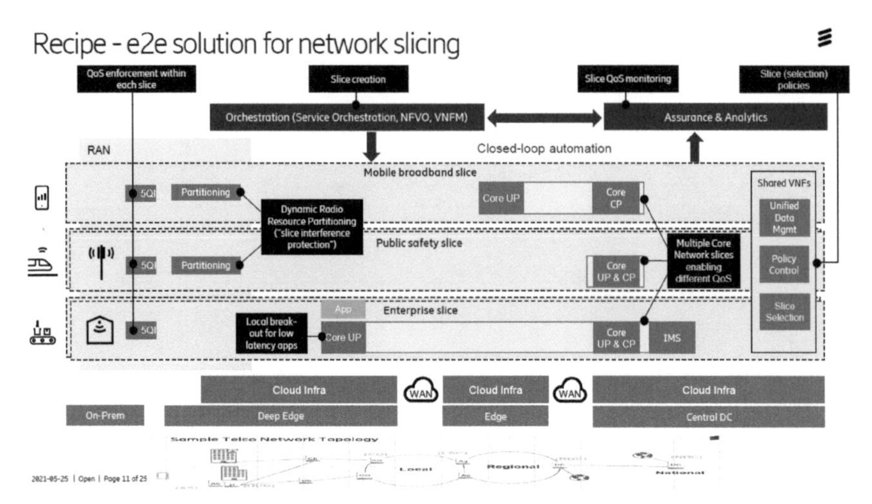

Figure 3.13 Recipe - E2E solution for network slicing

routed through the given logical network (impacts the latency) and by assigning more or fewer capacity resources to the individual slices (impacts the throughput). For services related to public safety, low latency services, or broadband IoT, there may be a need to introduce Edge computing to meet the resource demands for the services to work correctly.

Several design decisions need to be taken to design a slice, e.g., for trains. For the train case in Figure 3.13, the first decision is whether the coverage shall be provided by the entire macro network or only cells close to the train rails. The next decision is about the Quality Index (5QI – 5G QoS Identifier) and the quality demands of the service slice. Then each slice can decide to get a partitioning of the radio interface and how it shall be agreed against interference. On the core side, it shall be decided which level of isolation is required. It also needs to be decided if the User Plane shall be close to the UE or if all User Plane data can be transferred via the regional data centre. Finally, the orchestration and assurance are done outside the slice.

Each UE/device can simultaneously connect via (up to) 8 Network Slices, so different services at the device can get their slice with the characteristics needed for a given service. An example is a drone where the drone needs one slice for its command & control channel and one slice for the video feed from the camera to the servers in the cloud.

In principle, it is possible to make a slice for both different services and different areas. Figure 3.14 shows an example of four different slices

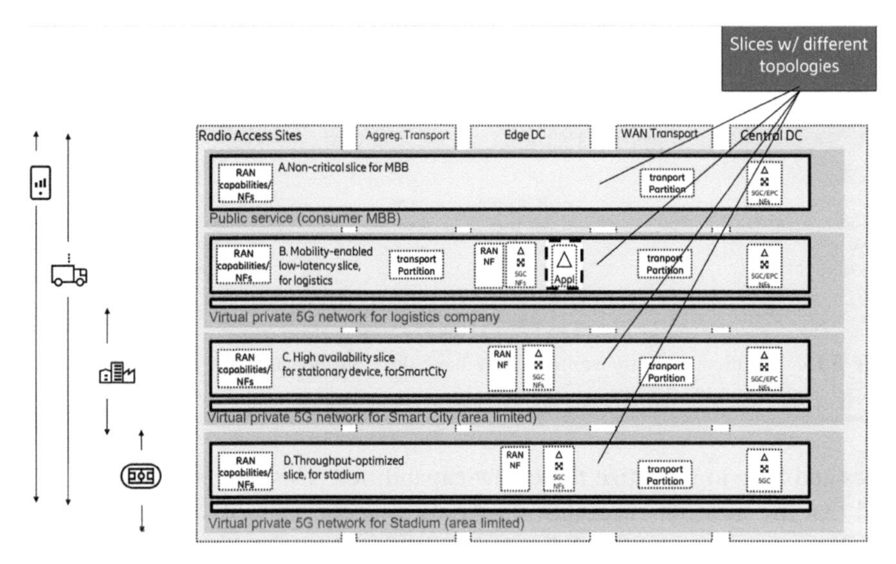

Figure 3.14 Example of slicing in wide-area/area-limited deployments

(A, B, C, and D), each slice covering a smaller or larger area. For example, slice D targets a stadium, or a hotspot location, where there is a need for high bandwidth and very short latency to provide the necessary capacity to offer exceptional services. The slice C is for high availability, mainly supporting stationary devices in, e.g., a smart factory. The slice B is a mobility-enabled low-latency slide for logistics, while slice A is for non-critical MBB service.

This means that Network Slices can be configured to support a given service type everywhere in the network or be dedicated to offering capacity or special characteristics to a limited area like a stadium or along metro tracks.

3.8 Conclusions

The 5G technology offers new capabilities that enable new use-cases and, thereby, new services to the consumer market and the industry verticals. Furthermore, the performance on 5G will be superior in all dimensions compared to the performance in 4G.

The total market potential for 5G in 2026 is expected to be 582 USDbn on the global market, which can only be realized through the careful introduction of the new features to the various industry verticals by showing how 5G can be a benefit in the digitalization processes. At the same time, the TSPs must

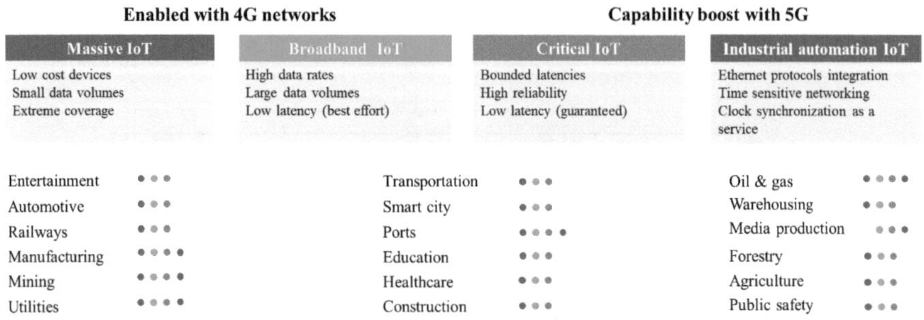

Enabled with 4G networks
Capability boost with 5G

Massive IoT
Low cost devices
Small data volumes
Extreme coverage

Broadband IoT
High data rates
Large data volumes
Low latency (best effort)

Critical IoT
Bounded latencies
High reliability
Low latency (guaranteed)

Industrial automation IoT
Ethernet protocols integration
Time sensitive networking
Clock synchronization as a service

Entertainment
Automotive
Railways
Manufacturing
Mining
Utilities
Transportation
Smart city
Ports
Education
Healthcare
Construction
Oil & gas
Warehousing
Media production
Forestry
Agriculture
Public safety

Figure 3.15 Different industry verticals may have different demands for critical IoT and industrial automation IoT services

understand how to monetize these new capabilities. This is a prerequisite to avoid that the new 5G capabilities are only used for capacity expansion to meet the increasing amount of data generated by Over Top players' applications and the continuous development of smartphones (larger screens, higher resolution).

With the introduction of 5G, new capabilities will be enabled to extend the IoT use-cases and include Critical IoT and Industrial automation IoT, use-cases that have not been possible with the 4G technology, see Figure 3.15. Alongside enabling Critical IoT and Industrial automation IoT, enhancements are also expected within Massive IoT and Broadband IoT to support the ongoing commercialization of services herein. Within Massive IoT, the technologies NB-IoT and LTE-M will evolve, while for Broadband IoT, 5G NR and 5GC will improve the performance compared to Broadband IoT solutions based on 4G.

As shown in Figure 3.15, every industry vertical has use-cases that benefit from technical capabilities in Critical IoT. However, the individual use-cases within may have different technical requirements that can still be quantified and understood. In this work, the Innovation Hubs can get a central role through the different pilots and provide feedback to the network infrastructure equipment vendors and the industry deploying the individual use-cases.

The initial use-cases in 5G do not require sub ms latency or a peak rate of 20 Gbps, but they require controllable performance, e.g. the SPOT Robot that needs a latency exceeds 300 ms or the drone that requires to be always reachable from the controlling platform.

The intention here has been to elaborate on the capabilities available in the near term 3GPP release, such as 3GPP Rel 16 and 17. In addition, it has been attempted to exemplify the capabilities through different use-case examples, e.g., the steering of the SPOT Robot and a typical mission for a drone.

5G may not be the only technology to enable these new use-cases. Furthermore, the 5G network infrastructure demands from the use-cases are heavily dependent on the practical implementation of the individual use-cases, e.g., whether the postprocessing of the captured data is expected to be done locally at the device or in the cloud. Another thing is the requirement for positioning accuracy, whether done via GPS at the device or by introducing accurate positioning from the cellular network.

In the monetization of the new 5G enabled services, the TSPs may be challenged that customers are used to buying access to the network, amount of voice minutes and SMS, today primarily free, and a certain amount of data, but they have limited experience of buying a guarantee for how the voice and how the data will be delivered. With the introduction of 5G and the ability to offer QoS based products and as a result of this supporting new use-case within Critical IoT and Industrial IoT, it becomes necessary for the customer and operators to agree on new business models where the customers pay for a guarantee, and the operator can document how well the services have been delivered with the agreed end-to-end scope. Even though today's networks have superior performance already, going after the support of new use cases and specifically beyond best effort will require new investments. These investments must be justified as part of enabling customers to apply new usages of the cellular technologies and for the TSPs to prepare the infrastructure for these new services.

It is hearting to see that many Industry Verticals, e.g., manufacturing with 5G ACIA [21], Automotive with 5GAA, Railway with 5GRAIL [22], and Media with 5G-RECORDS [23], unite to drive requirements in Cellular technologies via 3GPP. As described, Rel. 16/17 and 18 have much industrial focus. However, it is probably good advice for just starting work on the 6G journey, that besides new cellular enhancements in architecture and Radio technologies, the cross-industry focus is needed to capture new value beyond eMBB strategically.

The 3GPP community has been excellent historically to unite around and create global opportunities. The benefits that cellular technology provided for humanity with the facilitation of the Smartphone revolution are possible for the future machines. However, this will require further refocus from many pure architecture/cost efficiencies driven discussions to acute focus on end-user/vertical needs discussions ensuring that the future 6G architecture is also natively built with Industries in mind.

References

[1] Danish Energy Agency "National Telecommunication statistics", https://ens.dk/sites/ens.dk/files/Tele/mobil_baggrundsark_1h20.xlsx

[2] "Cisco Annual Internet Report - Cisco Annual Internet Report (2018–2023)", White Paper – Cisco, https://www.cisco.com/c/en/us/solutions/collateral/executive-perspectives/annual-internet-report/white-paper-c11-741490.html

[3] FAIST, "The Global Telecommunication Services Market: 2012-2019, https://www.faistgroup.com/news/the-global-telecommunications-services-market-2012-2019-data/

[4] Ericsson, "Switch on 5G for business", https://www.ericsson.com/en/5g/5g-for-business

[5] Reuters, "Who was first to launch 5G?", April 05/2019, https://www.reuters.com/article/us-telecoms-5g-idUSKCN1RH1V1

[6] Monika Byléhn, 5G Marketing Director at Ericsson, "The 5G BusinessPotential", 2017, https://www.terminsstarttelekom.se/upload/termin/pdf/pres475.pdf

[7] 5G Observatory, "National 5G Spectrum Assignment", https://5gobservatory.eu/5g-spectrum/national-5g-spectrum-assignment/

[8] Telecompaper – wireless, "TDC tests Boston Dynamics robot with 5G forfence inspection", March 5/2021, https://www.telecompaper.com/news/\tdc-tests-boston-dynamics-robot-with-5g-for-fence-inspection--1374875

[9] 5Groningen, https://www.5groningen.nl/en

[10] Gerry Christensen, Mind Commerce, "6G", https://searchnetworking.techtarget.com/definition/6G

[11] LTEto5G, Sanjay, "TDC reaches 1 Gbps through LTE Advanced Pro trial", June 12/2016, https://www.lteto5g.com/tdc-reaches-1-gbps-through-lte-advanced-pro-trial/

[12] GSMA, "Mobile IoT in the 5G Future", https://www.gsma.com/IoT/wp-content/uploads/2018/05/GSMAIoT_MobileIoT_5G_Future_May2018.pdf

[13] Ericsson, "Exponential data growth – constant CO2 footprints", https://www.ericsson.com/en/reports-and-papers/research-papers/the-future-carbon-footprint-of-the-ict-and-em-sectors

[14] STL Partners, "Curtailing Carbon Emissions – Can 5G Help?", October 2019, https://carrier.huawei.com/~/media/CNBGV2/download/program/Industries-5G/Curtailing-Carbon-Emissions-Can-5G-Help.pdf

[15] Juan Pedro Tomàs, "TDC completes 5G network deployment inDenmark", RCR Wireless, Dec 2/2020, https://www.rcrwireless.com/20201202/5g/tdc-completes-5g-network-deployment-denmark
[16] Vodafone Digital Innovation Hub Experience, https://www.platformgroup.co.uk/work/vodafone-digital-innovation-hub/
[17] TDC, "TDC annoncerer plan for 5G-net" (in Danish), 18-03-2019, https: tdcgroup.com/da/news-and-press/nyheder-og-pressemeddelelser/2019/3/tdc-annoncerer-plan-for-5g-net13570503
[18] Telia Division X, https://www.teliacompany.com/en/about-the-company/our-operations/division-x/
[19] Swisscom, "Greater bandwidth in trains," https://www.swisscom.ch/en/about/news/2020/10/21-mehr-bandbreite-im-zug.html
[20] Nicola Brittain, "27 innovative 5G use cases: we reveal what 5G is capable of", 5Gradar, May 2021, 5Gradar, https://www.5gradar.com/features/what-is-5g-these-use-cases-reveal-all
[21] 5G Alliance for Connected Industries and Automation, https://5g-acia.org/
[22] 5G-PPP, "5Grail: 5G for future RAILway mobile communication system", https://5g-ppp.eu/5grail/
[23] 5G-Records, "5G key technology enablers for emerging media content production services", https://www.5g-records.eu/

Biographies

Christian Kloch, PhD. & MBD, 5G Innovation Architect, TDC Net. Christian is one of Denmark's leading 5G experts and is part of the joint Ericsson/TDC Net strategic initiative Innovation Hub that focuses on new 5G business viable use-cases for the B2B market. As part of the engagement in the Innovation Hub, Christian focuses on how future capabilities of the 5G network can enable new use-cases and fulfil unmet needs in the different industrial verticals.

Before joining the Innovation Hub, Christian was Chief Network Architect for the Radio Access Network in TDC, where he was responsible for the Strategic Plan for the TDCs Mobile Access Network and hence defining the overall targets for "Denmark's Best Mobile Network."

Christian has more than 23 years of professional experience in the telecommunication industry working with mobile operators, mobile infrastructure vendors and a consultant for the local industry. On top of this, he has organized several technical events on telecommunication throughout his career, previously as Chairman of IEEE Denmark Section and now within IDA-Connect. Furthermore, he has published several peer-reviewed scientific papers on radio-wave propagation, mobile communication, and cloud-based services.

Thomas Uraz, Core Partnership Lead, 5G Innovation Hub, Ericsson. Thomas Uraz is a seasoned (20+ years) Wireless Telecom industry professional experienced in technology transformation, strategic product planning, and business development in partnership with multiple advanced service providers (AT&T, Verizon, TDC). He has proven successes in driving complex engagements resulting in value creation for all parties. In his current engagement, he is part of a strategic TDC Net/Ericsson initiative called Innovation Hub. The team focuses on new 5G opportunities that expand Cellular technology beyond Mobile Broadband. His mission is to help expand the cellular industry's value with mobile technology for humans, now onto machines, specifically into industrial, professional applications. Before taking a stunt in his home territory of Europe, he has spent most of his career in North America. He has worked with leading North American Operators, helping shape 5G and ensuring 3GPP technology's sustainable implementation into Operators' networks. Thomas's background is in Core Networks, but he always has an eye on the Radio side.

4

From 5G Technology to 6G Green Deals

Walter Konhäuser

Managing Director, Oktett64 GmbH and Associated Partner, Management Consulting Kastner GmbH & Co. KG
E-mail: walter.konhaeuser@gmx.de

We have seen tremendous societal changes and political policies to generate opportunities for new businesses and jobs addressing technological advancements, including Green Deal, Digitalization, Cyber Security, Industry 4.0, and Economy. PV-Systems on the roof or CHPs in buildings' cellars emerged as a critical Green Deal example to support the deployment of decentralized energy production. Digitalization among different verticals is foreseen as an essential aspect of future communication and business models with varying requirements. For instance, digitalization of Real-estate falls in the category of managed services demanding low-cost, low-rate, low-power, and long-range, the industrial applications require ultra-reliable low-latency communication (URLLC), whereas the entertainment services demand highspeed links. Radio technologies like Bluetooth LE, ZigBee, EnOcean, Z-Wave, and KNX-RF are used for local data generation and distribution using sensors, actuators, and meters. In contrast, local transmission gateway platforms with IoT radio technologies are used to deploy applications on the Cloud servers and enable data transmission. An efficient control, monitoring, and metering systems are essential for the digitization of electrical equipment and monitoring system state parameters in real-time. Thus, these applications require control, monitoring, and metering sensors, actuators, and meters capable of communication in distributed real-time computing platforms (gateways). Data from such sensors and meters serves as an input for many model predictive artificial intelligence (AI) algorithms to build real-time decisive/interactive future systems. This chapter proposes a gateway platform with some examples of future digital applications supported by future mobile communication systems [1].

4.1 Introduction

The evolving technologies are meant to serve humankind with ease of access and control. Figure 4.1 illustrates some of the recommended considerations to follow when one works towards developing new technologies and system concepts.

The world is moving toward a sustainable version to support a better quality of life. Loads of re-engineering are taking place in various sectors to reduce carbon emissions. Fossil fuels and nuclear power are the primary sources of generating electricity. Twenty-six percent of greenhouse gas emissions recorded worldwide are associated with electricity generation, which thus, makes it incompatible with sustainable development goals. It is essential to massive renewable energy expansion to achieve the 2-degree target to limit global warming. Varying climate changes in different parts of the world have resulted in high social and economic costs, such as rising sea levels and persistent droughts. Some disasterous changes include the events in Fukushima (2011) and Chornobyl (1986), which shows the significant and irreversible consequences of nuclear power on today's and all future generations. Also, the extraction of uranium and fossil fuels has high environmental impacts. Therefore, there is a dire need to switch to renewable energy-based electricity generation. The ecological risks and consequential costs are reduced with renewable energy expansion. Recently, we have found a drop in renewable power generation costs and a spike in the competition. The price of fossil

Figure 4.1 Considerations to support technology.

fuels and uranium has increased and is expected to increase further. Thus, renewable energy generation is emerging as a sustainable, cost-effective option in the foreseeable future.

The world markets for renewable energy technologies are booming, and a race has already broken out for the leading position in these markets. Countries that promote renewable energy early are in a favorable position [2]. The CO_2 emissions worldwide between 1960 and 2019 are illustrated in Figure 4.2, and the CO2 emissions from households are shown in Figure 4.3. The CO2 accounted from households is categorized primarily into living, traffic, and nutition. Living recorded about 36%, while the other two main categories accounted for 26,6% (traffic) and 12,3% (nutrition). Thus, we can conclude that humans contribute to about 75% of the private household's CO2 footprint. Other major contributors to the CO2 footprint are industry and energy. The industry is working hard to reduce its CO2 footprint, often in common concepts with Industry 4.0 deployment, and the energy footprint can be reduced most by using renewable energy production technologies. Reducing CO2 emissions in buildings and using buildings for local energy production must play an essential role in making our environment greener. For local energy production and energy management, a low-cost and efficient communication network inside and outside the building is necessary [1].

The evolution of mobile wireless communication systems from the second generation to the recent fifth generation has changed how people

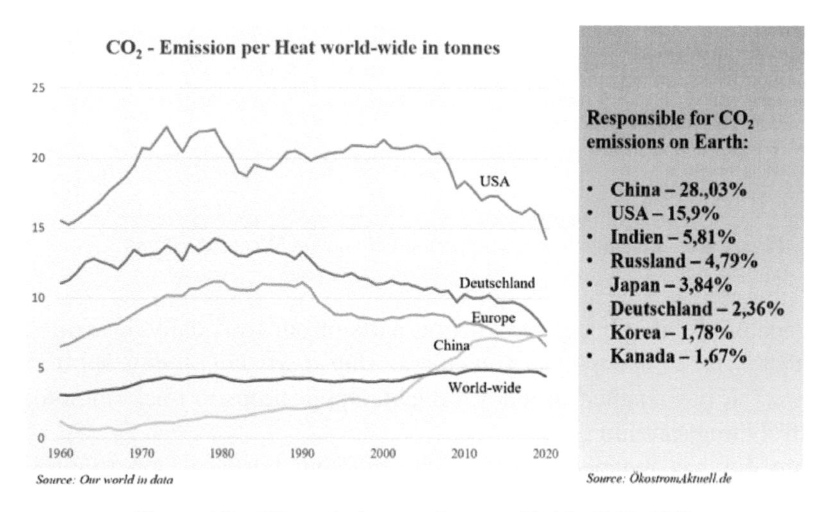

Figure 4.2 CO_2 emissions per heat worldwide 1960 -2019

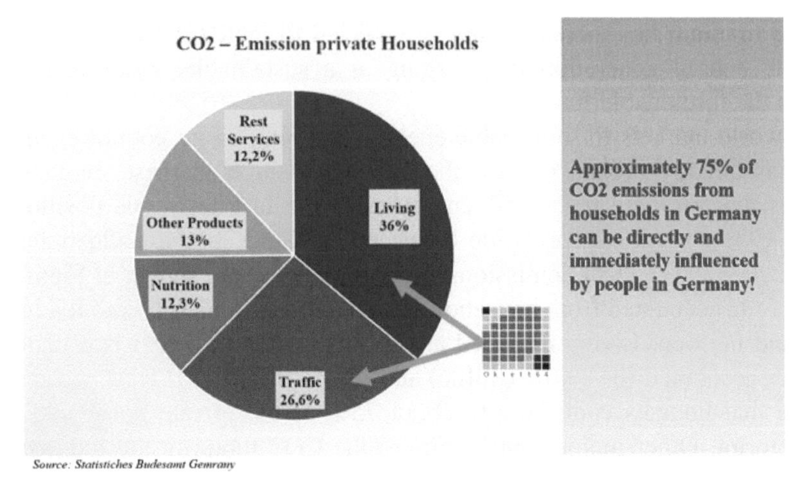

Figure 4.3 CO2 emissions per household

<table>
<tr><th>Business Opportunities</th><th>Operations & Management</th></tr>
<tr><td>•Core Infrastructure Sharing
•Infrastructure Crowdsourcing
•Enhanced Spectrum Sharing
•Guaranteed SLA Management
•Enhanced mobile Digitalization of Industry 4.0
•Vehicle to Anything Communication
•Mobile Digitalization in real estate
•Private managed Services
•Deploy of own private networks (campus network)
•Enable B2C, B2B, and B2B2X
•Enhanced tenant electricity concepts (produce, consume, sell)
•Operate as the virtual network operator
•Enhanced Business Models for contracting, operating, metering
•Cost-efficient Services</td><td>•Real-time Operation
•Enhanced Mobile Broadband
•Improved Network Scalability
•Operational Safety
•Massive Machine-Type Communication (mMTC)
•Ultra-Reliable Low-Latency Communication
•Long-range, low-rate, low-power, low-cost managed service
•Artificial Intelligence (AI)
•Blockchain Technology</td></tr>
</table>

Figure 4.4 Enabling business opportunities and O&M considerations

communicate. Mobile phones are integral parts of our lives daily. It is often quoted that "*with a smartphone, you can do almost anything today, but also make calls.*" It has resulted in many valuable applications to track, monitor and control things around us.

Figure 4.4 has mapped some of the enabling business opportunities with innovative services and new business models and the considerations to improve associated. Operations and management issues.

4.2 Applications to Reduce CO_2 in Buildings, Urban Districts, and Cities

Today, buildings in urban districts and cities produce a lot of CO_2 for a living (see Figure 4.5) and will have to be given more significant consideration in future emission targets. Therefore, using new technologies and providing appropriate communication networks are essential. Especially the installation of data networks in existing buildings is often a big hurdle. Therefore, internal and external mobile data networks are cost-effective solutions. For CO_2 reduction concepts in structures, digitalization plays an important role. New digital technologies, renewable local energy production systems, and communication networks must be deployed in real estate. In buildings, the following communication technologies should be available:

- Different mobile communication standards for data transmission,
- IoT networks (Figure 4.5) and
- Fixed networks standards (Figure 4.6).

The goal is to find out different architectures for a common Communication Platform (IoT, mobile, and fixed networks best merge) to manage applications for CO2 reduction within the building.

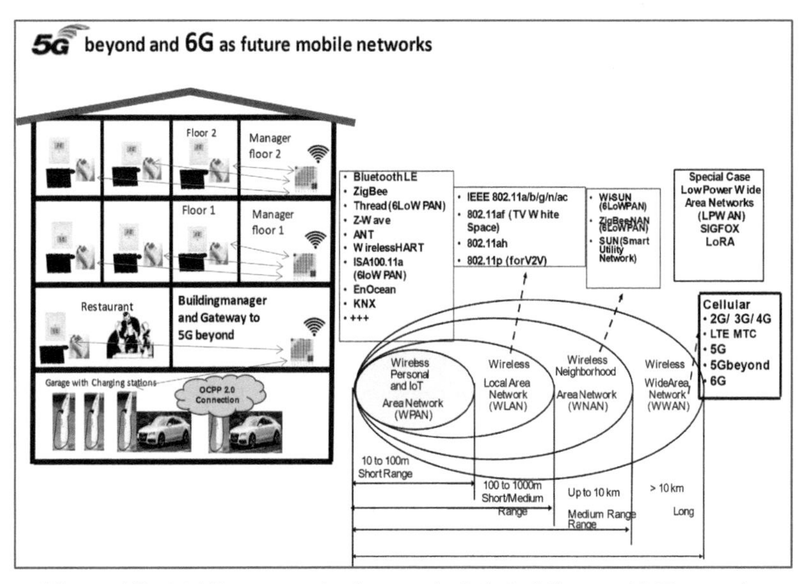

Figure 4.5 Mobile communication standards in buildings and IoT networks

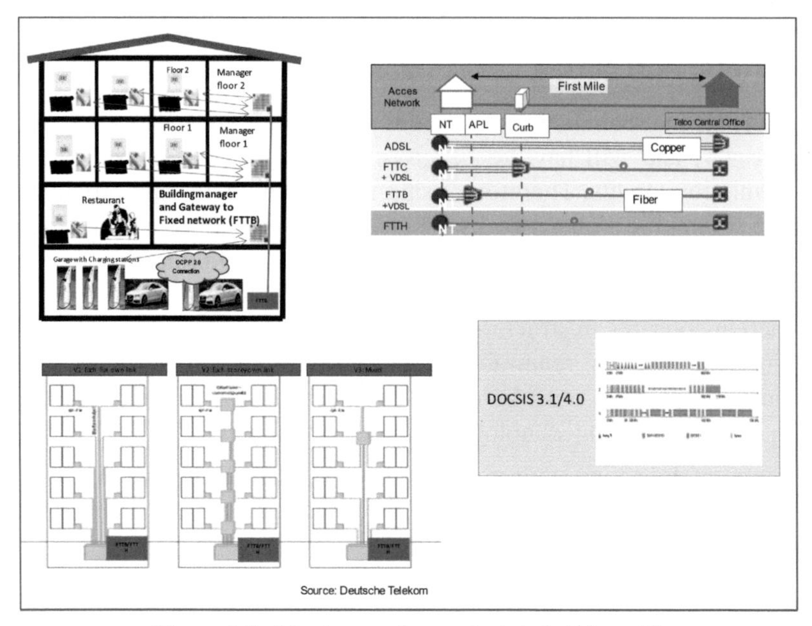

Figure 4.6 Fixed networks standards in buildings [5]

Figure 4.7 shows a combination of fixed, mobile, and IoT devices as a holistic communication network. The goal for the future and 6G should be to get all the services for the digitalization in buildings (e.g., energy management, monitoring) from one operator to a thrifty prize.

Local data generation in the Smart Home/buildings/urban districts/cities from sensors, actuators, and meters should use suitable radio technologies, e.g., Bluetooth LE, ZigBee, EnOcean, Z-Wave, KNX-RF. Data transmission to external clouds, where the applications could be deployed, should be managed by central transmission via gateways and mobile networks. Local date generation can be controlled by SW platforms [3] (Figure 4.8) with quick links to different IoT radio technologies. In the external cloud, the following services for the digitalization of buildings should be deployed based on SW solutions (Figure 4.9):

- Remote access and control of power generation and storage (Solar, CHP, batteries, etc.);
- Monitoring of technical equipment in buildings (e.g., CHP, DC/AC converter, etc.);
- Smart metering (electricity, heating, water, etc.);

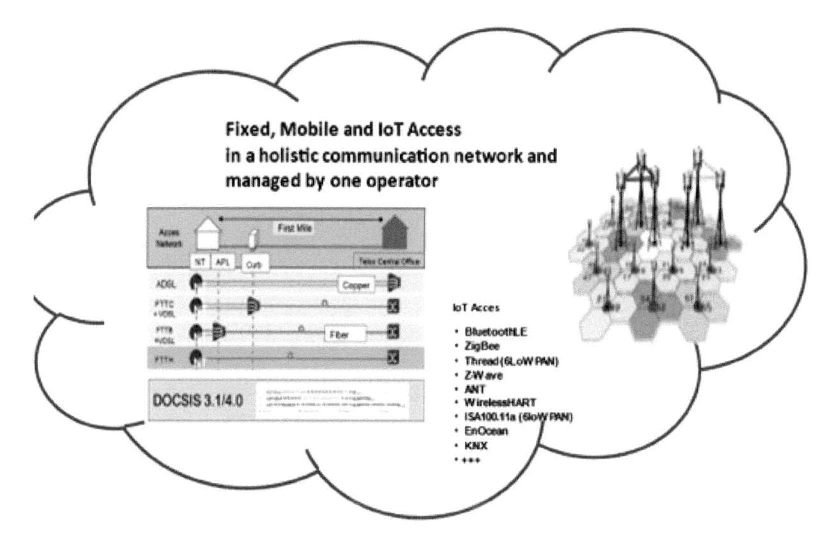

Figure 4.7 6G goal to offer the applications via external edge cloud server: holistic communication network for real estates

- Smart Home/Smart Building control with new operation concepts;
- Light control and sun protection;
- Device control: Remote control, scene control, reduction of Standby consumption;
- Maximization of self-consumption;
- HVAC control, including air quality monitoring and prevention;
- Network stabilization using buffers and control of local energy production;
- Integrated metering system: deliver data for billing the rent and extra costs;
- Smoke, fire, water hazard monitoring;
- Access Control and security;
- Mobility hub of a quarter: deliver billing data;
- Coordination mobility hub: calculate and regulate charging power for coordination of available network power, generation capacity, and user demand;
- Media control (FTTH, TV, Internet, phone): deliver billing data (if available from the operator).

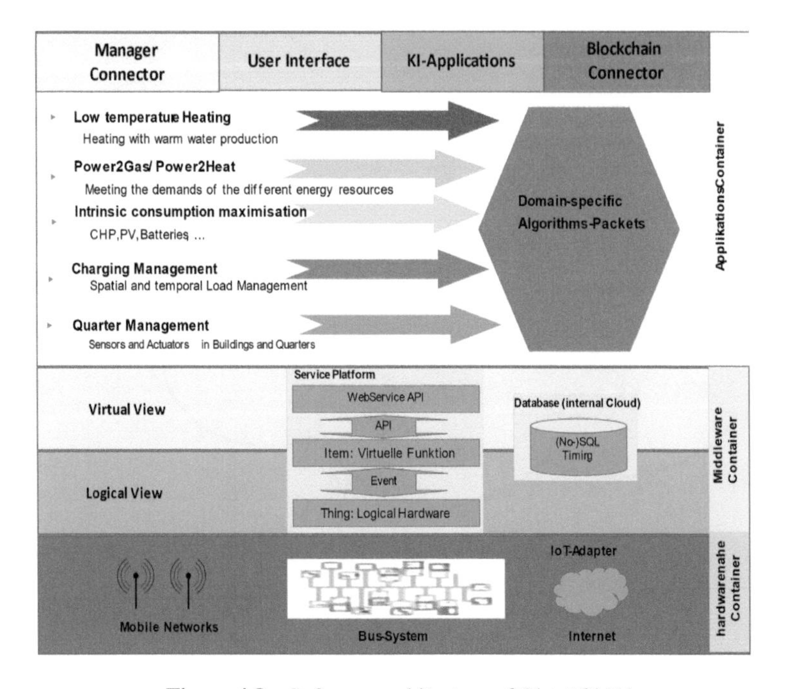

Figure 4.8 Software architecture of Oktett64 [1]

Many sensors, actuators, and meters with IoT - interfaces are available on the market. The critical point is to transfer these elements' data cost-effectively and with high security to the distributed real-time computer platforms. Communication technology will be crucial to monitor, control, and meter the various electrical equipment and system state parameters. This is critical for determining the current state of the Smart Home/Buildings/Urban districts/Cities applications and should be considered for new applications. This captured information will be input for many model predictive algorithms (e.g., AI) whose output supports decisions to achieve future goals. An appropriate control structure needs to organize a grid with an architecture based on a proper combination of central and decentralized control [4]. For predictive models and algorithms, improved computer-based models will be needed for the grid elements themselves, such as transmission and distribution lines, voltage and current transformers, flexible AC and DC elements, switches and breakers, protection equipment but also of all grid users, including generators, storage, consumer equipment, and behaviour.

Software architecture is needed to allow consumers and market players to compose new services like energy management for small units (offices, apartments, production facilities), monitoring and energy management for whole buildings and facilities, energy management, data measurement, and metering for quarters and industry parks. Moreover, supporting digital control for customer products and applications to satisfy own requirements related to energy services and products, thereby also using market interfaces and at the same time supporting the quality and security of supply of the grid-based electricity system [4]. Oktett64 as a platform (Figure 4.8) offers vendor-independent turnkey solutions based on standards and open-source frameworks, including security and operation concepts like DMZ, container, and virtualization, optimized for excellent performance at high efficiency. Solution components can be implemented without difficulty, and many different Hardware-systems can be used (including easy-to-distribute low-end hardware like SoC-Systems in the performance range of a Raspberry Pi). The Oktett64 platform is divided into a Hardware related container, a Middleware container with another division in Logical View and Virtual View, and an application container. The Hardware container connects mobile networks, bus systems, IoT adopters, and Internet access to the platform. There are also connections to other management systems, user interfaces, AI applications, and a connector to Blockchain Solutions (Figure 4.8).

An example of a holistic concept for two buildings is shown in Figure 4.10. CHP produces heating water, warm water, electricity, and PV on the roofs. Electrical energy can be stored in the battery systems. If the electrical power is insufficient, it can also be obtained from the distribution system operator (DSO). CHP is fed with gas. Apartment solutions are installed on all floors, and tenants can find out about the current energy consumption on a dashboard. In the cellar, a gateway (e.g., Oktett64) collects all data from sensors, actuators, and meters and passes this to an external cloud via the mobile network, where the data on the installed server platform (e.g., also Oktett64) processes the data and sends the results back to the building to control for example the CHP (Figure 4.9). If more electrical energy is generated in the building as it can be consumed and the battery is fully charged, then the excess energy can be sold to the distribution system operator (DSO). Electric cars can also be charged from the building with electrical power.

The buildings are also supplied by fibre optics (FTTB and FTTH) to the public communications network. This connection can also be used for data transfers to the external cloud. In the 6G standard, the Edge Cloud should also

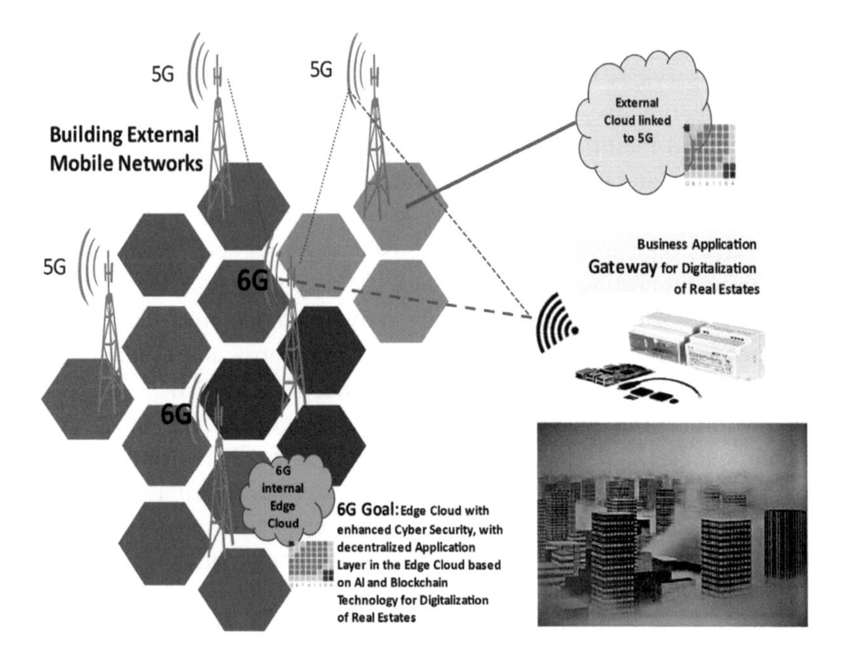

Figure 4.9 External application layer (5G in an external cloud, 6G in the edge cloud)

be extended for external applications like energy management in buildings (Figure 4.9). This has the advantage that there are short distances between execution and implementation and that several buildings can be supplied with digital applications by one Edge Cloud server implemented at one Node B. This new possibility within the access network of a mobile network creates a new business opportunity for 6G operators. This network opening via the radio interface also opens up the possibility for many other industries to outsource their applications to the edge cloud. Therefore, this opportunity should be introduced into the recommendations for the standardization of the 6th generation.

The following requirements by using external cloud solutions for Smart Home/building/urban districts/cities applications are requested:

- High network coverage in residential areas,
- High Number of subscribers,
- Long-range, low-rate, low-power, low-cost managed services,
- High transfer speed 50Mbit/s (4K compressed) for entertainment services,

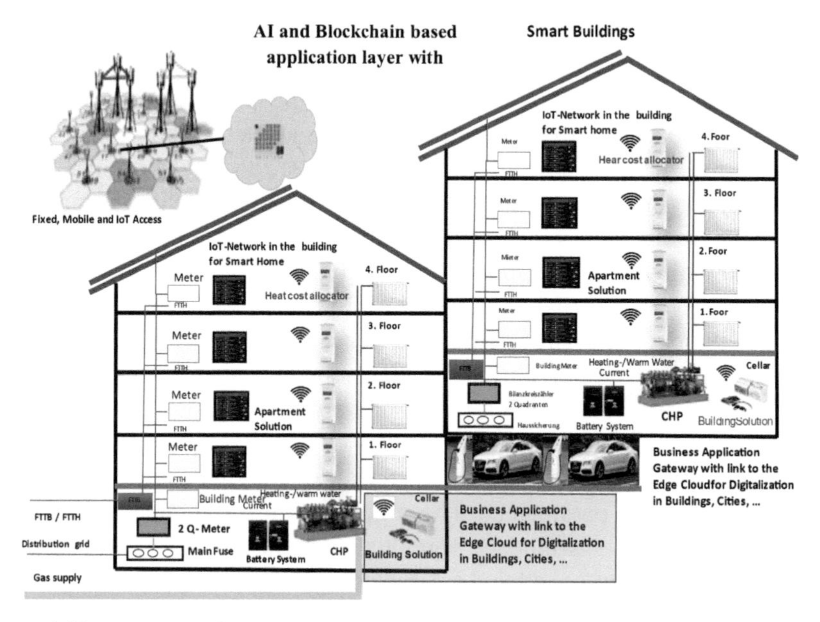

Figure 4.10 Example of a holistic solutions: energy supply for tenants with current and heating-supply in apartment houses, charging systems for EV, broadband communication supply and gateway to an external cloud solution [1]

- Fast Handover (< 2ms),
- Low latency (< 2ms) for the radio link to transmit entertainment services,
- High data security,
- Deploy of own 5G/6G private network (campus network) and
- Robust networks with less sensitivity contra interference like walls, cellars, halls, high buildings, trees, and weather conditions.

The goal is to find appropriate architectures for a common Communication Platform (IoT, mobile, and fixed networks best merge) to offer the applications via an external cloud server. This, in turn, is an excellent opportunity for the upcoming 6G networks (Figure 4.7) to offer a coherent solution that combines IoT access, mobile, and fixed communication network solutions managed by one operator. How quickly such external applications can be implemented depends crucially on the costs and security of such solutions. An intelligent urban district (quarter) can be developed to enable interoperability among devices or applications in a district environment. Massive IoT data, monitoring of all devices and meters, decentralized efficient energy generation, consumption, and distribution are the basis for optimizing the use of different energy resources. The level of complexity of this architecture is increased significantly by considering in the equation demand-side management approaches, different types of renewable energy sources, battery storage systems, electric vehicles, demand response, and dynamic pricing. The high density of metering nodes and the amount of data transmitted via low-cost endpoints can be easily translated via the business application gateway for digitalization in buildings which collects the amount of data from the IoT devices to the external Cloud server mMTC use case. Direct links to the mobile network are possible as well. A choice of services can be made at the deployed business application layer in the cloud server. Operations can be optimised by steering consumption power and production power at direct or implicit storage systems.

Using algorithms taking place, non-linear operations are steered by business- and demand parameters. The algorithms are dynamic tailorable or self-learning (prognosis-model). Remote maintenance for local energy production (CHP, PV, etc.) and local energy storage can be implemented, and metering for energy production and consumption and smart metering (current, heating/cooling, water, etc.). New operating concepts can be installed for heating/cooling control, light control, shutter control and equipment control

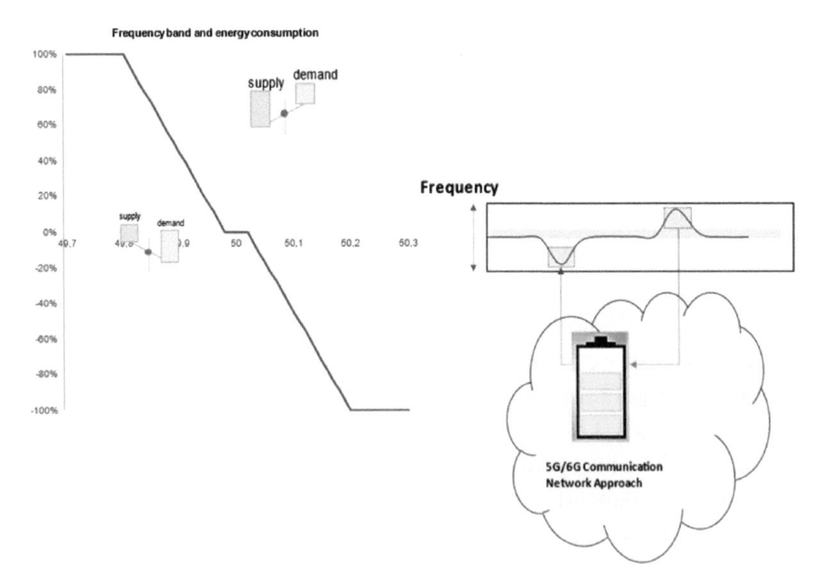

Figure 4.11 Balancing power controlled via 5G/6G

of different equipment, e.g., heating/cooling, light, shutter, indoor climate, air quality, mould avoidance, smoke and fire detection, water detection, admission control, and security. Tenant electricity for using the decentralized produced energy should be managed and integrated.

Another business opportunity for urban districts is the connection to balancing power supply (Figure 4.11) from battery systems and local energy production. The balancing power compensates as a reserve fluctuation in the power grid, more precisely, frequency. Using balancing energy, electricity can be fed into the grid and removed from the grid. The grid frequency control can be done via an energy management system like Oktett64.

4.3 Further Applications Supported by 6G

The Industry 4.0 applications offer innovative applications for 6G (Figure 4.13). The following requirements should be met:

- High network coverage in industry areas,
- A high number of subscribers: 10 subscribers per m2,
- High transfer speed 50Mbit/s (4K compressed),
- Fast Handover (<2ms),
- Low latency (<2ms) for the radio link,
- High data security,
- Deployment of own mobile private network (campus network) and
- Robust network with less sensitivity contra interference like electrical machines, walls, walls, halls, high buildings, and trees

Local data generation and distribution processes in industry areas from sensors with a suitable radio technology such as, e.g., Mesh, LoRa, SigFox and existing sensors like field busses, such as, e.g., OPC- UA, profinet/Profibus, Modbus EtherCAT is required. Data transmission to the 6G Edge cloud, as already suggested for the real estate industry where the applications are deployed, should be managed by central transmission due to costs from many local data via gateways or optional direct transmission from the machines via radio links. Applications should be deployed in the Edge Cloud-based on AI Technology to visualize data, condition monitoring, and process analysis about the whole production process automation, e.g., minimize downtime, increase efficiency, etc. If blockchain connectivity is available and Blockchain technology is used, then applications such as service data (run time, maintenance) and traceability and reproducibility (e.g., individual production, used machines) can be ordered.

Peak Shaving is another business opportunity for the industry with fluctuating power recalls. Any power consumer can generally use it. However, based on economic considerations, it is of particular interest to electricity consumers a recording power measurement, which records the power requirements every quarter of an hour. This can be done with digital electricity meters. In many power grids, network operators can apply applicable grid charges to the electricity consumption of the respective network area. For this purpose, network charges are levied in performance-based prices for electricity purchases. This performance-based component of the electricity bill refers to the maximum power call of a quarter of an hour interval of a period. With the help of Peak Shaving, this maximum power call can be reduced utilizing load shifting, load shedding, or via energy storage systems controlled by an energy management system. Depending on the service price, this can result in lucrative business models. Especially in the energy transition context and the associated increasing investments in the public electricity grid, rising performance prices and thus a rising incentive for peak shaving are expected (Figure 4.12).

Another innovative field of application for 6G is the control and monitoring of drones (also Unmanned Aerial Vehicles, UAVs). Therefore, drones are undergoing a boom as flying measurement systems with various sensors. Industries use well-tried approaches and tools from automotive engineering

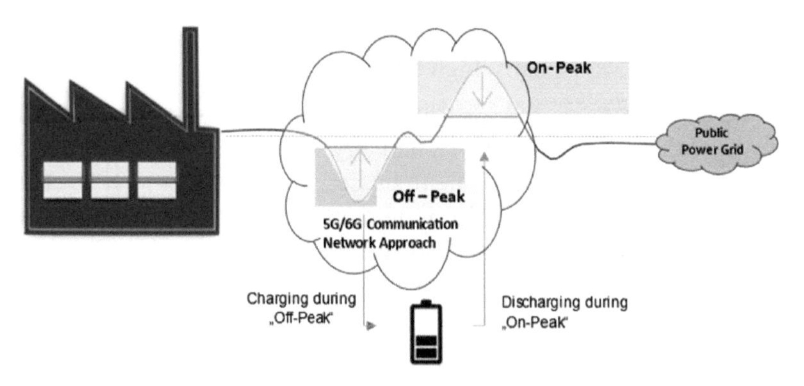

- **Costs for Power Peaks: 30 - 60% of the total electrical energy costs**
- **Reduction of maximum Energy Level: Reduction of energy costs without reduction of energy supply 20% - 50%**

Figure 4.12 Business opportunity peak-shaving

to open new scenarios for using remote-controlled aerial vehicles. Experts focus on controlling the UAVs by Wi-Fi, LTE, 5G, and 6G on evaluating sensor data and safety while in flight. This combination of readily available drone hardware and proven tools gives a system that is easy to use for a wide range of different activities [6]:

- Inspecting windmills and solar panels,
- Surveillance of parking areas,
- Analyzing the energy efficiency of buildings,
- Testing the performance of antennas,
- Measurement activities in automotive development,
- Supporting search and rescue activities,
- Supporting firefighters with thermography analytics,
- Urban planning, Cartography,
- Intra-logistics for industry, Package delivery,
- Event and film shooting,
- Agriculture.

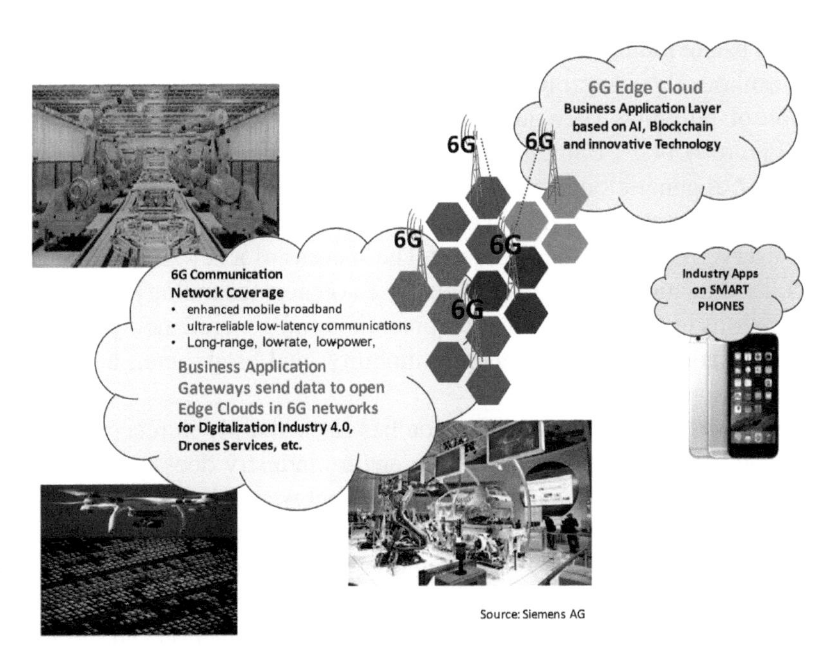

Figure 4.13 Further applications for 6 G: industry 4.0 and drones

4.4 Conclusions

Renewable energy is the key to a sustainable electricity supply and contributes to climate protection and a sustainable energy supply. Delays in the transformation of the energy system threaten economic and social development opportunities worldwide. Progressive climate change threatens the stability of global ecosystems and causes high economic and social costs. For example, the OECD estimates the cost of unchecked warming at 14 percent of global consumption by 2050, with the trend continuing to rise in the second half of the 21st century. The main criteria for a sustainable power supply are environmental and health compatibility, risk poverty and fault tolerance, resource conservation, and comprehensive economic efficiency considering external costs. Electricity generation costs of renewable energies in Germany are expected to be around 7.6 ct/kWh. Electricity from new natural gas and coal-fired power plants exceeds nine ct/kWh. This does not consider the costs of system integration. In many areas in Africa, India, Southeast Asia, and parts of the Middle East, the operation of photovoltaic systems is already economically comparable to a supply of diesel generators [2].

The real estate industry must also contribute to a sustainable CO_2 reduction to greenhouse gas neutrality in 2050. This will only be possible if the digitization of buildings (new and existing buildings) can be implemented quickly (Figure 4.14). Communication networks play a significant role in this, and the 6th generation of mobile networks can still offer many features contributing to climate neutrality. The 5G standard is already taking part in changing everyday life and the economy. The success of a new generation of mobile communication systems is mainly based on new technology, enabling new applications to address new market potentials and generate new jobs. 6G is expected to address industry, energy, mobility, real estate, etc., and will develop many applications.

Decentralized green energy production has risen sharply in recent years, primarily due to PV plants. Currently, the energy industry does not provide instruments for the direct exchange of electrical energy and the corresponding billing between prosumers, producers, and final consumers at the regional level. It is also not possible to label regionally generated electricity from local micro-grids. Future energy trading should aim to deploy within micro-grids a peer-to-peer energy market for local renewable production and storage units based on a blockchain that simultaneously clearly identifies the traded electricity.

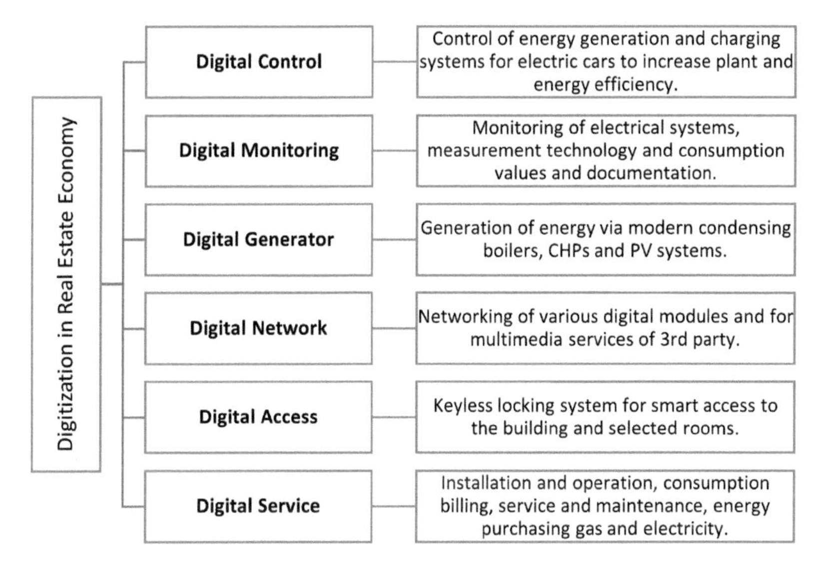

Figure 4.14 Examples for digitalization in the real estate economy [1][7]

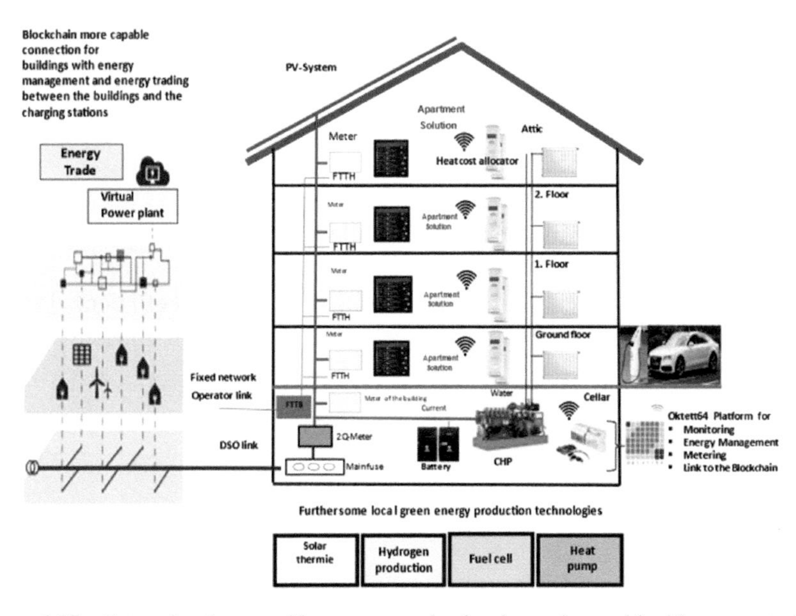

Figure 4.15 Future local renewable energy production in a micro-grid with energy trading by Blockchain technology.

The paper describes the import of new applications and a communication platform to generate new applications for the real estate industry and further applications, e.g., industry 4.0 (5.0) applications and drone applications, which become essential in the future, based on possible business association with future 6G networks. Limiting global warming to a maximum of 2 degrees above pre-industrial times can only be achieved if emerging and developing countries also participate in international efforts to reduce greenhouse gases and fall well short of the expected emissions trend. Therefore, 6G will be deployed in emerging and developing countries too.

References

[1] Severin Beucker, Walter Konhäuser, Ingo Schuck, Olaf Ziemann; Climate-neutral districts with decentralized energy production, E-Mobility and through the formation of an energy community exchange of electricity and heat; https://www.springer.com/series/15427.

[2] Andreas Burger, Benjamin Lünenbürger, Dirk Osiek. Nachhaltige Stromversorgung der Zukunft, Bundesumweltamt Pressestelle Wörlitzer Platz 1 06844 Dessau-Roßlau.

[3] Walter Konhäuser; The Next Mobile Communication Steps into New Application Areas; 5G Outlook-Innovations and applications; River Publishers; ISBN 978-87-93379-77-0.

[4] Decentralized Transactive Energy; White paper WEC (World Energy Consortium).

[5] Deutsche Telekom: Glasfaser-Inhouse-Verlegung.

[6] IAV; support from the air; remote controlled drones as flying measurement systems.

[7] Thomas Kastner, Walter Konhäuser, Ingo Schuck, Energieeffizienz dank effizienter Prozesse; Die Wohnungswirtschaft, Ausgabe 05/2021.

Biographies

Prof. Dr.-Ing. Habil. Walter Konhäuser is a CTO with strong strategy and technology competence, international business responsibility, and leadership experience. Walter Konhäuser, born on November 19, 1949, in Ruhpolding, is a promoted electrical engineer working as a professor at the Technical University of Berlin in mobile communication systems. He studied electrical engineering at the Technical University of Berlin; His doctorate and habilitation were in automation technology and networking.

For more than 25 years, he has been active in various business areas for SIEMENS AG, many years as CTO in mobile communication systems, and responsible for broadband data access. He has 40 years of professional and management experience in R&D and Marketing and Management with P&L responsibility. He has contributed to the Research and teaching in mobile communication networks, such as 5G at Technical University Berlin. His primary interests are Innovation Technology Development for energy areas (Batteries, Smart Grid, and Energy Management), Business development in renewable energies, Process Development for R&D, Marketing, Product Management, and Sales. He is an Advisory board member for a battery development project at Fraunhofer Institute IKTS and a former advisory board member at Fraunhofer Institute Fokus. He was the former head of the advisory board of a start-up company (Garderos). He participated in VDE Association in Germany.

5

Enhanced Massive Machine Type Communications for 6G Era

Razvan Craciunescu, Simona Halunga and Octavian Fratu

Telecommunication Department, University Politehnica of Bucharest, Romania
E-mail: razvan.craciunescu@upb.ro; simona.halunga@upb.ro; octavian.fratu@upb.ro

Massive machine type communications (mMTC) is one of the three pillars of 5G communication alongside enhanced mobile broadband (eMBB) and ultra-reliable low latency communications (URLLC). mMTC aims to optimize the resources in a 5G network to achieve high connectivity density, 1 million connections per square kilometre. 5G is only the beginning for mMTC use-cases. With the clear benefits of this 5G pillar, many challenges also arise, such as maintaining many devices and how long it will take to change the batter for all of them. In this chapter, we present some of the challenges of mMTC for the 5G communications and associated research activities. We provide a platform to advance knowledge beyond 5G or 6G communications.

5.1 Introduction

6G communications is a technology that is forecasted for 2030. Several drivers will influence the technology's development, like sustainability and zero energy devices, micro-networks for cities, enhanced future factories, autonomous driving, data as a service, enhanced experiences between different realities, Augmented/Virtual/mixed realities.

These drivers, mainly consumer drivers, will shape how enhanced mMTC (emMTC) will be developed. In this chapter, we are going to analyze some of the trends that are presented in the literature, such as

- Massive Connectivity – physical layer enhancement for massive connectivity;
- Zero Energy Internet of Things (IoT) – reduce the power consumption of sensors so that the lifespan will be around 40 years;
- Predictive resource allocation – traffic correlation based on neighbours' nodes transmissions;
- Low-Cost Authentication and Authorization – for example, group-based authentication;
- Swarm networking – a group of devices will be resolving a specific task;
- Personalized;
- Other specific industry-vertical specific trends such as massive twinning or automotive mobility;
- Satellite communication for IoT.

As one can observe, new business cases are being created alongside new applications and functions with each new generation of mobile technology. 6G network aims to be a ubiquitous network that will integrate both terrestrial and satellite and integrate more technologies for sensing, localization, holographic communications, human-machine interface applications, or multi-sensory experiences. Also, robots will be more present in our day-to-day life. In this context, emMTC will be even more present as an integrated part of the new ecosystem.

5.2 5G Systems

mMTC is one of the three use-cases that can co-exist in a 5G network alongside eMBB and URLLC. The main characteristics of these three use-cases are [1]:

- For eMBB – Bandwidth is the main parameter; hence the applications under this use-case need higher bandwidths;
- For URLLC – Latency is the main parameter; hence the applications under this use-case need low latency (1 ms) and high reliability;
- For mMTC – The number of connections is the main parameter; hence the applications under this use-case need a large number of connections, 1million connections per square kilometre, and this use-case is defined for IoT devices

Figure 5.1 explains the relation between the three use-cases and the applications under these use-cases.

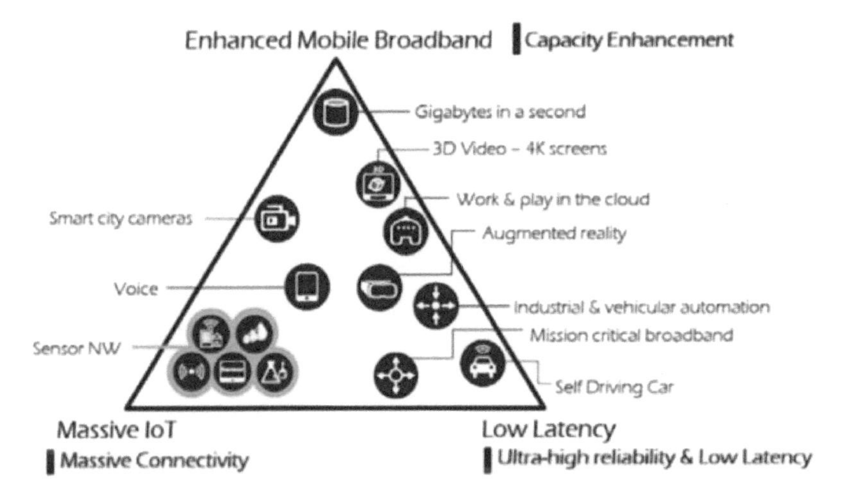

Figure 5.1 5G use-cases and applications (ETRI graphic, from ITU-R IMT 2020 requirements.

A concept that needs to be mentioned is the Open-Radio Access Network (Open-RAN) [2]. Open-RAN brings interoperability, security, and network slicing critical enablers in the 5G ecosystem. In terms of security, several innovations are brought, design frameworks and wireless technology combined to create a highly secure and resilient 5G network. Because the system becomes flexible and easy to split into separated blocks, every device or piece of software can be verified easily, and it can accept or deny it if it is malicious. Furthermore, network slicing can be made with Open-RAN and has a significant impact on the system architecture because it separates the network into smaller end-to-end pieces that can be managed individually. We are not going to detail the Open-RAN concept, but this one will be of interest for 6G networks in 2030. It is worth mentioning.

5.2.1 Network Slicing

For the 5G network to accommodate all the scenarios from Figure 5.1, the concept of network slicing is used.

Network slicing is one of the critical enablers and an architectural answer to the future communication system. Mobile operators provide all types of services in a single network, but with the implementation of network slicing, they will divide the entire network into different slices, each slice having its

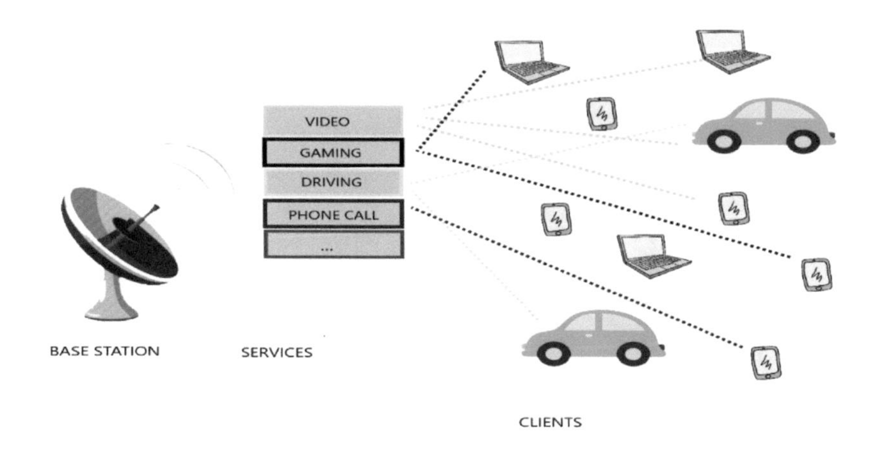

Figure 5.2 Network slicing made simple.

specifications and requirements in terms of QoS [3]. Thus, shortly, network slicing will be able to divide the physical network into multiple logical networks. This definition is, of course, a plastic one because the definition of this concept is still under heavy discussion.

By considering this concept's short definition, we can say that the *slice* is a set of programmable resources, isolated by the other slices, able to implement network functions, and even act like a complete end-to-end network, delivering some specific services for its customers. The delivery of the services requested by the users is done mainly in terms of the QoS of the slice, with every service having a predefined QoS class with an allocated slice, as one can observe from Figure 5.2.

Network slicing architecture works on a partially shared infrastructure as software defined networks (SDN) and network function virtualization (NFV) are embedded into 5G systems. The infrastructure contains two types of hardware: the dedicated hardware for the radio access network (RAN) and generic shared hardware for the NFV infrastructure resources. The network functions that work on shared hardware are customized based on the requirements of each slice and cannot be applied in the case where dedicated hardware is used.

In terms of deployment scenarios, two types of network slicing are present in the communication networks [3]:

- Slicing for QoS: creating slices when the quality of services parameter is considered regarding specific applications for end-users. This type of slicing is based on sharing services with a greater priority for the users, such as video streaming, cloud gaming, medical emergencies, etc.

- Slicing for infrastructure sharing: the ability to share the entire infrastructure, including the RAN level. For example, if an operator who owns a slice gives access and shares its infrastructure among its users, the slice could be easier to modify and adapt since the tenants can change its structure.

As NGMN Alliance describes, the network slicing concept consists of three functional layers:

- Service Instance Layer: represents services demanded by business companies or end-users. The operator can provide the services, or a service instance represents a third party and every service.
- Network Slice Instance Layer: provides network characteristics required by the Service Instance Layer. To create a Network Slice Instance, the operator uses a Network Slice Blueprint, a list of physical and logical requirements, and a complete description of the structure, configuration, and workflows to instantiate and control the Network Slice Instance. This layer may also be composed of none or multiple Sub-Network Instances created based on Sub-Network Slice Blueprints.
- Resource Layer: refers to the physical and virtual network functions used to implement a slice instance. The resource partitioning is based on NFV Orchestrator and application resource configuration at this layer.

To these three layers, the network management and orchestration (NMO) function is added to provide management and orchestration over the functions of the three layers. NMO should be able to manage every slice individually [3].

There are, of course, a lot of proposed network slicing architectures and models, see Figures 5.3 and 5.4, each having a different target for the delivered services and different requirements in terms of application, but in this work, we will discuss the generic architecture, the layout which is present at the core of the network slicing.

This type of architecture could be handled with the help of NFVs and SDNs by adding software-defined network functions and then virtualizing them to work at every point of the network, as in the following image.

At the resource layer, the physical and already virtual resources are passed through the virtualization layer to make them useful for virtual network functions. Next, the VNFs are created and contain the network functions that need to be used. Together with the SDN controller, they form the management

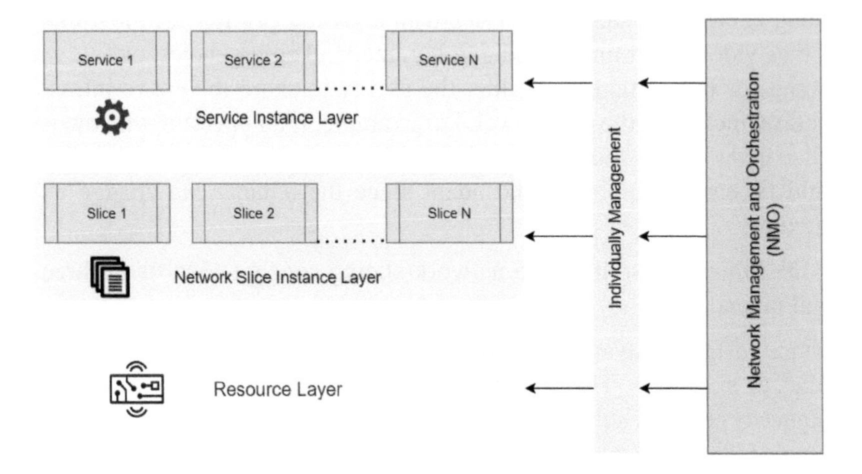

Figure 5.3 General network slicing architecture [3].

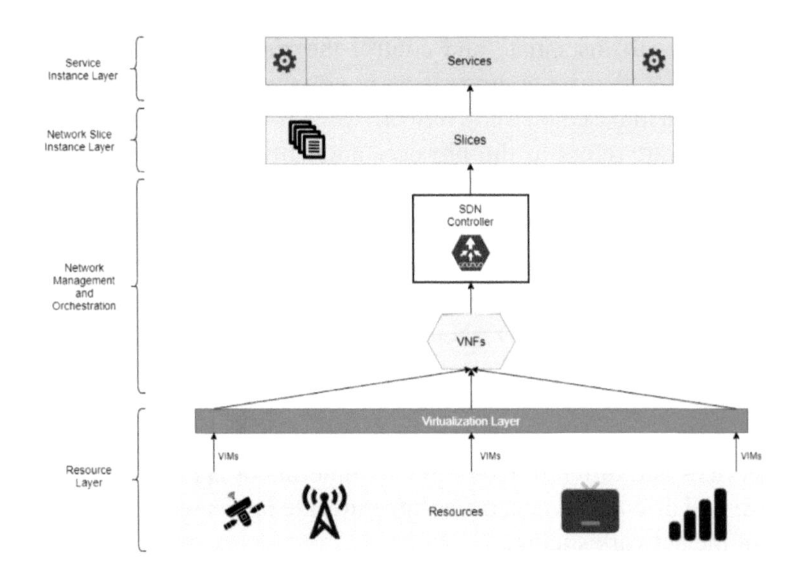

Figure 5.4 SDN and NFV network slicing architecture [3].

and orchestration of the network. The SDN controller manages the resources and data flow from the source to the correct destination. After a virtualized network function has a strongly defined purpose in the network, the slice is created at the Network Slice Instance Layer, then transmitted at the Service

Instance Layer where the user or company gets their requested service or application [3].

5.2.2 Massive Machine-type Communications

The air interface of 5G, the 5G new radio (NR), is defined in 3GPP Release 15 to optimize the resources for high-density connections and to have scalable and configurable numerology to support the ultra-low latency transmission. Furthermore, the next 3GPP releases added more features for the IoT ecosystem inside 5G systems, such as device-to-device communication, integrating narrowband IoT (NB- IoT), and Long-Term Evolution MTC (LTE-M) with 5G NR [4].

The mMTC characteristics can be observed in Figure 5.5 spider. The devices used under this use case are described by low cost, increased coverage, and enhanced battery life. mMTC needs to support a high density of devices in urban, extra-urban, and rural areas, where many sensors, meters, cameras, etc., are installed. These devices have the purpose of managing the lightning infrastructure in cities and buildings, measuring noise, pollution, and temperature levels, and controlling traffic, among many other applications. In addition, IoT devices have a small price and do not require great data rates or low latency since they stay idle for a long time, occasionally being activated [4].

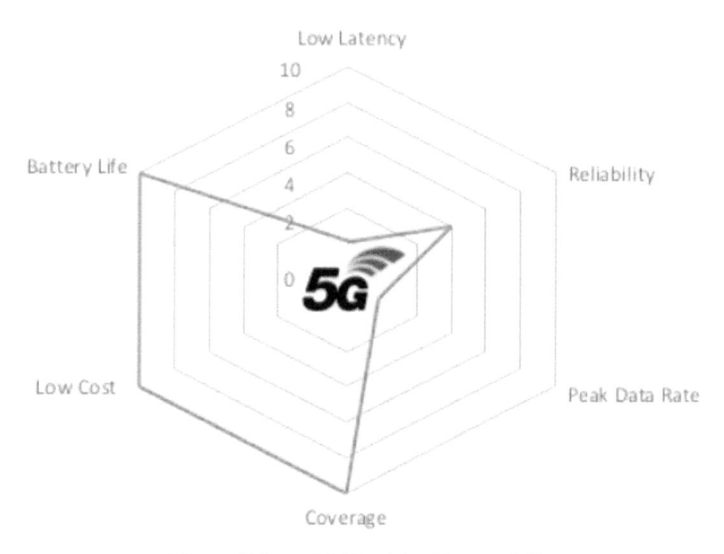

Figure 5.5 mMTC spider diagram [4].

5.3 Challenges for Massive Machine Type Communications in 5G Systems

Several challenges are previsioned for 5G mMTC that are tried to be addressed by the different releases. But some of them will be subject to further investigation and are assumed to enter the 2030 6G standard. For example, the characteristics of 5G mMTC are [5]:

- The transmitted packages are small, typically a few bytes;
- There is a considerable number of devices per cell;
- The uplink transmission is the dominant one;
- The user data rates should be around 10kb/s;
- The devices send data infrequently;
- The devices complexity and cost are low;
- Battery life should be optimal.

Besides the mMTC characteristics, there are also scenarios demanded by the market that add to the challenges. For example, mMTC can be a solution for a widespread of devices in a city to make it a smart city, but there is a need for other technologies such as Mobile Edge computing to offload the network backhaul processing part and to localize it as close as possible to the end-devices.

3GPP in [6] divided the mMTC into three classes of scenarios as follows;

A. Periodic and planned data delivery – suitable for sensors that wake up periodically to send data (e.g., water meters) but also periodic software updates are scheduled for that sensor
B. Initially, unplanned data delivery – devices can wake up when a large amount of data to be transmitted is accumulated. The software updates are sent only to the group of active devices
C. Initially, unplanned data delivery for critical data – if there is a bug update or a critical software update, the devices should be informed as soon as possible about an essential data delivery

All these characteristics and scenarios are creating challenges that are presented in the shape of research questions below:

- Should the access medium be Orthogonal or non-orthogonal?
- How do we decrease the control signalling and other mechanisms such as channel estimation or link adaptation complexity?
- How to increase the granularity as the amount of data is small?
- How to handle many devices that will access the network from time to time?
- How to increase device coverage?

- How to create devices that are self-powered from the environment?
- How to do fast authentication and authorization for a large number of devices?

Also, by combining mMTC with URLLC, we get the ultra-reliable mMTC use-cases where the 1 million per square kilometre connections are also ultra-reliable. These are some of the use-cases of the future releases of 3GPP.

All the presented challenges are being addressed as new 3GPP 5G releases are being rolled out and as research groups around the world are developing the beyond 5G systems as the 6G quest has already begun. In addition, the mMTC will be enhanced for new and more powerful use cases. Hence for 6G, a new enhanced mMTC (emMTC) use case will be developed.

5.4 6G Enablers and Trends for Enhanced Massive Machine Type Communications

There are several enablers for emMTC. These are technical enablers and socio-economic ones.

Hence according to a new 6G flagship project under the H2020 framework, the Hexa-X project [7], the new connectivity age should be in line with the Sustainable development Goal identified by the United Nations. These goals are illustrated in Figure 5.6.

From all these goals, the climate change issues will be embedded into the creation new technologies. Hence, new technologies, like 6G, should increase technological advancement, like data rates, but reduce the energy usage of networks to a zero-power consumption world.

Also, the 6G technology should be a technology for an open society, with a build-in trust and privacy, and security should be an essential requirement for creating such networks. Furthermore, artificial intelligence should also be implemented in these networks with ethical aspects in mind, reducing inequalities and increasing economic growth as evenly as possible.

Other enablers of the 6G networks, as identified by [5][7], are the regulatory trends. These trends will significantly impact how the technology will be developed. We can mention here the following:

- Spectrum and operational aspects – as the future networks will be created with spectrum sharing flexibility and with a dense radio access network (RAN) devices from macrocells to small cells, the regulatory framework should be adapted.

Figure 5.6 UN 17 sustainable development goal.

- Ethical aspects such as privacy as the future networks will increase the number of sensors and the amount of data collected about cities/cars/infrastructure and humans.
- Environmental aspects – regulation regarding the carbon footprint decrease will impact the deployment of future networks and the vast number of sensors.

Besides the socio-economic trends, there are also technological trends that will enhance the emMTC for future networks. As presented by [5], these trends are fueled by some drivers like:

- Autonomous mobility - mMTC will be used to make anything that moves autonomously; as the number of sensors increases, the edge processing will increase
- Connected living – having everything connected to everything means a lot of sensors and connections that will transform, for example, cities in connected cities.
- Factories of the future – MTC plays an essential role in creating the factories of the future, and in 6G, this will be enhanced by having personalized products for factories combining sensing with actuation and haptic scenarios as well as implementing the factories' digital twin on a larger scale

- Digital reality as frontier technology – emMTC will help enhance the augmented/virtual and mixed reality as the user experience will be complete by adding more sensors/actuators or haptic feedback.
- Zero world – emMTC will have to deliver high performances with a low energy consumption alongside zero latency and zero error for critical scenarios
- Data as the new oil- the data generated by many deployed sensors will have an enormous business value if combined with analytics tools and machine learning/AI algorithms.

All these trends are creating use-cases for emMTC for a 6G era. In [5, 7, 8], the authors propose several use-cases for emMTC. For example:

- Swarm networking use-case – autonomous vehicles, either terrestrial or aerial, can connect using sensors to perform a specific task or with the sensors from the infrastructure to navigate through different environments. These use-cases will be used in emergencies or logistics/supply chains.
- Personalized factories – as the number of sensors/actuators in factories is increasing, these will enable an agile approach to optimize factories by analyzing the massive amount of data that the sensors are producing.
- Massive twining - with the deployment of emMTC, we can have a digital representation of a part of the environment (e.g., a factory) and our entire environment (e.g., entertainment, social interactions, digital health).
- Merged cyber-physical world – adding to mixed reality and holographic telepresence, a large number of sensors that will also transmit senses will merge the cyber world with the physical one and, for example, one can have an immersive sports event experience.
- Micro-networks for smart-city – in a massive sensor deployment scenario, the network management should be divided into small networks based on different aspects such as public vs private networks, data rates requirements, and so on, and in these use-cases, the extensive overall networks should be divided into smaller one, having in mind also the benefits of network slicing.

All these use-cases and trends are driving research for enhancing the technology advancement so that some of them will be feasible to be tested and validated. And in the next chapter, we will present some of these technological enhancements.

5.5 6G Technologies for Enhanced Massive Machine Type Communications

According to [9], the authors divided the enhancement into six areas: Efficient massive connectivity; Multi-access edge computing as an enabler; Security for MTC; vertical-specific solutions; Powered by AI and machine learning (ML); Energy Efficiency.

5.5.1 Efficient Massive Connectivity

For example, predictive resource allocation and scheduling techniques should be developed to enable efficient massive connectivity. Hence, in a factory where we have a production line with multiple robots connected to the 6G network, to process the information from each robot, there should be a schedule so that the time difference between robots on the production line is reflected in the network. Also, a Machine Learning algorithm should be used to create a predictive resource allocation based on the environment. Non-orthogonal multiple access (NOMA) will also be a key enabler as the research will begin to focus on improving transmitter and receiver side processing. [10, 11]. Besides the multiple access schemes, coding schemes for MTC should be revised as we deal with short packages sent from time to time. Some research objectives here should be the construction of error detection capabilities without the overhead of CRC code [12]. Moreover, the authors in [13] present the advantages of list decoding for short packages with unknown channel state information (CSI). The proposed principle applies to polar codes and short binary linear block codes and uses short pilot fields only to derive an initial channel estimate.

As the number of devices increases, new MAC layer approaches should be developed. For example, the authors in [14] propose a novel unsourced (every user employs the same codebook) random-access communication. However, this new method is used because the pilot signalling user identification techniques are not efficient in an emMTC scenario, and a grant-less scenario is preferable, and all the users should use the same codebook.

5.5.2 Multi-access Edge Computing as an Enabler

Multi-access edge computing (MEC) can offload tasks performed by MTC devices and ensure low latencies due to the proximity of the computing facilities to the MTC point of attachment. Having these characteristics, MEC will help the MTC localize the data analysis and deliver it faster so that for critical MTC scenarios, the information latency is as low as possible.

Also another critical characteristic of MEC in the context of mMTC is that analyzing data as close as possible to the source will distribute the load of filtering unnecessary data from the user data. Thus, the data that reaches the cloud is already filtered and preprocessed by MEC. MEC also introduces a security layer, as the data is processed near the attach point and only preprocessed data is sent to be stored in the cloud and adding energy efficiency to the 6G systems [15]

5.5.3 Security for emMTC

The number of devices will increase even more in a 6G scenario. There are device constraints due to the low complexity, low memory, memory storage, and low data rates in an MTC scenario. Hence, one must develop efficient authentication/authorization (AA) and security systems for emMTC [16].

Security loopholes will be exploited as the systems evolve and developers patch the most obvious ones. For emMTC, the AA process for millions of connected devices will evolve to signalling congestions. Therefore, 6G group-based AA will be deployed, and lightweight physical authentication or integration of authentication in the access protocol will be researched [17].

Subscriber identity modules (SIM) are not scalable, and other user identification mechanisms are being developed.

As mentioned before, MEC will add a supplementary security layer as the data will be processed near the attachment point as possible. These will be translated into local trust zones for both humans and MTC devices,

We already presented network slicing in 1.1.1. Network slicing can also enable security mechanisms so that isolation between different slicing is enabled and different AA protocols can be implemented per slice and per MEC node.

5.5.4 Vertical-specific Solutions

5G was different from other G technologies because the specifications were vertical-specific compared to user-specific. Verticals, as expected, have different needs and different requirements, from the Automotive industry to the entertainment industry or public protection. Therefore, they will require different levels of support in taking their applications and services from concepts to prototypes; and, finally, products. However, it also becomes more and more frequent that services in different verticals need to exchange data. And these data exchanges will be more predominant in a 6G emMTC scenario. This data sharing will enable compatibility in a multi-operator system, including service provisioning, accounting, and authorization across different systems with different stakeholders. These requirements will be even more critical in a critical MTC scenario where the application will require ultra-reliability and low latency [9].

Also, connectivity solutions are personalized based on vertical-specific requirements as the Software-defined networks (SDN), and network function virtualization (NFV) will enable the same core network for different verticals.

5.5.5 Powered by AI and Machine Learning

The 6G systems will start building AI as a Service (AIaaS) [7]. MTC devices will consume these services in different scenarios, such as finding your friends from a high-resolution stadium image. The AI algorithms will be located in the network and be called to process the image and deliver the results. This will also be the case for MTC devices that are low-power, low-complexity but that will need machine learning techniques to process the collected data.

Besides the specific verticals of AIaaS, other machine learning techniques will be used for the system itself, such as decluttering networks by accessing adjacent cells in advance or increasing the QoS for a certain slice in advance. Hence, the emMTC will be dynamic and self-organizing networks, learn from usage patterns and security concerns, and use AIaaS for dynamic slicing and resource allocation [18, 19].

AIaaS can also be used to optimize the energy consumption and to make the MTC devices grouping for AA in an efficient manner.

5.5.6 Energy Efficiency

Energy efficiency is one of the hot topics in every regulatory, economic, or technological meeting, as the primary goal of the countries is to reduce the carbon footprint as much as possible. Of course, MTC devices, by nature, are small, low-complexity, low energy devices, but when you look at a million of them deployed in a square kilometre, the carbon footprint will be significant. Therefore, a zero-energy consumption policy will be developed for 6G. Also, new architecture for efficient energy devices are foreseen as one of the critical aspects.

One of the main parts of zero-energy is how energy harvesting can be done from different sources from the received signal designated to the MTC device from natural sources like solar, vibration, human-powered (e.g., moving), and others.

Another technique for zero-energy devices is backscatter communications, which were usually used in radio frequency identification (RFID) tags to report an ID to an inquiry. For emMTC, this can be enhanced so that the sensors can communicate using ambient signals (from Base Stations), extracting the necessary power from them, and using this power to communicate with other MTC devices in their proximity. These types of devices can be called the Internet of Tags. The tags will be present everywhere and help collect information, track, monitor or act on environmental stimuli [20].

Also, as in 5G, different techniques for sleep and wake-up signals were deployed as power saving mode (PSM) or wake up signal (WUS). These efforts will be further enhanced in 6G as in the sleep mode, the devices will consume even less power, and the WUS signals will be optimized for different use-cases and verticals.

5.6 Business Impact of the 6G Technologies for emMTC

Section 5.5 discussed some of the technological enhancements for MTC use-cases. In this chapter, we are going to try to capture the business impact.

There are several business impact areas that we can discuss. One of them is the economic growth, others are the number of new jobs created by adopting the tech, and the CO2 avoided emissions. For example, by 2030, 5G/5G+ is estimated to generate a +8 trillion USD for the global GDP, according to a forecast by Nokia [21].

Also, in terms of job creation, 5G/5G+ will contribute to more than 1 million jobs and reduce CO2 levels by 33 million Tones [22].

In Romania, for example, the 5G+ impact in 2030 will be as follows [22]:

- 20 billion in 5G enabled sales
- 69 000 more jobs
- 2Mt CO2 reduction

The biggest impact of 5G+ will be in Manufacturing, where, due to the emMTC devices, the overall sales of the factories of the future will be around 5 billion euro. Also, 5G+ enabled factories will reduce their carbon footprint by 24% in Romania, followed by transportation and energy [22].

In terms of workplace transformation, [22] identified that the desk workers and the companies that have desk workers would not invest a lot in 5G. However, workers in peripatetic, field, and industrial roles that can work from home or work from anywhere, that can have a nomadic workplace or that need to travel for the job but also need to have real-time data intelligence and context-appropriate collaboration tools, these workers will be the ones that can move the 5G era. 5G effect on the workforce will be 1.03 million additional jobs in Romania, Belgium, France, or Poland by 2030, 0.6% of the population and 1.3% of total employment [22].

5G/5G+ value chain is diverse, spanning end-users (consumers, enterprises, and other organizations), telecom operators, infrastructure vendors, device manufacturers, component providers, software and platform providers, content and application providers, etc.

According to a report from Research & Markets 6G will achieve initial commercialization by 2028-2030, and it will have the following consequences [23]:

- By 2030 the first phases of the 6G infrastructure and testbeds market will reach $4.83 billion by 2030
- The overall 6G market will have a cap of 1773 Billion USD by 2035
- The most prolific investors in 6G will be countries from the Asia Pacific region, followed by the US and European Union.

Also, regarding the emMTC market, according to Transforma Insights [24], the number of devices will grow by 2030 to 24 billion, and the number of connections will grow to 4.7 billion.

These economic impacts are created by the research done in the emMTC 6G field and will reach the first maturity milestones by 2030.

5.7 Conclusions

In this chapter, a way to see the future of communications is proposed by creating a framework on how to view the emMTC devices in the context of 6G. 5G features were presented, like network slicing and how the mMTC use-case is integrated into the technology.

The authors presented several challenges that 5G mMTC has, such as security, energy consumption, network access, and so on, and framed several research questions for the future 6G networks to investigate. Also, key elements driving the emMTC 6G development were presented. And we presented the 17 UN sustainability goals and how some of them will impact the new technology, like climate change and energy efficiency, ethical algorithms, and sustainable cities.

Besides the socio-economic trends that enable research on the 6G front, technology trends were presented, such as Autonomous mobility, Connected living; Factories of the future; Digital reality as frontier technology; Zero worlds; or Data as the new oil. All these tech trends have particularities for emMTC that were discussed, and based on these particularities' requirements for 6G are created.

Moreover, use-cases for emMTC were presented and how these use-cases, alongside the tech and socio-economic trends, are creating research questions and are creating the beyond state of the art in the emMTC 6G field.

These scenarios are divided into six categories as follows:

- Efficient massive connectivity,
- Multi-access edge computing as an enabler,
- Security for MTC,
- Vertical-specific solutions,
- Powered by AI and machine learning (ML),
- Energy Efficiency.

Lastly, several business cases and the economic impact of 5G+ by 2030, 6g, and emMTC were presented. When presented, the economic impact was divided by revenue or GDP growth, the number of new jobs, or the carbon footprint reduction. For example, for Romania, the impact of 5G+/6G is translated into 20 billion dollars of 5G enabled sales, 69 000 more jobs and a 2Mt CO2 reduction. Also, Romania manufacturing factories will reduce their carbon footprint by 24%, followed by the transportation and energy sectors.

The chapter is an overview of what will be emMTC in 2030 from different standpoints. As the first commercially available 6G network is predicted for 2030 and as several research flagship projects have already started, such as the H2020 Hexa-X project, the Oulu 6G flagship project, several others in Japan, the USA and China, we expect to see a lot of research papers and more research projects that will answer some of the proposed challenges and some of the research advancements such as new architectures for IoT devices in a zero-energy world or new architectures for enabling AI and ML algorithms in edge computing.

Acknowledgements

The work for this chapter has been done with the support of the H2020-MSCA-RISE project nr. 872857 "RECOMBINE (REsearch COllaboration and Mobility for Beyond 5G future wIreless Networks)" funded by European Commission and the Romanian PN3 project no. 33PCCDI/2018 "Multi-MonD2 (MultiAgent Intelligent Platform for Environment and Water Quality Monitoring along with Romanian Sector of the Danube and Danube Delta)", funded by UEFISCDI. Also, this work has been performed at the Future Networks 5GLab, an advanced experimental facility created in partnership between Orange Romania and CAMPUS Research Institute of University POLITEHNICA of Bucharest.

References

[1] A. Ghosh, A. Maeder, M. Baker and D. Chandramouli, "5G Evolution: A View on 5G Cellular Technology Beyond 3GPP Release 15", in *IEEE Access*, vol. 7, pp. 127639-127651, 2019, doi: 10.1109/ACCESS.2019.2939938.

[2] S. Kukliński, L. Tomaszewski and R. Kołakowski, "On O-RAN, MEC, SON and Network Slicing integration", *2020 IEEE Globecom Workshops (GC Wkshps*, 2020, pp. 1-6, doi: 10.1109/GCWkshps50303.2020.9367527.

[3] Mohammad Asif Habibi, Bin Han, Hans D. Schotten, Network Slicing in 5G Mobile Communication Architecture, Profit Modeling, and Challenges, arXiv 2017

[4] ***, New services & applications with 5G Ultra-reliable Low Latency Communications, 5G Americas, Nov 2018

[5] Nurul Huda Mahmood et al., White Paper on Critical and Massive Machine Type Communication Towards 6G, arXiv 2020

[6] 3GPP 26.850 V15.1.2 Technical Specification Group Services and System AspectsMBMS for IoT. 2018.

[7] ***, D1.1 6G Vision, use cases and key societal values, Hexa-X project consortium

[8] P. Ahokangas, S. Yrjölä, M. Matinmikko-Blue, V. Seppänen and T. Koivumäki, "Antecedents of Future 6G Mobile Ecosystems", *2020 2nd 6G Wireless Summit (6G SUMMIT)*, 2020, pp. 1-5, doi: 10.1109/6GSUMMIT49458.2020.9083756.

[9] N. H. Mahmood, H. Alves, O. A. López, M. Shehab, D. P. M. Osorio and M. Latva-Aho, "Six Key Features of Machine Type Communication in 6G", *2020 2nd 6G Wireless Summit (6G SUMMIT)*, 2020, pp. 1-5, doi: 10.1109/6GSUMMIT49458.2020.9083794.

[10] L. Liu, E. G. Larsson, W. Yu, P. Popovski, C. Stefanovic and E. de Carvalho, "Sparse Signal Processing for Grant-Free Massive Connectivity: A Future Paradigm for Random Access Protocols in the Internet of Things", in *IEEE Signal Processing Magazine*, vol. 35, no. 5, pp. 88-99, Sept. 2018, doi: 10.1109/MSP.2018.2844952.

[11] Y. Chen *et al.*, "Toward the Standardization of Non-Orthogonal Multiple Access for Next Generation Wireless Networks", in *IEEE Communications Magazine*, vol. 56, no. 3, pp. 19-27, March 2018, doi: 10.1109/MCOM.2018.1700845.

[12] Z. Yuan, Y. Hu, W. Li and J. Dai, "Blind Multi-User Detection for Autonomous Grant-Free High-Overloading Multiple-Access without Reference Signal", *2018 IEEE 87th Vehicular Technology Conference (VTC Spring)*, 2018, pp. 1-7, doi: 10.1109/VTCSpring.2018.8417836.

[13] M. Xhemrishi, M. C. Coşkun, G. Liva, J. Östmanl and G. Durisi, "List Decoding of Short Codes for Communication over Unknown Fading Channels", *2019 53rd Asilomar Conference on Signals, Systems, and Computers*, 2019, pp. 810-814, doi: 10.1109/IEEECONF44664.2019.9048806.

[14] Alexander Fengler and Peter Jung and Giuseppe Caire, SPARCs for Unsourced Random Access, arXiv 1901.06234,2020

[15] M. Ejaz, T. Kumar, M. Ylianttila and E. Harjula, "Performance and Efficiency Optimization of Multi-layer IoT Edge Architecture", *2020 2nd 6G Wireless Summit (6G SUMMIT)*, 2020, pp. 1-5, doi: 10.1109/6GSUMMIT49458.2020.9083896.

[16] B. Mao, Y. Kawamoto and N. Kato, "AI-Based Joint Optimization of QoS and Security for 6G Energy Harvesting Internet of Things", in *IEEE Internet of Things Journal*, vol. 7, no. 8, pp. 7032-7042, Aug. 2020, doi: 10.1109/JIOT.2020.2982417.

[17] Nuno K. Pratas and Sarath Pattathil and Cedomir Stefanovic and Petar Popovski, Massive Machine-Type Communication (mMTC) Access with Integrated Authentication, arXiv 1610.09849, 2017

[18] J. Zhu, M. Zhao, S. Zhang and W. Zhou, "Exploring the road to 6G: ABC — foundation for intelligent mobile networks", in *China Communications*, vol. 17, no. 6, pp. 51-67, June 2020, doi: 10.23919/JCC.2020.06.005.

[19] H. Song, J. Bai, Y. Yi, J. Wu and L. Liu, "Artificial Intelligence Enabled Internet of Things: Network Architecture and Spectrum Access", in *IEEE Computational Intelligence Magazine*, vol. 15, no. 1, pp. 44-51, Feb. 2020, doi: 10.1109/MCI.2019.2954643.

[20] X. Li *et al.*, "Physical Layer Security of Cognitive Ambient Backscatter Communications for Green Internet-of-Things", in *IEEE Transactions on Green Communications and Networking*, doi: 10.1109/TGCN.2021.3062060.

[21] ***, 5G redineess report, Nokia Corporation

[22] ***, 5G Impact 2030, OMDIA group \
[23] ***, Sixth Generation Wireless by 6G Technology Development (Investment, R&D and Testing) and 6G Market Commercialization (Infrastructure, Deployment, Apps and Services), Use Cases and Industry Verticals 2021 – 2030, ResearchAndMarkets
[24] ***, Total Addressable Market for IoT, Transforma Insights .

Biographies

Razvan Craciunescu received the Engineer's degree, Master's degree and PhD degree in electronic engineering and telecommunications from the University POLITEHNICA of Bucharest, Bucharest, Romania, in 2010, 2012 and 2018. He is currently leading the first 5G Lab in Romania, the Future Networks 5G Lab at CAMPUS Research Institute, University Politehnica of Bucharest. He is an Assistant Professor and a post-doc researcher at the same university in communication systems, the internet of things, and wireless sensors networks. Between 2015-and 2016, he was a Researcher at the Centre for TeleInfrastruktur, Aalborg University, Denmark, in eHealth and digital incubation for startups. In 2018 he was an invited professor at the University of Rochester, USA, in emerging technologies and entrepreneurship. His publications include more than 40 papers published in recognized international journals and presented at international conferences or book chapters. He is a reviewer for several top journals and international conferences. He participated as a researcher in over 15 national and international R&D projects and as a project manager in 4 international R&D projects.

Simona Halunga received her M.S. degree in electronics and telecommunications in 1988 and her PhD degree in communications from the University POLITEHNICA of Bucharest, Bucharest, Romania, in 1996. Between 1996–1997 she followed post-graduate courses in Management and Marketing, organized by the Romanian Trade and Industry Chamber and University POLITEHNICA of Bucharest, in collaboration with Technical Hochschule Darmstadt, Germany, and 2008- postgraduate courses in Project Management—Regional Centre for Continuous Education for Public Local Administration, Bucharest She has been Assistant Professor (1991–1996), Lecturer (1997–2001), Associate Professor (2001–2005) and from 2006 she is a full professor at in University POLITEHNICA of Bucharest, Electronics, Telecommunications and Information Theory Faculty, Telecommunications Department. Between 1997–and 1999, she has been a Visiting Assistant Professor at the Electrical and Computer Engineering Department, the University of Colorado at Colorado Springs, USA. Her domains of interest are Multiple Access Systems & Techniques, Satellite Communications, Digital Signal Processing for Telecommunication, Digital Communications - Radio Data Transmissions, Analogue and Digital Transmission Systems.

Octavian Fratu received a PhD degree in Electronics and Telecommunications from the University "Politehnica" of Bucharest, Romania, in 1997. He achieved postdoctoral stage as a senior researcher in 3rd generation mobile communication systems, based on a research contract between CNET-France, ENS de Cachan - France and Universite Marne la Vallee – France. He is currently a professor in Electronics, Telecommunications and Information Theory Faculty of University Politehnica of Bucharest. His research interests include Digital Mobile Communications, Radio Data Transmissions, Mobile Communications, Wireless Sensor Networks and Communication Security and Cyber-Security. His publications include more than 200 papers published in national or international scientific journals or presented at international conferences. He participated as director or collaborator in many Romanian national funded research projects and international research projects in the areas mentioned above.

6

6G Multinetwork Convergence

R. Chandramouli[1], S. Srikanth[2], K.P. Subbalakshmi[3] and Vidya Sagar[4]

[1,3]Stevens Institute of Technology,
[2]Nanocell Networks,
[4]Spectronn Inc.
E-mail: mouli@ieee.org; sribommi@gmail.com; ksubbala@stevens.edu; vidya@spectronn.com

In this chapter, we discuss network convergence in a 6G world. 6G will consist of many networks with diverse data rates, latencies, and coverage ranges. Managing these networks to satisfy application requirements is a key challenge. The evolution of a convergence architecture is presented, and the relevant technical challenges are discussed.

6.1 Introduction

6G networks will be more diverse than 5G. It will encompass terrestrial and non-terrestrial networks (e.g., low earth orbit satellites and unmanned aerial vehicles). 6G applications will require access to significantly higher bandwidth than 5G applications. Therefore, frequencies between 100 GHz and 1 THz gain importance. A range of frequency bands from sub-6 GHz to 1 THz can be expected to be accessed.

New cellular and WiFi [1] networks combined with space-based wireless networks necessitate multinetwork convergence in 6G. A converged 6G architecture will be able to provide on-demand bandwidth and network access to support a multitude of data rate and latency requirements. Holographic presence, real-time mission-critical networks, and broadband connectivity leveraging satellite clouds are a few 6G application examples.

A common theme in 6G is multinetwork convergence, i.e., combining disparate networks with their own PHY/MAC layers governed by different standards into seamless network behaviour. Applications will feel like they operate on a single network, whereas the underlying network will consist of different networks with their packet error rate, latency, bandwidth, coverage, capacity, etc.

Distributed hybrid edge-cloud computing will be more prevalent in 6G. Multipath connectivity to these computing resources will ensure resilience, security, and quality of service. But the converged architecture must be designed, deployed, and managed carefully to achieve this goal.

Applications will have different priorities. For example, a mission-critical public safety application may preempt other applications from accessing network and computing resources during a major emergency. At the network level, this will require very fast handoff, predictive load balancing, bandwidth aggregation, spectrum bonding, predictive computational offloading, predictive distributed computing, etc. New dynamic spectrum sharing/access policies and policy-based radio behaviour control that are anticipatory will also be required. Reactive radio behaviour, network control, and current sensing-based dynamic spectrum sharing or access policies may not satisfy the bandwidth and latency requirements. Machine learning algorithms play a natural role in enabling these capabilities.

6.2 Multinetwork Convergence

6.2.1 6G Network Convergence Architecture

While there are many possibilities for 6G network convergence, our vision is presented in Figure 6.1. Unlike the current network convergence and control architectures that are reactive in nature, the proposed architecture is predictive. Telemetry data from PHY/MAC, IP, Transport, and Application layers are leveraged to predict future values. Two types of predictive control mechanisms can be anticipated:

1. **Per-layer predictive control**: time-series data from each network stack layer drives predictor-control algorithms for only that layer. For example, at the PHY/MAC layer, a dynamic interference model is computed with past measurements. The model then drives the choice of the spectrum, spectrum aggregation/bonding, transmit power, spectrum access scheduler, etc.

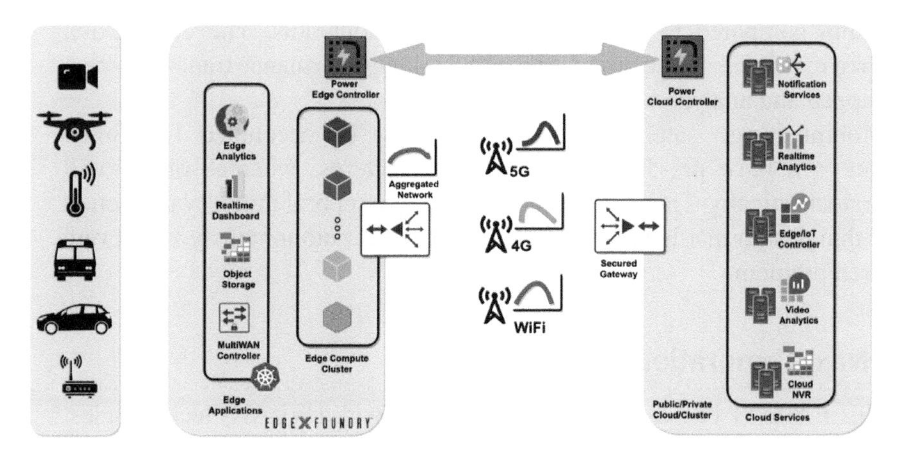

Figure 6.1 6G network convergence architecture.

2. **Cross-layer predictive control**: Here, data from across the layers are combined for cross-layer prediction and control. This approach is especially applicable for end-to-end predictive optimization and robust and stable network convergence control.

Note in Figure 6.1 that in 6G, both computing and networking will be optimized jointly. Such joint optimization will maximize network efficiency while satisfying application parameters such as latency, storage, memory, battery power, etc.

Software-defined networking will continue to expand and play an even more significant in 6G network convergence. Distributed computing software platforms will be embedded with predictive intelligence for offloading computations between different edge and cloud servers. New data exchange protocols must be developed to transition between cross-vendor edge and cloud computing platforms seamlessly.

Machine learning algorithms capable of forecasting data at multiple time scales will be essential. Notice that the PHY/MAC layer operates at a significantly shorter time scale compared to the Application layer. The forecasted data will be unreliable and unstable if forecasting algorithms are not carefully designed and analyzed for multi-time scale operations.

Prediction and control operations will be deployed at the edge and the cloud. This means that machine learning algorithms must be optimized for heterogeneous devices. For example, a resource-poor edge device (e.g., wearable sensor) would require low-complexity forecasting and control algorithm compared to a resource-rich cloud apparatus. The complexity, resource requirements, accuracy, and application performance trade-offs must be mapped and analyzed.

Mobility (e.g., satellites and UAVs) must be accounted for. Some devices will have pre-determined mobility patterns, whereas others will move stochastically. Therefore, dynamic Spatio-temporal mobility prediction algorithm design that learns to predict and correct autonomously is a severe research problem.

6.3 Next Generation WiFi – WiFi 7

The IEEE 802.11 based wireless local area network (WLAN) technologies, popularly referred to as WiFi, have a tremendous impact on people worldwide. While cellular technology has had an impressive march in generations, Wi-Fi has also grown in technology and user base in the last few decades. It is also poised to evolve its next generation to serve users with higher data rates and lower latencies. Figure 6.1 provides a snapshot of the development of 802.11 based Wi-Fi generations.

In the initial days of Wi-Fi, the objective was to provide an extension to Ethernet in enterprise environments. Over time, Wi-Fi has become prevalent in many indoor spaces, and the recent work from home (WFH) scenario during the pandemic has made it a crucial part of our work, education, and entertainment accessing solutions. The most recent evolution to WiFi 6 did not significantly impact peak rates as the focus was more on improving high-density performance. However, the recent announcements worldwide of the availability of a 6 GHz spectrum for unlicensed usage have given more choices to augment the standards with higher bandwidth possibilities. The IEEE 802.11be standards, which is likely to be branded as WiFi 7 by the WiFi Alliance (WFA), are pursuing some key technologies which are likely to be the technology of choice for most applications indoors and in low mobility environment from 2025 onwards. The motivation for 802.11be is to provide a high-speed indoor solution working in unlicensed bands capable of meeting the needs of applications like virtual/augmented reality and serving traditional voice and broadband needs. While we talk about indoor environments, WiFi is also the technology of choice for many closed environments like cars, trains, planes, etc.

We shall discuss some of the proposed key technology approaches for achieving the objectives in the next generation of WiFi standards.

6.3.1 Large Bandwidth Channels

The IEEE 802.11 standards are designed to operate in unlicensed bands. Recently, many regions have opened the 6 GHz band to unlicensed technologies like WiFi. The attraction of the new allocation is that it adds upwards of 500 MHz of spectrum with more than 1 GHz of spectrum in countries like the US and Canada. This is an opportunity for the standards to utilize the spectrum to provide higher bit rates. In 802.11be, a new bandwidth possibility of 320 MHz is added to the existing choices, as shown in Figure 6.2. The 320 MHz channels shall help double the peak rates leading to better end-to-end throughputs for higher bandwidth applications. In addition, the 6 GHz spectrum also enables better utilization of 160 MHz channels as there are enough non-overlapping choices available.

Table 6.1 Large bandwidth possibilities in WiFi 7.

<table>
<tr><th colspan="6">UNII5</th><th>UNII6</th><th>UNII6 /7</th><th colspan="3">UNII7</th><th>UNII7 /8</th><th colspan="2">UNII8</th></tr>
<tr><td>80</td><td>80</td><td>80</td><td>80</td><td>80</td><td>80</td><td>80</td><td>80</td><td>80</td><td>80</td><td>80</td><td>80</td><td>80</td><td>80</td></tr>
<tr><td colspan="2">160</td><td colspan="2">160</td><td colspan="2">160</td><td colspan="2">160</td><td colspan="2">160</td><td colspan="2">160</td><td colspan="2">160</td></tr>
<tr><td colspan="4">320</td><td colspan="4">320</td><td colspan="4">320</td><td></td><td></td></tr>
<tr><td></td><td></td><td colspan="4">320</td><td colspan="4">320</td><td colspan="4">320</td></tr>
</table>

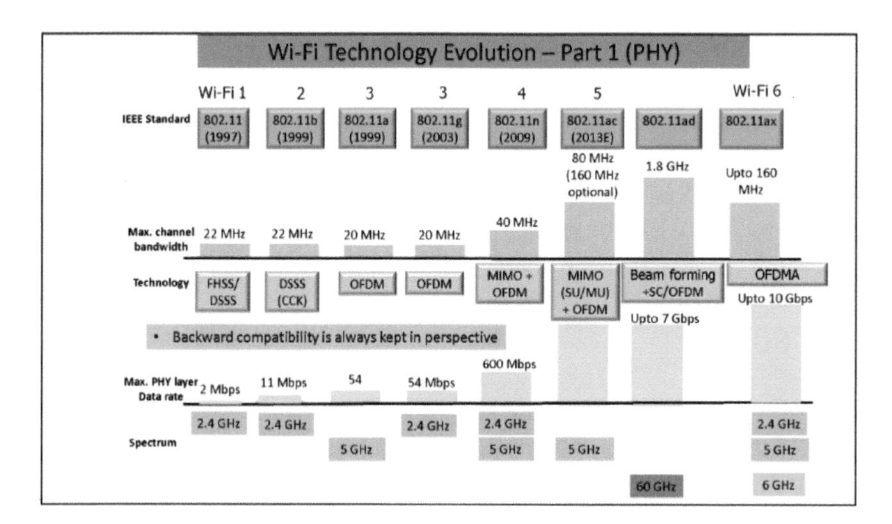

Figure 6.2 WiFi technology evolution.

6.3.2 Flexible Bandwidth Usage Using Enhanced OFDMA

The IEEE 802.11ax standard, also known as WiFi 6, specified the use of OFDMA to solve some of the challenges in high-density implementations. Both 4G LTE and 5G NR also use OFDMA, and the convergence of OFDMA as a PHY/MAC solution for the dominant wireless access technologies bode well for the development of chipsets that combine these technologies in products like smartphones and internet gateways for homes, offices and other places.

The next generation 802.11be/WiFi 7 specifications add more flexibility to the OFDMA operations by allowing more flexibility in resource scheduling to a user by enabling the combination of resource units to a single user. In addition, a technique called preamble puncturing (see, Figure 6.3) proposed for both single and multi-user transmissions allows operation in dense WiFi environments with other networks occupying intermediate parts of a large bandwidth channel.

In most evolutions of the WiFi generations, the legacy devices had to be admitted and served using the legacy specifications. Usually, during the initial years of the next generation WiFi equipment, the older generation devices are dominant, and the benefits of the new generation features are limited.

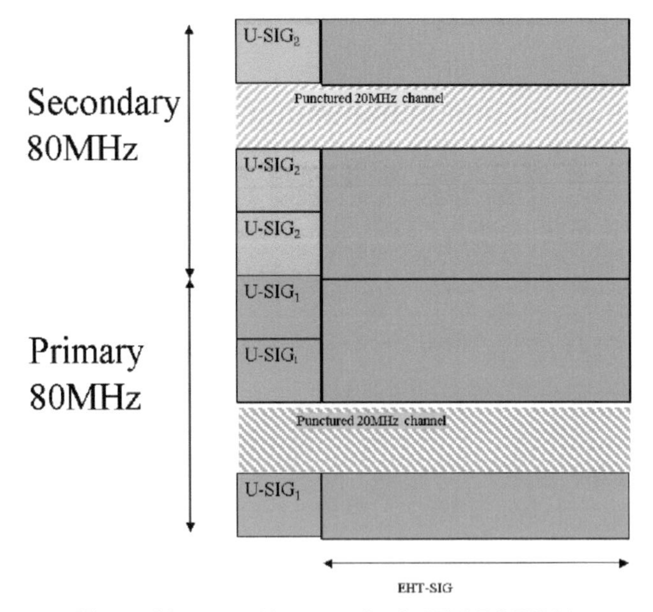

Figure 6.3 Preamble puncturing in WiFi 7 OFDMA.

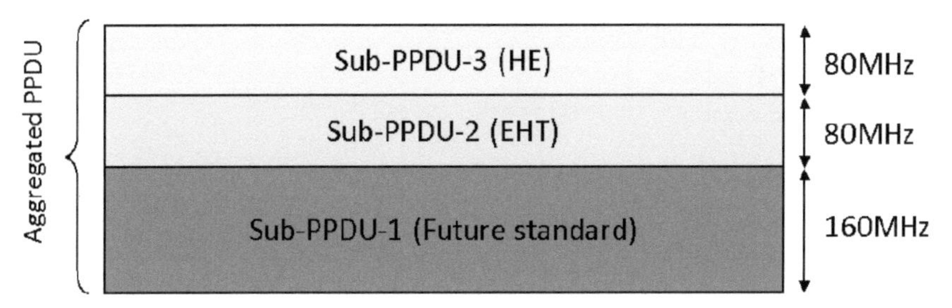

Figure 6.4 Legacy-friendly operations in WiFi 7.

In the 802.11 be standards, specifications are proposed to accommodate the immediate earlier generation standard's waveform as a part of the new generation transmission, i.e., a WiFi 7 and WiFi 6 transmission could happen simultaneously in a network. This is due to the common OFDMA PHY/MAC layer used in these standards. A future possibility due to the developments in 802.11be is illustrated in Figure 6.4 as an example.

6.3.3 Enhanced MIMO and MU-MIMO Operations

MIMO has been especially important in the WIFi evolution, starting with the 802.11n specification. Most real-life access points and many data accessing client devices like laptops, tablets, TVs, and smartphones incorporate MIMO to obtain better throughputs in practical conditions. Recently, MU-MIMO has also been introduced as a part of WiFi 5 with enhancements in the 802.11ax based WiFi 6 specifications. Downlink (DL) and uplink (UL) MU-MIMO schemes are being made mandatory for clients to support, and we expect high-end APs to support this as a feature enabling better throughputs for multiple users simultaneously accessing the channel for their services. We have seen during the pandemic times that many people in a single home are accessing/sharing high-bandwidth content as a part of their work/education/entertainment needs. This is one of the scenarios where MU-MIMO technologies could become very handy. The enhancements planned for the 802.11be standards increase the MIMO spatial stream capability. In addition, the feedback mechanisms necessary for downlink MU-MIMO to be effective are being enhanced to allow increased efficiencies. In summary, we can expect to see better support for MU-MIMO in real-life products and a more efficient feedback mechanism resulting in better performance numbers in 802.11be/WiFi 7 based products. The potential of the enhanced MU-MIMO capabilities is shown in Figure 6.5.

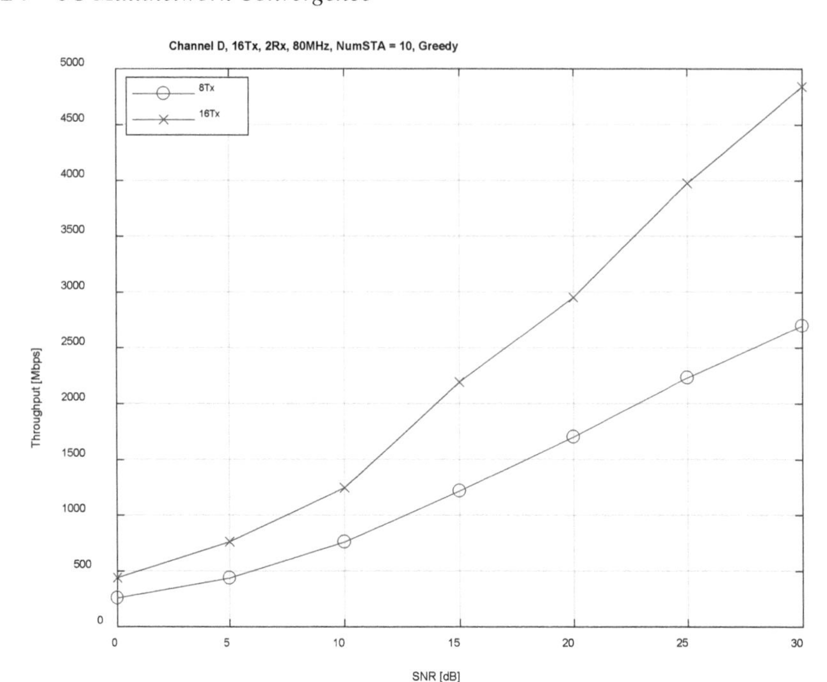

Figure 6.5 MU-MIMO potential in WiFi 7.

6.3.4 Multi-link Operations

Many WiFi products support more than one band of operation, and many of these products can operate in multiple bands simultaneously. With the advent of the new 6 GHz spectrum, we expect many products to support tri-band operations with multiple radios capable of working simultaneously. In such a scenario, the 802.11be specifications propose techniques to combine the multi-band support available in products for better throughput, latency, and reliability, amongst other advantages like load balancing. The multiple links in different channels/bands between two devices is an essential new feature likely to harness the multi-band support in WiFi products. Two prominent use cases mentioned as a part of the standards activity are dynamic load balancing across the bands supported between two devices and the aggregation of the links across the different bands. These are illustrated in Figure 6.6 and Figure 6.7, respectively. While the aggregation use case can enable higher possible rates between two devices, the load balancing features can allow handling scenarios wherein a higher quality of service (QoS) class of traffic can be carried in a quieter spectrum, e.g., the 6 GHz band. Such abilities

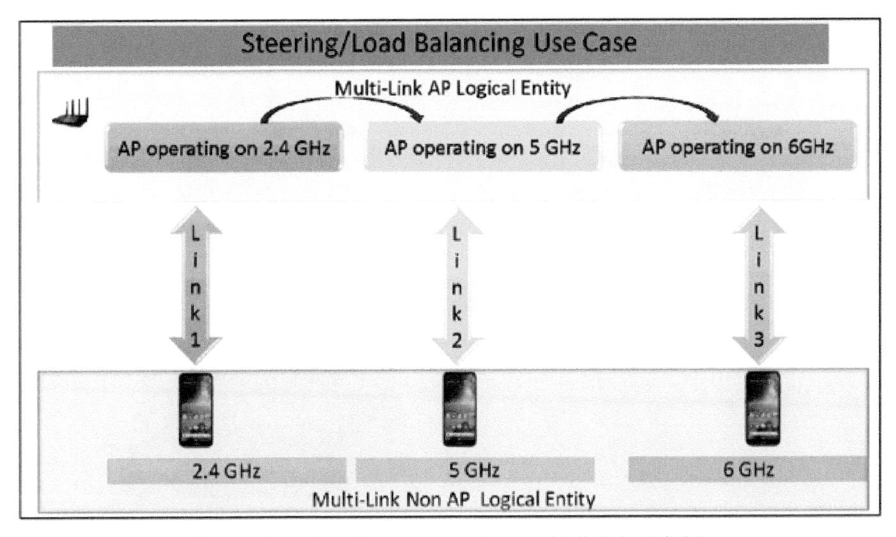

Figure 6.6 Load balancing using multi-link in WiFi 7.

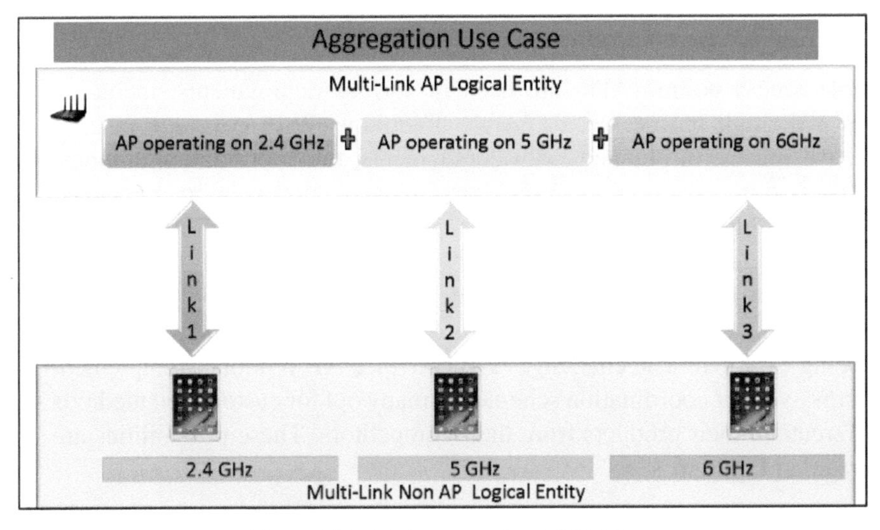

Figure 6.7 Link aggregation in WiFi 7.

to fine-tune traffic to a link are possible in the multi-link specifications proposed in the 802.11be standards. As this feature becomes widespread, it is possible that even the 60 GHz radio could become a part of the WiFi product, especially for short-range applications that can enable highly high bandwidth applications like virtual reality and hologram communications.

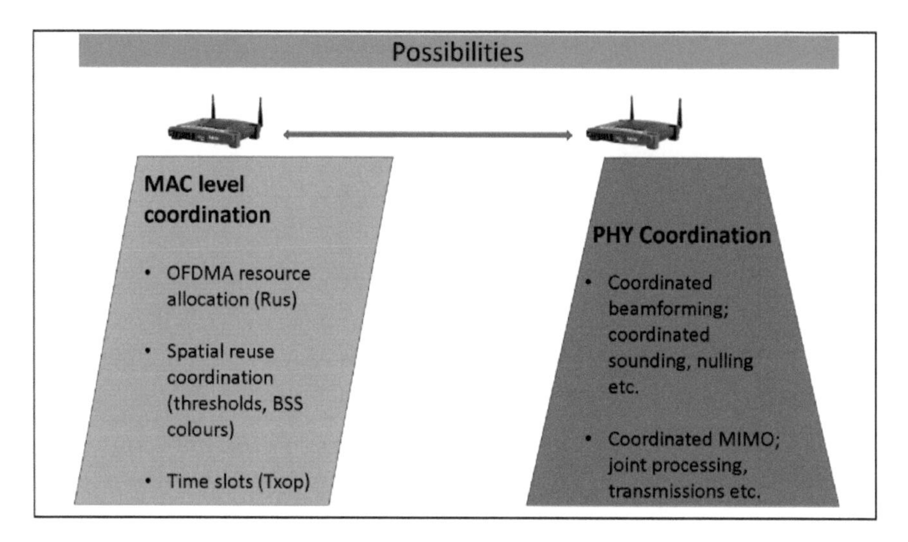

Figure 6.8 Multi-AP coordination in WiFi 7.

6.3.5 Multi-AP Coordination

Multiple access points (APs) are used in various deployments, including homes and small businesses, to enable ubiquitous WiFi coverage in their respective spaces. In planned WiFi deployments, the coordination amongst the APs is done in a vendor-proprietary manner. However, in unplanned deployments like in homes and offices, there is scope for coordination amongst APs to provide additional benefits. This is one of the proposed techniques in 802.11be. At the PHY level, we have coordinated OFDMA and MIMO techniques being discussed, while resource sharing at the time level is also being explored. The challenge is to convince AP vendors to implement such cross-vendor coordination schemes as many opt for customized methods to differentiate their products from their competitors. These possibilities are illustrated in Figure 6.8.

6.4 Conclusions

It is an exciting time for the whole connectivity industry spurred on by the challenges of the pandemic, which has increased the demands for next-generation wireless network technology. Work, education, entertainment, and many other activities are now being conducted at home, and the need for an improved network experience is required in most indoor and closed spaces. The evolution of the 5G standards into 6G for satisfying many of these demands has led to standards activities such as 802.11be for delivering the WiFi 7 experience. The proposed technologies are designed to address the increased demands of throughput, latency, reliability, and other related parameters. This should help the user experience wireless connectivity in various scenarios for satisfying the next-generation applications.

References

[1] https://www.wi-fi.org/

Biographies

R. Chandramouli is the Thomas Hattrick Chair Professor of Information Systems in the Electrical and Computer Engineering Department at Stevens Institute of Technology. His research expertise covers cognitive radio networking, wireless spectrum issues, and machine learning. He is an NSF CAREER awardee, IEEE COMSOC Distinguished Lecturer, and keynote speaker and serves on numerous journals' advisory and editorial boards.

Srikanth Subramanian obtained his B.E. degree from the College of Engineering, Anna University, Chennai, and M.A.Sc, and Ph. D. degrees from the University of Victoria, British Columbia, Canada. He is currently the Chief Knowledge Officer for Nanocell Networks Pvt. Ltd. and consults for various companies.

Srikanth began his career as a research associate at the University of Victoria, British Columbia, Canada, working in DSL and CDMA Systems. After his Ph. D., he joined Harris Corporation and worked on baseband algorithms for various wireless standards, including IS-136 and 1S-95 systems.

He then joined the KBC Research Foundation/AU-KBC Research Centre, and he has focused on OFDM systems for powerline and wireless applications. From 2004-to 2007, he was awarded a Young Scientist Fellowship by the Government of India to work on technologies related to upgrades on 802.11 and 802.16 standards. He has been following 802.11 and 4G/5G standards and delivers several courses to corporate clients in this area.

He has consulted on various areas of OFDM systems and has also been involved in setting up a test lab for 802.11. He is also involved with the organization of Wi-Fi and 5G knowledge summits to help develop a community to exchange technical information in these areas.

K.P. (Suba) Subbalakshmi is the Founding Director of the Stevens Institute for Artificial Intelligence and a **Professor** in the Department of ECE at Stevens Institute of Technology. She is a Fellow of the National Academy of Inventors and a Member of the *National Academy of Science Engineering and Medicines' Intelligence Science and Technology Experts Group.* She has Co-founded several technology start-up companies.

Her current research areas are in Artificial Intelligence and Machine Learning, emphasising mental health, cyber safety/security, and cognitive radio networking. Specific areas are Mental Health: early-stage cognitive impairment detection (including Alzheimer's' disease), depression and other mental health conditions; Cyber security/safety — explainable AI for fake news detection, rumour dynamics on social media, deception detection from text content, image/video steganography and steganalysis, information confusion in social media; NLP problems**:** keyphrase extraction; Cognitive radio networks and security – algorithms and design; attacks, countermeasures and systems-level impact; Cognitive mobile cloud computing – spectrum and radio resource-aware optimal mobile computing.

Vidya Sagar is the Chief Technology Officer at Spectronn, Inc. He received a PhD from Stevens Institute of Technology and a Masters's from the IIIT-Bangalore. He was an R&D engineer at Motorola, working on early versions of LTE.

Vidya Sagar received a New Jersey Inventor Hall of Fame Graduate Student Award for his research on software-defined cognitive networking. He holds several patents in wireless networking and has published in peer-reviewed conferences and journals.

7

6G QoE for Video and TV Applications

Paulo Sergio Rufino Henrique[1] and Ramjee Prasad[2]

[1]Paulo S. Rufino Henrique, CTIF Global Capsule (CGC) / Aarhus University/ Spideo, Paris, France
[2]CGC, Aarhus University-Denmark
E-mail: rufino@spideo.tv; ramjee@btech.au.dk

The behavioural shift in video and TV services consumption by viewers in the past ten years was triggered by fixed and mobile broadband network infrastructure improvements such as 3G, 4G, and 5G. Another major contributor to revolutionizing how broadcasters, internet service providers (ISPs), and content providers (CP) engaged with viewers was reinforced by creating a more democratic concept of IP content distribution, defined as TV Everywhere. This new service evolved, creating a new audience that could decide what, when, and where to watch or listen to a video or audio streaming based on the over the top (OTT) content services. OTT business model is sitting on the networks' application layer. It does not rely on any particular ISP network infrastructure, as it uses cloud services provider (CSP) and a content delivery network (CDN). CDNs specialize in bringing the content much closer to the network's edge. Such architecture reduced video and audio impairments, improved the content quality distribution over the network infrastructure, and refined the user experience (UX). However, ISPs, mobile network operators (MNOs) focused until now on investing in enhancing the quality of services (QoS) by utilizing traditional metrics in the network architecture to indirectly improve the user's quality of experience (QoE) as 6G is defined as a future human-centric network, designing a methodology to guarantee excellent QoE for consumers. This chapter proposes a 6G QoE based on artificial intelligence (AI) and quantum machine learning (QML).

7.1 Introduction

The technological evolution offered by mobile broadband generations allowed the deployment of a better quality of video and TV services for mobile subscribers. Then, ISPs, broadcasters, CPs, and OTT providers benefitted from enhanced wireless technologies integrated with Cloud Services Providers, known as Hyperscalers and the Content Delivery Networks. In summary, the technological evolution of fixed and wireless broadband networks permitted new applications for video communication services, particularly mobile broadband services. Such investment granted a superior level of quality control for broadcasters, ISPs, and OTT providers to leverage their user's engagement via a better user experience. With these service improvements in wireless architecture, video and live TV streaming over the Internet has popularized during the past decades. Video is now dominating the Internet data traffic for consumers around the globe. This trend can be noticed in both fixed and cellular networks. Worldwide, the growth of video streaming data traffic is more prominent. Currently, the global mobile traffic for the consumer is comprised of 66% of video data [1]. A recent report [2] shows an expected 48.7% global mobile video traffic rise from 2021 to 2022.

The improvements delivered by mobile broadband generations, such as universal mobile telecommunication systems (UMTS), Long Term Evolution -Advanced (LTE-A), and 5G, created a new business opportunity for the media industry and increased the portfolio of products with better user experience (UX) for video and tv streaming. Furthermore, another major contributor to revolutionizing how broadcasters and content providers (CP) engage with viewers has reinforced the innovation in the traditional TV, based on digital terrestrial broadcast (DTTB) to the innovative concept of TV Everywhere and video on demand (VoD). TV Everywhere, along with a service orchestration propelled by CDNs, forged the success of consumers' engagement with VoD and Live TV streaming. In this case, the name streaming is co-related to the capacity to deliver the audio-visual content using a process similar to streaming a fluid. The quality of the content is related to bandwidth and other QoS metrics. However, theoretically speaking, if the bandwidth is more comprehensive, the quality of audio-visual content can be transmitted better. In case of a smaller bandwidth or traffic congestions, the content can be delivered but in degraded quality for continuity in the UX. Such adjustment in the video quality is possible thanks to dynamic adaptive streaming over HTTP (DASH) [3] and the adaptive bit

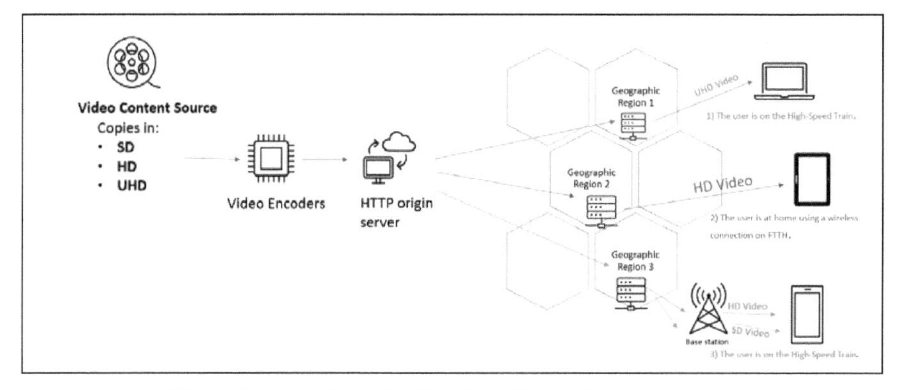

Figure 7.1 DASH and ABR video distribution architecture.

rate (ABR) [4] deployed in the CDN. Figure 7.1 shows the content streaming distribution workflow based on the earlier process.

As shown above, the Figure 7.1, the first user is in the office utilizing a corporate network with optimal QoS, then this user can watch content in ultra high definition (UHD). User 2 is at home, but the user's network is a brownfield network with a smaller bandwidth, allowing only the maximum high definition (HD) video quality. Finally, user 3 is on the move, in a high-speed train and sometimes served by a three network in total capacity, and another time a 4G network is operating. Therefore, the catch servers at geographic region 3 control the video format according to the quality of the wireless channel. Thus, the user's experience changes based on the HD video format downgraded to the SD video format. This co-occurs in the high-speed train as the user needs to change to different cells and distinct QoS in each new allocated Base Station.

The TV everywhere concept and the VoD business model radically changed how users are used to consuming content. With these two models, the users were freer from a rigid electronic program guide (EPG) to select their preferred content whatever and whenever they wanted. In addition, with the enabled technologies of CDNs, CSP and the OTT providers found a new niche to explore. This process posed a threat to the original and well-established status quo of the Broadcasting Industry. As the user engagement with the traditional TV started declining, the rise of OTT platforms started gaining traction amongst customers. This new disruptive digital content distribution business model was the VoD. Along with this model, the different types of film quality increased in variety from standard definition (SD) to full high definition (HD), and the ultimate commercial film quality is used the ultra high definition (UHD) video. There is a specific bit rate for these video qualities to guarantee the QoE.

Table 7.1 below describes how TV Everywhere and VoD Techno-Business Concept changed the traditional broadcasters' business landscape.

Equally relevant to the user's user experience in video streaming is associated with various video quality formats and technical forms. The evolution of encoding techniques and the development of superior video formats have also contributed to the success of video streaming.

Therefore, the OTT providers utilizing the mobile broadband infrastructure aligned with the Cloud Service Providers could optimize the video traffic. A new class of services (CoS) and MNOs utilized a combination of subjective and objective metrics to improve video and audio streaming quality. One of these subjective metrics used in the industry is the mean opinion score (MOS) based on the International Telecommunication Union recommendation ITU-P.800.1 [5]. MOS is a standardized subjective key performance index (KPI) to evaluate the audio and video quality of experience. The MOS

Table 7.1 TV everywhere and VoD techno-business concept.

Concept		Techno-Business Description
TV Everywhere		Digital IP TV model – protocols: broadcast, multicast, unicast - medium: fixed or wireless network
Derivatives xVoD		
VoD (Video on Demand)		Video Streaming on Demand for users. Users can choose what to watch in a video library.
SVoD (Subscribed Video On Demand)		Video Streaming is based on the user's subscription. The user needs to pay a fee to watch content, which unlocks his access to the OTT Provider content catalog.
TVoD (Transactional Video On Demand)		Video Streaming is based on purchasing or renting any content available in the OTT catalog. In this case, any content asset will have an exact price, and it can vary according to the release date or film's quality format SD, HD, or UHD.
AdVoD (Advertising Video On Demand)		Auto target advertising on the xVoD services.

Table 7.2 MOS ACR.

MOS Absolute Category Rating (ACR)	
Rating	**Label**
5	Excellent
4	Good
3	Fair
2	Poor
1	Bad

can be calculated following the given formula [6] and Table 7.2.

$$MOS = \frac{\Sigma_{n=1}^{N} R_n}{N}$$

The MOS calculation is based on the user's experience with audiovisual data streaming. When looking at the formula, Rn is the user's unique ratings for several inducements N.

The MOS still being a very subject way of measuring the QoE for users. QoE is still subject to the user's perceptions, the human capacity to retain information, and the QoS metrics. The ISPs and MNOs aimed to enhance the QoS on conventional KPIs for the fixed and wireless networks to improve the user's quality of experience QoE indirectly. Also, mobile generations did not yet have a QoE KPI defined on top of this. The QoE KPIs currently are very subjective in terms of UX. Improvement is necessary for this field, especially when new robust video standards and formats requiring better user experience are gearing up to dominate the market. Provide a few examples of video application that needs better QoS on the network level to achieve a good user experience are:

- Augmented Reality (AR)
- Cross Reality (XR)
- 3D Videos
- 8K Video
- Holographic type communications (HTC)

In summary, this book chapter is organized into subchapters. First, section 2 describes the evolution of mobile network infrastructure for multimedia applications and the innovation delivered by 4G and 5G. Then, section 3 describes Quantum Computing, the subset of AI, Machine Learning, and finally presents quantum machine learning (QML) as a solution to improve and guarantee QoE for 6G subscribers.

7.2 Current Status of the Wireless Network Infrastructure for Multimedia QoS and QoE

7.2.1 2G and 3G inaugurate the Multimedia Services over Cellular Radio Network

Before the launch of OTT services, the Second Generation of Mobile Services, 2G, or global system mobile (GSM), has released the short-message-service (SMS). However, SMS only carries digital text messages. Thus, the innovative concept of OTT services to distribute video and audio content via IP connectivities to end users located in the fixed or wireless broadband networks, including Wi-Fi, is adequately established in the 3G Networks. Multimedia messaging services (MMS) [8] for non-real-time unified applications as standard service such as video, text, image, and audio in the 3G specifications was a big jump to allow the cellular broadband network to create a new market based on different types of heterogeneous media. Thus, 3G Networks' architecture enabled mobile video calling to become a reality. It also permitted ISPs to bundle services to attract subscribers based on QuadPlay's business strategy, which combined packages with Home Broadband Internet, Mobile Services, Fixed Phone Lines, and IP Television (IPTV). However, due to 3G limitations, most video streaming on universal mobile telecommunication systems (UMTS) was based on standard definition (SD) videos. Therefore, fewer high definition (HD) videos could stream in this architecture as it would depend on the 3G optimal capacity and optimized QoS delivered to the end-user.

The specificities of 3G enabling video streaming are based on the (3GPP) IP multimedia subsystem (IMS) [9] that is the key enabler of IP multimedia services in the 3G architecture, including the usage of the Session Initiation Protocol (SIP) [10]. SIP protocol enabled controlling an end-to-end Internet session amongst computers for an application established via hypertext transfer protocol (HTTP). Below, Figure 7.2 presents a high-level overview of the MMS transmission based on the 3G Architecture and its protocol framework. In this figure, the MMS protocol is introduced following the workflow:

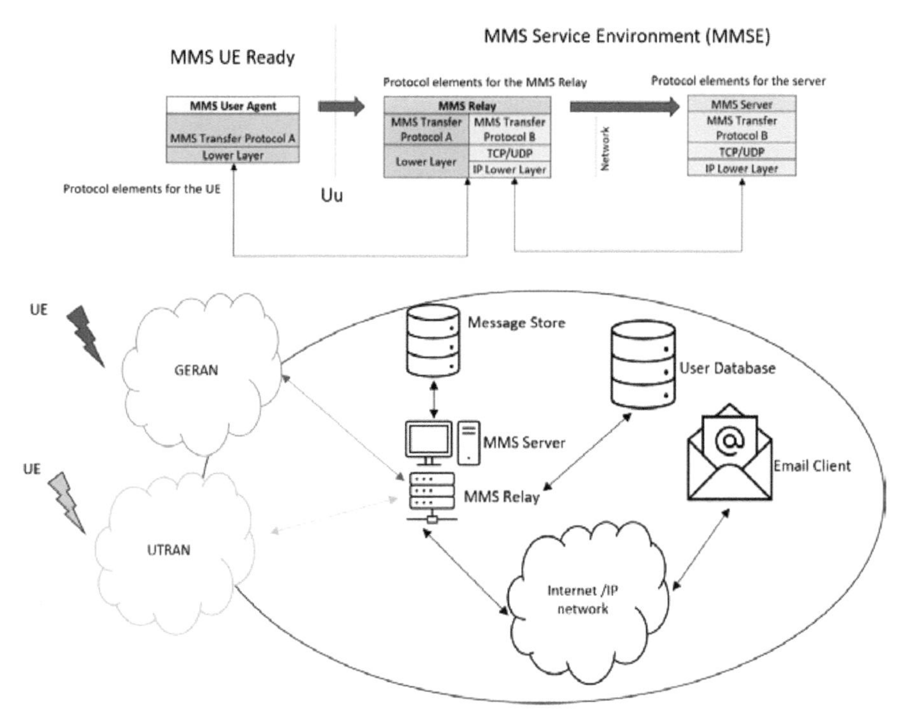

Figure 7.2 MMS framework and the 2G – 3G architecture based on MMS.

1. The MMS Agent (Application Layer) sends the MMS msg to the MMS Relay via UMTS air interface (Uu).
2. The MMS Relay forwards the MMS msg to the MMS Server.
3. The area which occurs the MMS exchange message is considered MMS Service Environment.

On the other hand, to initiate the streaming of data real-time streaming protocol (RTSP) is utilized, returning the transport control protocol (TCP) and user datagram protocol (UDP). The protocols are selected following the media service in use, and all transactions are exchanged via IP Network.

After All, the figure above shows the high-level overview of streaming communications considering the MMSE and the two distinguished Radios, the GSM EDGE radio access network (GERAN), representing the 2G Networks following the 3G Radio designated as UMTS terrestrial radio access network (UTRAN). Ultimately the UMTS Radio Bearer [11] service located in the Layer 2 and Layer 2 of the OSI Network will establish the QoS for the MMS. In this approach, one can envisage why peak signal to noise-ratio (PSNR) is selected as one of the first QoS metrics to influence the audio-visual quality of an MMS streaming transmission over the cellular networks.

7.2.2 4G Innovates the LTE-Broadcasting for Multicast Services

With the advent of the 4G networks, the TV everywhere concept via OTT providers was consolidated and became prevalent. It enabled new ways for viewers to watch TV and film at the anytime-anywhere (AA) concept, and it unleashed business opportunities for ADVoD services. The power of LTE-enabled Full HD video transmission and certain types of UHD video streaming in its best possible network condition. Additionally, to this success, Broadcasters and OTT platforms can use better ways of streaming content with the QoS and control based on the long-term evolution broadcast (LTE-B) standard. LTE-B considerably improved the UX for mobile video consumers. This was possible by implementing multiple-input and multiple-output (MIMO) antennas and evolved multimedia broadcast multicast service (eMBMS) as advanced features in the 4G ecosystem. Figure 7.3 displays the architecture of LTE-B along with the eMBMS.

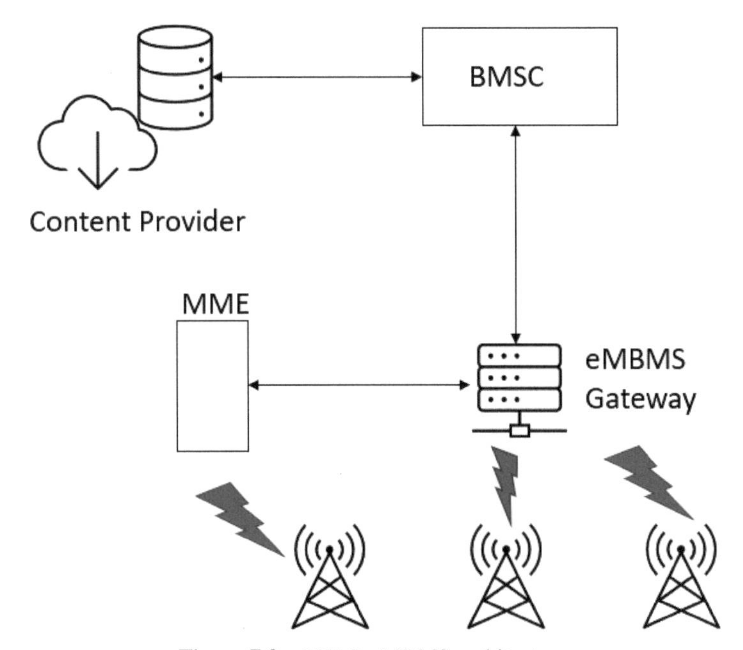

Figure 7.3 LTE-B eMBMS architecture.

As shown in Figure 7.3 above, eMBMS can enable advanced communication on the LTE Radio bearer, enabling communication-based in point-to-multipoint communications [12], something not available in the previous cellular networks. First, the CP sends the content to the broadcast multicast service center (BMSC). Then, the MBMS Gateway transmits the multicast signal to several cell sites. The entity responsible for controlling UE sessions is the mobile management entity (MME), which works parallel with the eMBMS gateway. Furthermore, this facilitates the ISPs and Broastacaster to save bandwidth, enabling multicast transmissions over unicast, finally optimizing network traffic and saving bandwidth [13]. The advantages of using eMBMS are:

- Authorizes LTE UE to switch from unicast to multicast
- Improve network effectiveness in the LTE Radio while numerous UEs are connected to LTE radio bearers and start receiving the same content streaming.
- Creates opportunities to scale up broadcasting streaming content for UEs.

The LTE-B concept also is based on DASH technology to enable CDNs to handle different file sizes and share them in different segments. It also reduces further latency based on the moving picture experts group (MPEG) standards.

LTE-B democratizes the content distribution creating new business model transactions for mobile network operators (MNO), CPs, and ISPs. Extra services like multiple-cameras viewer point and build a bridge with the next TV generation, standardized as ATSC 3.0. Content Streaming is the new disruptive digital service competing with Live TV Broadcasting using the digital terrestrial TV broadcast (DTTB) distribution. This novel architecture unleashed the full potential of streaming digital video content based on different strategies such as unicasting, multicasting, and broadcasting services. The MNO can dynamically adjust the network by allocating the right resource to users and increasing the QoE. Looking beyond the 4G Networks, LTE-B can be part of 5G Network Slicing, offering premium multimedia services for high-end users.

Figure 7.4 shows the OTT concept, including VoD business transactions underpinned by MNOs, CSPs, and CDNs.

On the other hand, mobile broadband generations authorized the deployment of a better level of quality control for content to ISPs, broadcasters, CPs, and OTT providers, increasing the QoS and QoE for multimedia applications

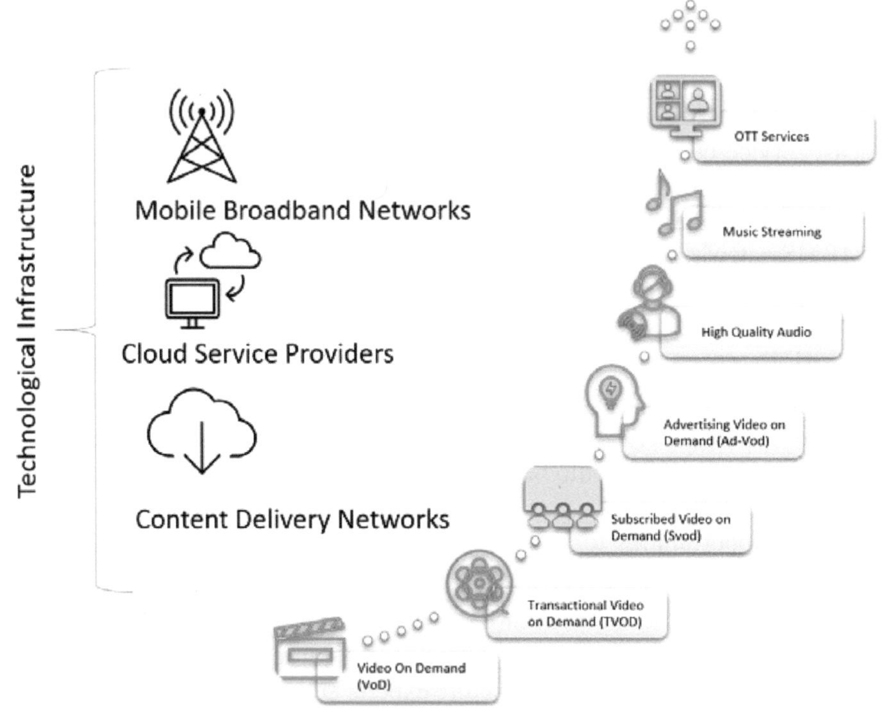

Figure 7.4 OTT and VOD disruptive services.

at the cloud radio access network (C-RAN) Level. To catch up with the disruption created by the OTT service, the advanced television systems committee (ATSC) and the TV industry created the latest model of Digital TV, defined as Next Gen TV, standardized as ATSC 3.0.

7.2.3 5G Networks and Media Streaming with Higher QoS

The 5G network architecture brought innovation to the media and broadcasting industries. These innovations are in the beginning being explored. 5G innovations are based on features designed in the 5G New Radio (NR). The new features will also impact the QoE for users, including services that could not be further explored in the 4G like 360° Video Calling, eXtended reality (XR), and 4K videos. For instance, 5G XR Radio is planned under the 3GGP specification TR 26.928 [14], allocated under the 5G Release 16. In this specification, there is a preparation for 5G to handle all applications related to extended reality for all industry verticals like education, health, entertainment, military, emergency response, etc. The XR services expected to gain with 5G XR Radio are:

- Augmented Reality (AR)
- Virtual Reality (VR)
- Mix Reality (MR)

The user's experiences target to be delivered in these specifications are:

- 3DoF (3 Degrees of Freedom)
- 6DoF (6 Degrees of Freedom)

The degrees of freedom are part of the UX for the consumers of extended reality. It represents the possibilities of movement that the user can have while interacting with the XR image. Below in Table 7.3 is a description of some use cases envisioned for the 5G XR Radio, which will require supporting high data rates and low latency type of communications to achieve a satisfactory QoE.

The table below displays some 5G XR use cases envisaged on the 5G specifications. Table 7.3 also brings the expected user experience for each use case, including the area in which the QoS will need to apply to guarantee the QoE for the expected devices rendering XR images.

Likewise, 5G video and TV streaming will have enhanced capabilities using Non-Public Networks (NPN) based on 5G architecture to deliver content in an end-to-end IP private network. Also, as CPs convert their production environment system to be fully digitized and IP conversant, 5G arrives with extra gears to deliver novel solutions to the media industry, leveraging Live TV productions' versatility. NPN is designed under 5G Release 16. It allows

Table 7.3 5G XR uses cases and user experiences.

Use Case	TYPE of Service	User Experience Expected	QoS KPIs	Device
3D Image Massaging	AR	3DoF+, 6DoF	DL, UL improvements	Phone
Immersive Game Spectator	VR	6DoF	Streaming	2D Screen, HMD
360-degree reference meeting	AR, MR, VR	3DoF	Conversational	Mobile/Laptop
Police Critical Mission with AR	AR, VR	3DoF to 6DoF	Local Streaming, Interactive, Group Communications	5G AR, Glasses, Helmet, VR camera, 5G geopositioning

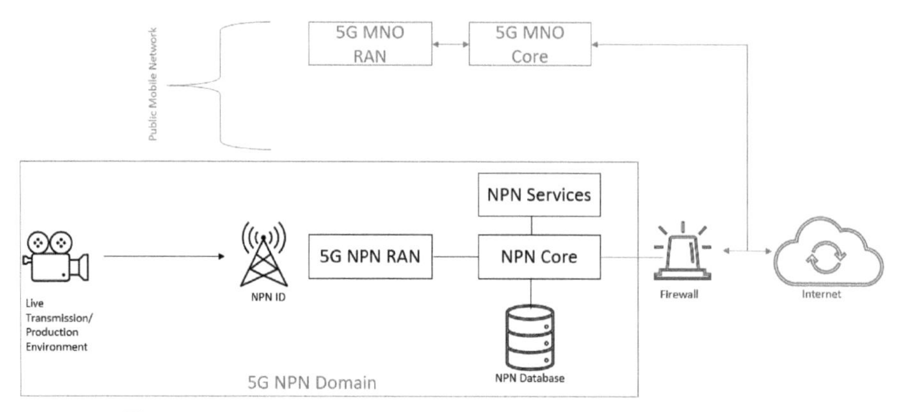

Figure 7.5 5G NPN for live transmission or production environment.

CPs to create private networks fully dedicated to the media streaming, reducing latency, eliminating traffic congestions, and increasing bandwidth allocated for the content streaming and the UEs. It is also important to mention the architecture of LTE Based 5G Terrestrial Broadcast (5G-TB). 5G-TB also incorporates the functionalities of LTE-B, as previously mentioned. With these added functionalities, the 5G architecture can utilize Single Frequency Networks (SFN), which enables transmitting a single content via multiple cells synchronized in a different time interval, maximizing the gain and the content quality at the UE. Figure 7.5 describes the 5G Private Network for Live Broadcasting [15], also known as Non-Public Networks.

However, 5G architecture based on Network Slicing offers new tactics to improve the user's perception of the video delivery in the traditional Live Streaming or XR, there still needs a review of a new methodology to parse and process the big data at the edge of the network specially at the Cloud RAN to increase the traffic prioritization and to ensure the QoE (ML) for mobile broadband subscribers. Studies propose using Machine Learning in the 5G networks to handle this laborious and meticulous task [16].

7.3 6G Quantum Machine Learning for QoE

With the 6G research ongoing for creating a future human-centric network, it is essential to plan and design a mechanism to ensure a higher QoE for all video streaming applications over the next generation of mobile networks (NGMN). Therefore, this chapter proposes a 6G QoE process for the future users of this network. The QoE process will be based on Artificial

Intelligence, Machine Learning, and Quantum Computing. As 5G Networks are currently being implemented worldwide, there is a need to solve the foreseen growth of heterogeneous data at the cloud radio access network (CRAN) to manage QoS effectively.

Multi-Access edge computing network (MEC) technologies are also being explored to enable Cloud Services to become available to the edge of the cellular network. In this aspect, RAN and MEC complements each other [17] as technologies and undoubted will be part of the 6G evolved CRAN (6G-eCRAN).

With these combined forces of MEC and CRAN, the future wireless network, 6G, will enable new services once fully integrated with Quantum computing and Artificial Intelligence. The prediction made here is related to the fact that quantum computing (QC) will probably be deployed first in a Hyperscaler and then be available to the 6G architecture via 6G MEC to power the computing resources of Big Data being exchanged in the CRAN.

6G will initiate the novel concept of Society 5.0, intensifying the integration of Communication, Navigation, Sensing, and Services (CONASENSE) and the Knowledge Human Bond Communications Beyond 2050 (Knowledge Home). Lastly, models like the Internet of Things (IoT), Industrial Internet of Things (IIoT), and Internet of Nano Things (IoT) will converge to the Internet of Beings (IoB) to offer complete assistance to human beings in all sectors from Smart Cities to Space Explorations. To name just a few areas to be transformed by 6G. Therefore, technological investments in Ultra-Massive MiMO (UM-MIMO), Holographic Radio, and Quantum Communications will be vital to architect 6G as a critical enabler of society's advancements beyond 2030. Figure 7.6 presents these deliverables that 6G achieve.

7.3.1 Quantum Computing (QC)

What is important to mention is that Quantum Computing surpasses classic computation based on quantum mechanics theories. Quantum computers, instead of considering only 0 (zero) or 1 (one) as in the classical computers binary code known as **bit**. It opens the ability to simultaneously have any value between 0 to 1, known as superposition. Thus, the quantum equivalency for a bit is the quantum bit or simply **qubit.** It is possible to obtain a unique value of a qubit by adding **quantum interference**. Quantum interference is when a qubit can *collapse* [18] to an established value of 1 or 0.

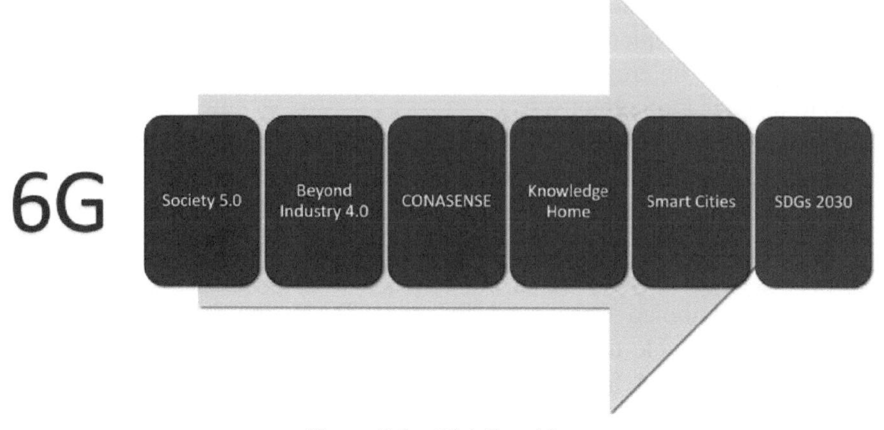

Figure 7.6 6G deliverables.

The power of Quantum Supremacy [19] is already creating opportunities and challenges at all levels of our society. The popularization of a QC shortly will change human life drastically and how data is transmitted on the Internet. It is predicted that QC technology will enable novel use cases, for instance, the next generation of the Internet, the Web 5.0, underpinned by quantum communication. Quantum communications is another field of studies of the QC. Quantum communications focus on building a safer network based on quantum key distribution (QKD). This latter was recently subject to national security policy [20], establishing a new technological security stack for protecting data. QC can break the most secure public encryption keys in a few hours against the years taken by a classical computer.

The challenges to deploying QC and quantum communication networks are vast, but the scientific community progressively announces new solutions to the current technical obstacles. Though, QC is still a challenge to be overcome, as there is no defined architecture to construct a quantum computer yet. Investments are being made in this sector, especially by Hyperscalers like Google, IBM, and Microsoft. Hyperscalers will probably enter the quantum computing market first and offer its services as Quantum as a Service (QaaS) for future mobile operators. As known, quantum states are sensitive to noise. Therefore, few theoretical quantum algorithms could work to overcome this obstacle. The most famous quantum algorithm is Shor's Algorithm, created by Peter Shor [21]. Peter Shor created an algorithm cable to measure a quantum noise with error correction without destroying the quantum information. Many research and development (R&D) programs are dedicated to advanced

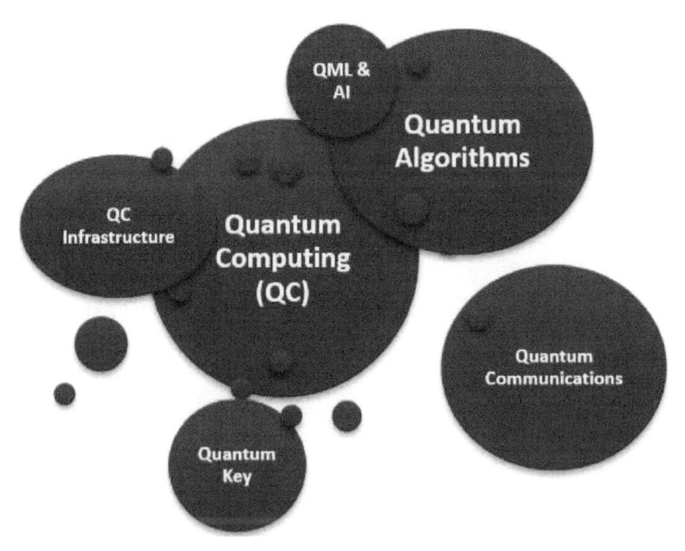

Figure 7.7 Quantum computing technologies research areas.

knowledge in QC and QKD. Then this is the reason to believe that QC will be ready to merge with AI and increase the computing power of handling Big Data in the edge and core of the network exponentially and also control data traffic and the network resources allocated in the Network Slicing attending the expected service level agreements (SLA). Figure 7.7 illustrates the areas under R&D for Quantum Computing, including quantum machine learning (QML).

7.3.2 Machine Learning (ML)

Moreover, the growth of machine learning (ML) research topics to improve Edge Computing is gaining traction, as many published papers reveal. ML is a division area within the AI framework. The main objectives of ML are to emulate the human learning process and progressively enhance its correctness [22]. ML also has applied algorithms for specific tasks in which it is employed. The ML methodologies and algorithms are nowadays well defined, and it is still evolving. Starting from the ML methodology, there are three main learning processes, and they are:

- **Supervised Machine Learning** – this is the ML method cable of learning from a labelled dataset. It is also primarily used in neural networks (NN), Linear Regression, etc.

Table 7.4 Machine learning method and algorithms.

Supervised ML Learning	Unsupervised ML Learning	Semi-Supervised ML Learning
• Labeled Dataset • Neural Networks Algo. • Linear Regression Algo. • Support Vector Machine Algo. • Logistic Regression Algo. • Naïve Bayes Algo.	• Unlabeled Dataset • Neural Networks Algo. • K-means clustering Algo. • Anomaly Detection Algo. • Hierarchical clustering Algo.	• Scarce/Minimum labeled dataset • Generative Adversarial Network (GAN) Algo. • Semi-Supervised GAN Algo.

- **Unsupervised Machine Learning** – this methodology is used to classify unlabelled datasets. It is well explored to classify the image and recognize patterns.
- **Semi-Supervised Machine Learning** – It is a mixture of the previous methodology, and it is deployed when there is an insufficient number of the labelled dataset.

There are many ML algorithms to be mentioned. They are somehow interconnected to the Machine Learning strategy methodology to solve a problem or improve a task. Support vector machines (SVM), K-Nearest Neighbour, Decision Tree, Logistic Regression, Random Forest, etc. are some of the ML algorithms. Below there is a description of some of these principles. Table 7.4 shows a summary of the different types of ML processes.

Also, many papers and designed technologies apply Machine Learning at the MEC to improve the QoS for the cellular network and increase the QoE for multimedia applications [23]. These papers and solutions are currently focused on solving the challenges of guaranteeing SLAs for the 5G Networks, including the entities of the 5G Release 17. 5G Release 17 is consisted of:

- eMBB (enhanced Mobile Broadband)
- Ultra-Reliable Low Latency Communications (URLLC)
- Massive Machine Type Communications (mMTC)

ML can support these three core entities in the 5G Networks utilizing different methodology and algorithms to increase the QoS and QoE for advanced multimedia solutions from health to cross-reality videos. The same would benefit 6G Networks and the 6G evolved version of the eMBB, URLLC, and mMTC.

7.3.3 6G Quantum Machine Learning for Multimedia QoE

Quantum computing and AI's proposed synergy is not new [24]. Other papers have now published the importance of looking at it, focusing on the subject of the NGMN. However, this chapter focuses on the present specific use of quantum machine learning (QML) [25] to improve the quality of experience for the future 6G users while interacting with multimedia video streaming applications eligible to take off commercially during the 6G era. Considering this point of view, below is a description of these future multimedia services, as shown in Tables 7.5 and 7.6.

Table 7.5 above shows the standard 2D video formatting and their respective resolution, bit rate, and latency tolerance. 8K TV formatting is in the embryonary stages of evolution. Its roadmap is not fully consolidated yet, as slowly, the UHD TV and 4K cinematic formatting are catching up with consumers. More and more content based on these formats is becoming available for users. On the other hand, there are new video streaming services in 3D, including 3D Video calling, Cross-Reality, and holographic types of communications (HTC).

These applications require a higher data throughput and are extremely sensitive to latency, rebuffering, playback error, and startup time. Thus, an intelligent network such as 6G must predict to prevent those issues and immediately offer to correct the network by redistributing resources and

Table 7.5 2D video/TV/cinema format.

2D Video/TV/Cinema Format	Pixels	BitRate	Latency
UHD and 4K	8,294,400	20-25Mbs	15-35ms
8K	33,177,600	100Mbs	15-35ms

Table 7.6 Beyond conventional video streaming.

Another Display format	bit rate	latency
XR	10-60Mbps	5-30ms
3D	50-200Mbps	5-20ms
Holographic Comm	4-10Tbps	sub milliseconds

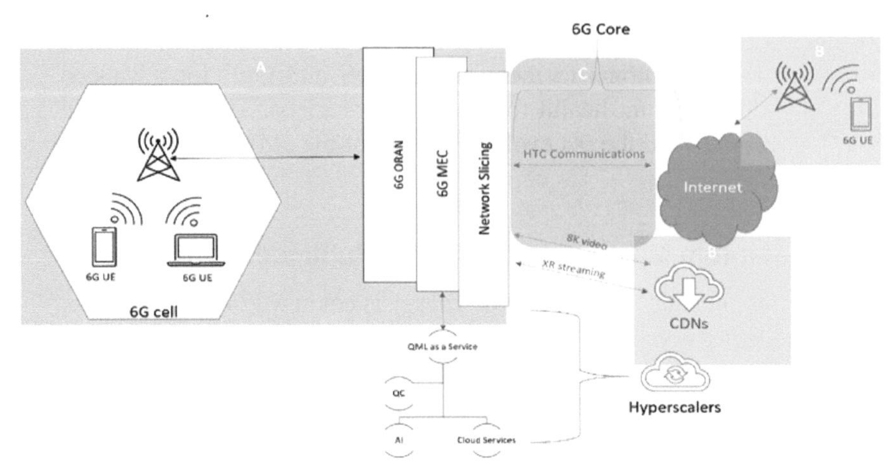

Figure 7.8 QML architecture for 6G networks.

delivering the optimal QoE for users. Consequently, nothing is better than joining quantum computation and Machine learning to tackle these issues to deliver this intelligence to the network. The QML architecture for managing QoE for video streaming focus on 6G networks is described in the following figure. Figure 7.8 show in part A the 6G cell that handles the 6G users' equipment (UE), including the 6G Open-Cloud RAN (6G-OCRAN), the 6G MEC responsible for delivering the Hyperscalers services of Quantum Machine Learning as a Service (QMLaaS), and then the Network Slicing part dedicated to receiving the prioritized traffic decided by the QML. In part C, the 6G Core and the future Internet-based on Quantum Communications, followed by the B part, comprises 6G cell B and the CDNs for 8K and XR streaming.

7.4 Conclusions

The success of 6G Networks will depend on being an intelligent network to serve the future society [26]. Therefore, it is essential for the 6G Open Ran architecture to embrace the next generation of Multi-Access Edge Computing interlaced with AI and Quantum Computing. If these promises were met by 2030, the improvements delivered to the user's experience based on video streaming and other multimedia services would be excellent. However, the challenges are significant first, to improve the Machine Learning algorithms to expand them further, and secondly, to create a commercial Quantum Computing able to be merged with AI and create the Quantum Machine Learning (QML) as an actual network entity.

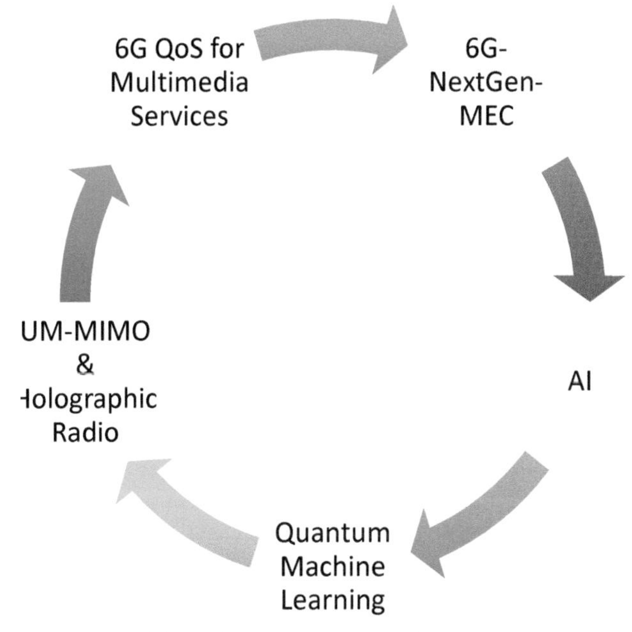

Figure 7.9 6G core areas for R&D.

On the other hand, there is a need further to advance the AI Quantum Computing and Quantum Communications studies. These will be the key areas to secure the network and scale up the current computing power. Henceforward, the authors of these book chapters will continue delving into these areas in the upcoming months and years. The figure below states some of the main areas of study for the 6G architecture, which will explore new technological services from the physical to the application layer.

Thus, as shown in Figure 7.9 above, the areas of 6G that will require investments are QML, Ultra Massive MiMO, Holographic Radio, 6G QoS, and QoE, and finally,, visible light communications and advanced artificial intelligence.

References

[1] Ericsson, "Mobile data traffic outlook," Ericsson Mobility Report data and forecasts. [Online]. Available: https://www.ericsson.com/en/mobility-report/dataforecasts/mobile-traffic-forecast. [Accessed: 12-Jun-2021].

[2] J. Clement, "Global mobile video traffic 2022," Statista, 03-Mar-2020. [Online]. Available: https://www.statista.com/statistics/252853/global-mobile-video-traffic-forecast/. [Accessed: 12-Jun-2021].
[3] T. Stockhammer, "Dynamic Adaptive Streaming over HTTP – Design Principles and Standards" W3C, 2010. [Online]. Available: http://www.w3.org/2010/11/web-and-tv/papers/webtv2_submission_64.pdf. [Accessed: 13-Jun-2021].
[4] K. Graham, "Adaptive Bitrate Streaming (ABR): Optimizing the User Experience," Dacast, 01-Apr-2021. [Online]. Available: https://www.dacast.com/blog/adaptive-bitrate-streaming/. [Accessed: 13-Jun-2021].
[5] ITU. Tsbmail, P.800.1 : Mean opinion score (MOS) terminology, 27-Nov-2019. [Online]. Available: https://www.itu.int/rec/T-REC-P.800.1-201607-I/en. [Accessed: 12-Jun-2021].
[6] Arimas, "MOS Mean Opinion Score," arimas, 23-Jun-2020. [Online]. Available: https://arimas.com/2017/08/01/mos-mean-opinion-score/. [Accessed: 13-Jun-2021].
[7] GSMA, "Network 2020 - GSMA," GSMA. [Online]. Available: https://www.gsma.com/futurenetworks/wp-content/uploads/2017/03/Network_2020_The_broadcasting_opportunity.pdf. [Accessed: 12-Jun-2021].
[8] 3GPP, "3GPP TS 23.140 V6.16 - ARIB." [Online]. Available: http://www.arib.or.jp/english/html/overview/doc/STD-T63v9_20/5_Appendix/Rel6/23/23140-6g0.pdf. [Accessed: 15-Jun-2021].
[9] 3GPP, "TS 123 278 - V10.0.0 - Digital cellular telecommunications system (Phase 2+); Universal Mobile Telecommunications System (UMTS); Customised Applications for Mobile network Enhanced Logic (CAMEL) Phase 4; Stage 2; IM CN Interworking (3GPP TS 23.278 version 10.0.0 Release 10)," ETSI, May-2011. [Online]. Available: https://www.etsi.org/deliver/etsi_ts/123200_123299/123278/10.00.00_60/ts_123278v100000p.pdf. [Accessed: 15-Jun-2021].
[10] Syngress, "SIP Architecture." [Online]. Available: https://cdn.ttgtmedia.com/searchVoIP/downloads/Building_a_VoIP_Network_Ch[1]._8.pdf. [Accessed: 15-Jun-2021].
[11] M. Rupp and M. Ries, "Video streaming test bed for UMTS network," Jun-2016. [Online]. Available: https://publik.tuwien.ac.at/files/pub-et_11905.pdf. [Accessed: 15-Jun-2021].
[12] GSMA, "Network 2020: The 4G Broadcasting Opportunity," Future Networks, 10-Jan-2020. [Online]. Available: https://www.gsma.com/futurenetworks/resources/network-2020-the-4g-broadcasting-opportunity/. [Accessed: 15-Jun-2021].

[13] F. Lachance, "Unicast, multicast, or both: what's right for you? - Genetec." [Online]. Available: https://www.genetec.com/blog/products/unicast-multicast-or-both-whats-right-for-you. [Accessed: 15-Jun-2021].

[14] 5G-MAG, "5G-MAG Explainer - Non-Public 5G Networks for Content Production.pdf," Nov-2020. [Online]. Available: https://drive.google.com/file/d/1MkBjZkVG30wKqwCMOm8GAmEYJu9hb4fE/view?usp=sharing. [Accessed: 15-Jun-2021].

[15] 5G-MAG, "5G-MAG Workshop' Media Production over 5G NPN,'" MEDIA ACTION GROUP, 19-Apr-2021. [Online]. Available: https://www.5g-mag.com/post/5g-mag-workshop-media-production-over-5g-npn. [Accessed: 18-Jun-2021].

[16] J. Kaur, M. A. Khan, M. Iftikhar, M. Imran and Q. Emad Ul Haq, "Machine Learning Techniques for 5G and Beyond," in IEEE Access, vol. 9, pp. 23472-23488, 2021, doi: 10.1109/ACCESS.2021.3051557.

[17] IBM Cloud Education, "What is Machine Learning?," IBM. [Online]. Available: https://www.ibm.com/cloud/learn/machine-learning. [Accessed: 18-Jun-2021].

[18] "Shor's algorithm," IBM Quantum. [Online]. Available: https://quantum-computing.ibm.com/composer/docs/iqx/guide/shors-algorithm. [Accessed: 18-Jun-2021].

[19] F. Tavares, "Google and NASA Achieve Quantum Supremacy," NASA, 23-Oct-2019. [Online]. Available: https://www.nasa.gov/feature/ames/quantum-supremacy/. [Accessed: 18-Jun-2021].

[20] Lochi.orr@nist.gov, "Quantum Communications and Networks," NIST, 02-Oct-2020. [Online]. Available: https://www.nist.gov/programs-projects/quantum-communications-and-networks. [Accessed: 18-Jun-2021].

[21] D. Castelvecchi, "Quantum-computing pioneer warns of complacency over Internet security," Nature News, 30-Oct-2020. [Online]. Available: https://www.nature.com/articles/d41586-020-03068-9. [Accessed: 18-Jun-2021].

[22] "2.4 What is reinforcement learning? Grokking Machine Learning MEAP V14," liveBook · Manning. [Online]. Available: https://livebook.manning.com/book/grokking-machine-learning/2-4-what-is-reinforcement-learning-/v-4/1. [Accessed: 18-Jun-2021].

[23] P. S. Rufino Henrique and R. Prasad, "The Road for 6G Multimedia Applications," 2020 23rd International Symposium on Wireless

Personal Multimedia Communications (WPMC), 2020, pp. 1-6, doi: 10.1109/WPMC50192.2020.9309478.

[24] J. Biamonte, P. Wittek, N. Pancotti, P. Rebentrost, N. Wiebe, and S. Lloyd, "Quantum machine learning," Nature News, 14-Sep-2017. [Online]. Available: https://www.nature.com/articles/nature23474/. [Accessed: 18-Jun-2021].

[25] A. Sag, "5G and AI: Complementary Technologies Now and Into The Future," Forbes, 25-Nov-2020. [Online]. Available: https://www.forbes.com/sites/moorinsights/2020/11/25/5g-and-ai-complementary-technologies-now-and-into-the-future/. [Accessed: 18-Jun-2021].

[26] P. S. Rufino Henrique and R. Prasad, "6G The Road to the Future Wireless Technologies 2030," River Publishers: Professional Books, 31-Mar-2021. [Online]. Available: https://www.riverpublishers.com/book_details.php?book_id=920. [Accessed: 18-Jun-2021].

Biographies

Paulo Sergio Rufino Henrique (Spideo-Paris, France) CTIF Global Capsule, Department of Business Development and Technology, Aarhus University, Herning, Denmark. Paulo S. Rufino Henrique holds more than 20 years of experience working in telecommunications. His career began as a field engineer at UNISYS in Brazil, where he was born. There, Paulo worked for almost nine years in the Service Operations, repairing and installing corporative servers and networks before joining British Telecom (BT) Brazil. At BT Brazil, Paulo worked for five years managing MPLS networks, satellites (V-SAT), and IP-Telephony for Tier 1 network operations. During that period, he became the Global Service Operations Manager overseeing BT operations in EMEA, Americas, India, South Korea, South Africa, and China. After a successful career in Brazil, Paulo got transferred to the BT headquarters in London, where he worked for six and a half years as a service manager for Consumers Broadband in the UK and IPTV manager for BT TV Sports channel.

Additionally, during his tenure as IPTV Ops manager for BT, Paulo also participated in the BT project of launching the first UHD (4K) TV channel in the UK. He then joined Vodafone UK as a quality manager for Home Broadband Services and OTT platforms and worked for almost two years. During his stay in London, Paulo completed a Post-graduation Degree at Brunel London University. His thesis was entitled 'TV Everywhere and the Streaming of UHD TV over 5G Networks & Performance Analysis'. Presently, Paulo Henrique holds the Head of Delivery and Operations position at Spideo, Paris, France. He is responsible for integrating the Spideo recommendation platform on the OTT and IPTV providers. He is also a PhD student under the supervision of Professor Ramjee Prasad at Global CTIF Capsule, Department of Business at Aarhus University, Denmark. His research field is 6G Networks - Performance Analysis for Mobile Multimedia Services for Future Wireless Technologies.

Dr Ramjee Prasad is the Founder and President of CTIF Global Capsule (CGC) and the Founding Chairman of the Global ICT Standardization Forum for India. Dr Prasad is also a Fellow of IEEE, USA; IETE India; IET, UK; a member of the Netherlands Electronics and Radio Society (NERG); and the Danish Engineering Society (IDA). He is a Professor of Future Technologies for Business Ecosystem Innovation (FT4BI) in the Department of Business Development and Technology, Aarhus University, Herning, Denmark. He was honoured by the University of Rome "Tor Vergata", Italy as a Distinguished Professor of the Department of Clinical Sciences and Translational Medicine on March 15, 2016. He is an Honorary Professor at the University of Cape Town, South Africa, and KwaZulu-Natal, South Africa. He received the Ridderkorset of Dannebrogordenen (Knight of the Dannebrog) in 2010 from the Danish Queen for the internationalization of top-class telecommunication research and education. He has received several international awards, such as the IEEE Communications Society Wireless Communications Technical Committee Recognition Award in 2003 for contributing to the field of "Personal, Wireless and Mobile Systems and Networks", Telenor's Research Award in 2005 for outstanding merits, both academic and organizational within the field of wireless and personal communication, 2014 IEEE AESS Outstanding Organizational Leadership Award for: "Organizational Leadership in developing and globalizing the CTIF (Center for TeleInFrastruktur) Research Network", and so on. He has been the Project Coordinator of several EC projects, namely, MAGNET, MAGNET Beyond, and eWALL. He has published more than 50 books, 1000 plus journal and conference publications, more than 15 patents, and over 150 PhD. Graduates and a more significant number of Masters (over 250). Several of his students are today worldwide telecommunication leaders themselves.

Ramjee Prasad is a member of the Steering, Advisory, and Technical Program committees of many renowned annual international conferences, e.g., Wireless Personal Multimedia Communications Symposium (WPMC); Wireless VITAE, etc.

8

Honey, I Flung the RAN to Space!

Rajarshi Sanyal

67, Route De Bouillon, Arlon,6700, Belgium
E-mail: rajarshi.sanyal@proximus.lu

Dedicated to the memory of my two dear friends and telecom wizards, David Berrocal Bustos (Madrid) and Imtiaz Ahmed (Shenzhen/Calcutta)

With an upsurge in the proliferation of mobile devices, the service providers are continuously adding up cells and bolstering their network footprint. A standard cell implies a physical radio tower unless it is a small cell. Adding up the cells instantly may not be possible beyond a limit due to infrastructure cost, spectrum cost, and threshold limits of frequency reuse factor for a given network. Hence, radio infrastructure and spectrum sharing between multiple operators are quite common nowadays, especially in the era of 4G and 5G. Sun-setting legacy 3GPP access technologies (2G & 3G) facilitate spectrum re-farming to fill the gap further. Some satellite operators are releasing a portion of their C band spectrum, which coincides with the 5G spectrum in the range of 3.7 GHz. However, the question is: what more can be done to optimise the network (and the use of infrastructure and spectrum). In this paper/chapter, we present one such technology where the satellite to 5G/6G UE access can be potentially exploited to address use cases for IoT and human-centric devices for 5G and 6G networks at a massive scale. We broadly cover the implementation aspects, design choices, technological and regulatory constraints, and the solutions around this idea. Few use cases around E-MBB, URLLC, and m-MTC have been exemplified to deliberate on these three pillars' opportunity, strengths, weaknesses, and risks. This will be a key ingredient for network evolution towards 6G.

8.1 Introduction

At times, friends may prove to be incredibly resourceful and living portals to the realm of information. Sometimes, from those friendly blathers, we come to know better what's happening around the world and us. In one such breakfast gossip on a slothful Sunday couple of years back with a friend, we conversed on 'tree BTS' (tree- Cellular Base Station). Queer idea indeed. The intention is to eliminate those unsightly cell towers in our neighbourhood and conceal the BTS in artificial trees. Of course, it is not as simple as it sounds as they had to contemplate means to retain the characteristics of the signal and to ensure that the propagation loss, signal attenuation, the physical effects of fading, reflection, and dispersion remain unaltered compared to traditional BTS.

He had subsequently spawned up a startup, though I do not know how successful it was, how many cell towers he could replace, and how many birds he could lure to his trees.

However, during a technology scan, another novel idea pertaining to a similar theme popped up in recent times. There is an endeavour of late to beam the cellular signal from the sky directly to the smartphone (and vice versa). Eventually, if all goes well, the satellites may replace the cell towers before we realise. Is it just wishful thinking? No, things may indeed move fast, as a couple of heavy-weight protagonists are behind this idea. But, the impediments around the technological and regulatory domains are bountiful, and they need to be addressed before this can be made operable.

Building up a massive overlay of satellite networks for global mobile communication may not be solely for environmental beautification. It may be pertinent to recall that satellite phone services were never as affordable as cellular services due to the inordinate expense of such non-terrestrial transport networks. The devices were expensive compared to a cellular smartphones. The cost of scaling up the network was manifold compared to a traditional terrestrial 5G network (say).

However, it has been anticipated that cellular communication on the satellite will be one of the key propellants for 6G networks. Infrastructure costs globally are on the rise. We are striving towards reducing our carbon footprint. Building the RAN in space powered by solar energy may sound quite logical to alleviate environmental and signal pollution. The sharp beams from space are likely to reduce terrestrial radio pollution due to lesser chances of terrestrial scattering and reflection, at least for the LOS use cases. This is also synergised with the cloud RAN /cloud Core initiatives and RAN/Spectrum sharing. The core network may be hosted anywhere in the cloud while the RAN (or part of RAN) may be in space. This renders excellent flexibility to operators to commence operation rapidly without implementing any physical equipment at the RAN and the core.

It created quite a stir when China sent the first 6G satellite in November (2020). They had announced that the satellites could operate in the terahertz frequency that could transceive data several times the speed of 5G. However, water vapour could easily absorb terahertz waves, which poses a challenge for using such carrier frequency and questions the efficacy of such a satellite. Other prominent players based in Europe and the US are building up a constellation of low orbit satellites presently with 5G capability and upgradable to 6G. To name a few (non-exhaustive), we have SES, Intelsat, Starlink, ESA, Amazon, and Iridium. Off late, an American satellite company AST is enabling a chunk of mobile users of Rakuten, Japan, to be directly served by a 5G radio network conceived by satellites. These users have no or little access to terrestrial RANs.

Meanwhile, there are rumours that iPhone 13 will have native Satcom capability to enable mobile communication via LEO satellites using customised Qualcomm X60 chipsets. However, we need to wait for a formal communication from Apple on this.

So, what can be the motivation factors for building a celestial mobile network? Are the reasons simply obvious? What can be the design challenges, regulatory constraints or environmental constraints? Can this technology be a game-changer?

The rest of the chapter is arranged as follows. In section 8.2, we explored the type of satellites that can be utilised to build this future mobile network. In the second section, we delved into the system architecture. We have brought in the architectural alternatives, discussed network design, and attempted to address pertinent questions mainly related to handset characteristics, coverage and cell/tracking area planning, and interference. Section 8.3 exemplified a few use cases on URLLC, m-MTC, and E-MBB. In section 8.4, we had put up a few regulatory bottlenecks and deliberated on some possible solutions. Finally, conclusions have been drawn in section 8.5.

8.2 Classes of Communication Satellites

The four types of communication satellites that we have today in space and their application in telecommunications are as follows [2] as depicted in Figure 8.1:

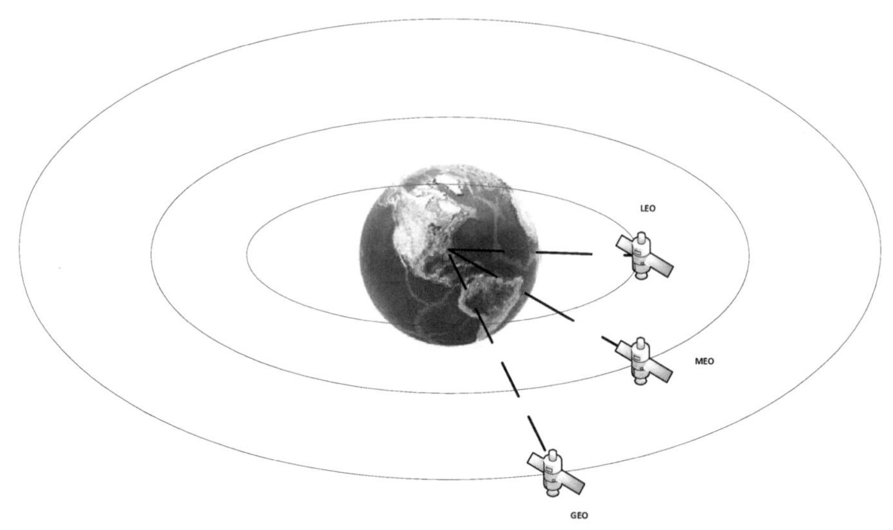

Figure 8.1 LEO, MEO and GEO satellites.

LEO (Low Earth Orbit) satellite: These satellites are in orbit between 500 to 1500 Kilometers from the earth. Typical signal latency is between 20 to 50 milliseconds. Iridium uses these satellites for voice and SMS communication and achieves a data rate of 2.4 kbps. To use LEO satellites for 5G at the height of above 1000 kilometers, some NAS and core network timers may need to be extended as the present timers cannot tolerate the latency. The core network should be more resilient to re-transmissions. A constellation of LEO satellites communicating between themselves on 'Inter Satellite Crosslinks' can form DSSN, which can focus high-density beams to earth to render coverage/mobile service to the high number of users (comparable to user density in urban cellular networks).

MEO (Medium Earth Orbit satellite: These satellites are in orbit between 8000 to 18000 Kilometers from the earth. Typical signal latency is between 100 to 150 milliseconds. Thuraya and Immarsat use these satellites for communication and achieve a modest throughput between 250 and 400 KBPS.

O3b mPOWER communications satellites owned by SES would provide broadband connectivity for cellular backhaul to remote rural locations and IP trunking applications. The higher the satellite's altitude, the lower the beam density would be. Hence the number of users / Kilometer^2 would decrease. However, this can be compensated by the state of the art beam forming, beam spotting and beam steering technologies discussed in the latter part of this chapter.

To use MEO satellites for 5G, some NAS and core network timers may need to be extended as the present timers cannot tolerate the latency. The core network should be more resilient to re-transmissions.

GEO (Geo Stationary Earth Orbit) satellite: These satellites orbit around 36000 Kilometers from the earth. Typical signal latency is between 250 to 350 milliseconds. With the recent advances in satellite technologies, the type HTS (High Throughput Satellite) GEO satellites can provide high bandwidth per user. So the applications where high bandwidth is the key but low latency is not the prerequisite. Some examples are SOTA (Software Over The Air) and FOTA (Firmware Over The Air). A single GEO satellite has the complete visibility of the earth, which is a significant advantage.

NAS timers should be increased to cope with the higher latency and round-trip delays.

8.3 System Architecture

GPP has proposed some architectural options in Rel 17 [1] to engage the satellite network at different legs of the communication, be it at the access network layer (between the device and the RAN), between RAN to Core network, or as a backhaul from the core network to data network, between two foreign PLMNs at the shared border, or as a relay node, etc. However, the implementation scenario and the type of satellite to be used depends on the use cases (URLCC, E-MBB, M-MTC), capability of the devices, mobility state, and the roaming landscape. Below is a snapshot of the implementation choices.

8.3.1 Satellites Directly used for Communication between UE and RAN/Core on 3GPP Defined Interface

The device shall have a satellite modem with 3GPP and non-3GPP satellite access support, communicating directly with the satellite in full-duplex mode. 3GPP release 17 has emphasized this aspect and included the different options for 3GPP and non-3GPP satellite access and integration choices with the 5G core network

There are a couple of possibilities with this mode:

a) Option1: The satellite has a G-NodeB component and communicates with the smartphone device on the Uu interface over one leg and communicates with the core network (AMF / UPF) at the other end on SRI (Satellite Radio Interface), simulating the N2 and N3 interfaces.

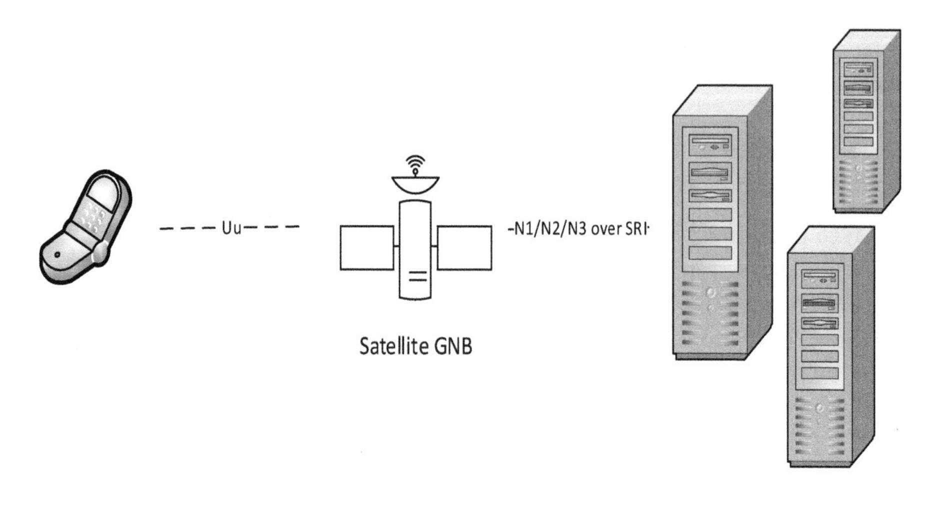

Figure 8.2 UE communicates with onboard satellite GNodeB.

The disadvantage of this model (as in Figure 8.2) is that the satellite will need to implement onboard electronics, the necessary software/firmware of a full-scale Gnode B, implement a MIMO antenna array with beam-forming and beam-steering capability, support multiple 5G interfaces on SRI, all of which can make the satellite expensive and challenging to maintain. For handovers, G-NodeBs of satellites can communicate on the Xn interface. But for that, the satellites need to establish an Xn mesh.

b) Option 2: The satellite has a G-NodeB- distributed unit (DU) component while the G-NodeB- centralised unit (CU) component remains terrestrial and is interfaced to the CU on the SRI interface (as in Figure 8.3). The satellite has only the DU component, so the design is more superficial. It is also more straightforward to maintain the connectivity to the core network as the N2 and N3 are terrestrial interfaces. As per the standard NG-RAN CU/DU distributed model, multiple DU (Satellites) could connect to a single terrestrial CU unit.
c) Option 3: The satellite will act as a relay bent between the device and the RAN (G-NodeB) as in Figure 8.4.

In this case, the satellite acts as a layer two bent. It alters the frequency between the two legs and amplifies the signal in uplink and downlink. However, it retains the content of the signal. This is the simplest and the cheapest mode in this category, as the satellite does not have to bear the

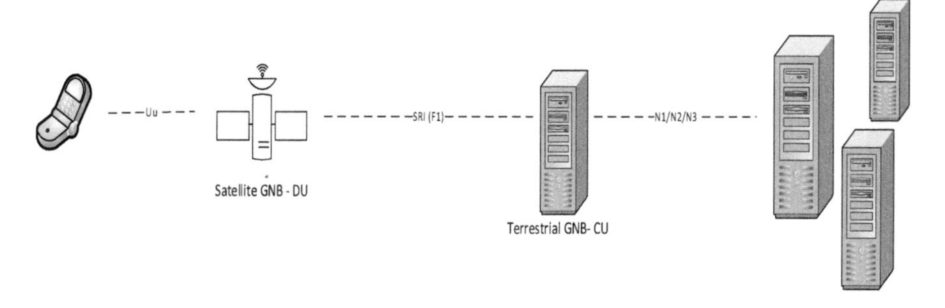

Figure 8.3 UE communicates with onboard satellite GNodeB distributed unit.

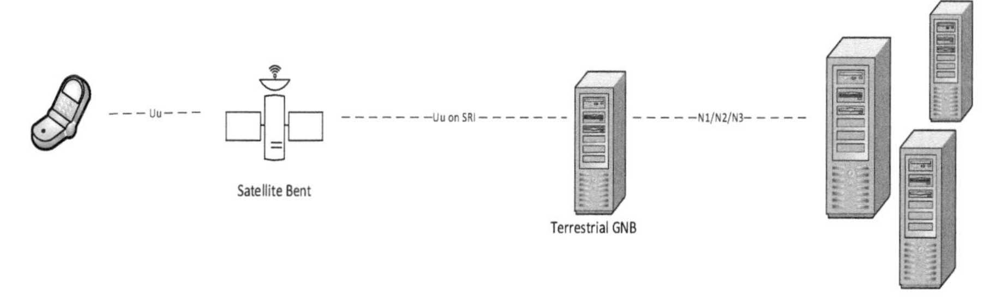

Figure 8.4 UE communicates with satellite relay.

onboard GNodeB. Instead, there should be an onboard repeater present. Also, the satellites do not need to establish the Xn interface (for handover). Xn will be a terrestrial interface in this case.

d) Option 4: Satellite communication between UE and the Core network using the non-3GPP link.

3GPP 5G specification introduced a network function called N3IWF (Non-3GPP Interworking Function), which acts somewhat like E-PDG and TWAG of an LTE network to enable a mobile device to connect to the mobile core network and avail various services, like Voice, SMS, VAS. The 5G core network can utilise this network element to enable the device to communicate with the core network using a non-3GPP satellite link. The radio access technology that can be exploited depends on the satellite's support. DVB-S2 for the forward link and DVB-RCS for the reverse link can be good options. Please refer to Figure 8.5 for details.

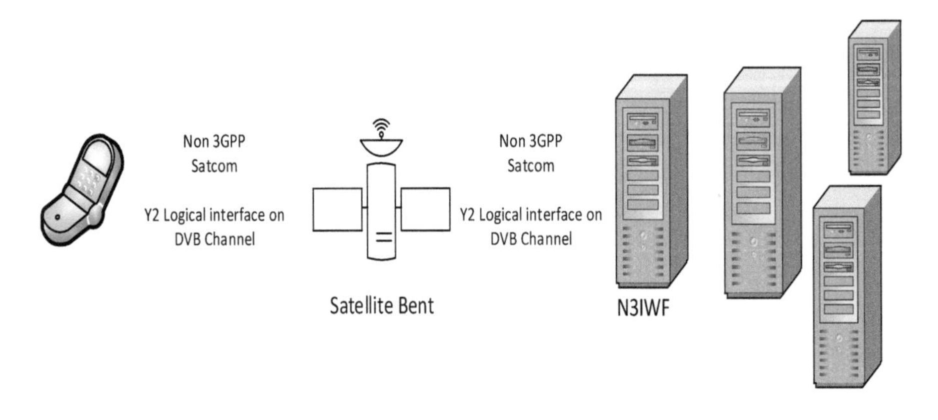

Figure 8.5 UE communicates on the non-3GPP channel on DVB using satellite relay.

8.3.2 Satellites Link as Backhaul to the Core Network for Control and User Plane Traffic

In case the edge (slice) comprising the radio head, GNodeB, and the UPF is geographically isolated from the core network and is in a remote location, then the N4 control channel that uses PFCP protocol between the UPF and the SMF can be backhauled on SRI interface on the satellite link. In addition, the same satellite link could convey the N2 signalling between the GNodeB and the AMF. Please refer to Figure 8.6 for the end to end architecture.

Typical use cases are offshore oil rigs and cruise ships where the GNodeB and UPF can be hosted in a single slice and connected to the on-shore core network (AMF and SMF).

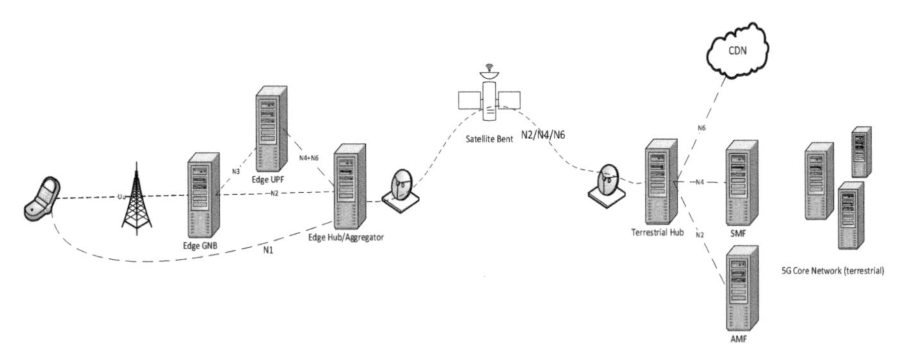

Figure 8.6 Backhaul traffic from edge via satellites.

If local breakout to the content delivery network or the internet is unavailable, the UPF can backhaul the traffic via the satellite link to the CDN on the N6 link.

8.4 Network Design and Planning: Key Constraints, Challenges and Solutions

New technologies invoke questions. Some of them have been addressed below.

a) How many cells can a single satellite cover if the technology is cellular-based?
b) How does the network manage the mobility of users in a fixed PLMN when the coverage of a satellite is moving or when the beam area covers more than one country?
c) How do we guarantee consistent coverage and ensure reliability?
d) How can the satellite address the ultra-low latency requirements and enhanced bandwidth per user?
e) Can the existing access layer technologies like MIMO and beamforming be used to achieve the 5G KPIs?
f) How can the satellite network be integrated with the 5G core?
g) What can be regulatory constraints and challenges for deploying a satellite-based 5G+/ 6G network?

Before we delve into details to address these questions/topics, let us first identify and investigate the key stumbling points

a. Cell planning where satellites cover large areas or else the coverage is dynamic.
b. Interference between satellite GNodeB, terrestrial GNodeB, satellite capable UE, and non-sat cellular UE can affect SNR, SINR, and CINR.
c. High round trip time due to propagation delay of radio signal
d. Spectrum availability

8.4.1 Cell Size and Type, Tracking Area, and Coverage

The cell size and the tracking area can be realised depending on a couple of factors:

- Nature of the satellite: LEO, MEO or GEO.
- Throughput supported by the satellite: depends on whether it is a High Throughput satellite with MIMO support, capable of beamforming and steering antenna array.

- Types of deployment: whether the satellite is directly communicating with the UE (either acting as a GNodeB or a repeater) or whether the satellite is used for backhaul
- Number of PLMNs supported by the satellite: single or multiple
- Frequency bands: C band or V band (for example)

The maximum radius of the cell can be 100 Kilometers considering timing advance, while the nearest LEO satellite is 500 Kilometers away. The larger the cell radius, the lesser the number of users that can be served with the same available spectrum. Typically, in city areas with standard demography, an 8 to 11 Kilometer cell radius could be ideal, although there can be exceptions for densely populated areas where we may need smaller cells, say with a radius between 1 Kilometer to 2 Kilometers.

A couple of 5G satellite communication providers attempt to realise a radio network of such nomenclature with their high throughput (HTS). LEO satellites are equipped with MIMO and beamforming capability, thanks to onboard GNodeB, or GNodeB DU unit. A cluster of HTS satellites communicating between themselves on Inter Satellite Crosslinks forms the DSSN. The beam includes a sort of cone, and the beam's coverage area increases with the height of the satellite. But beam density becomes lesser with the altitude of satellites. In LEO satellites, typically, where the beam width ranges from 30° to 45°, the signal coverage range is around 10 to 50 Kilometers, perfect for moderately sized cells.

Two beam coverage schemes are used, namely spot beam and hybrid wide spot beams. Each satellite provides multiple high-power spot beams to render coverage in a wide area [5]. But the beams move along with the satellite orbit. For example, SES O3b mPower MEO satellites can focus 5000 spot beams on earth.

With the hybrid comprehensive spot beam approach, the satellites beam low power wide beams, which are digitally steered to achieve fixed coverage and high density.

The user density (users / Kilometer^2) may decrease dramatically for MEO satellites. Typically, a cell of 100 Kilometer radius can cater to a user density of 5/ Kilometer^2. In case moderate cells need to be conceived on MEO satellites, the system must be highly dependent on beamforming and beam steering actuated by the MIMO antenna array.

The beams form multiple cells, and the cell clusters form the tracking area (Figure 8.7). If the cell size is well within the tolerable limit, no further restriction or complication is caused by the timing advance parameter.

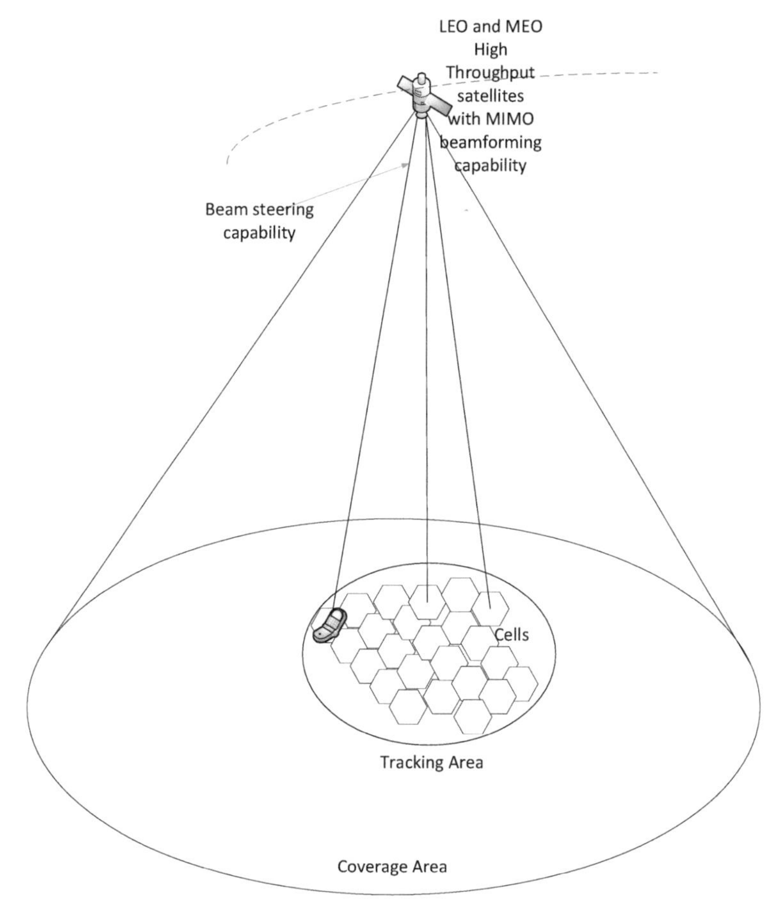

Figure 8.7 Formulation of cell and tracking area by satellite beams.

The number of cells that can be realised by one such satellite depends on the capability of the satellite to handle the number of beams about ARFCNs, the type of satellite (LEO, MEO or GEO), the coverage area and the number of PLMNs each satellite can handle. However, having said so, it seems that we do not have much room to extend the size of the antenna array due to technology limitations and cost. Typically, a satellite that can function as a complete GNodeB can be very complex and weigh up to a one-ton kilogram.

When a device moves from one TA to another TA, which is handled by another satellite, a handover procedure is initiated using the Xn mesh between satellites. It is assumed that the UE has the means to determine its position within the TA. For reliability, at least two satellites serve the cells and TAs. This reduces the chances of glitches and session discontinuity during handover. If a satellite cannot serve a cell by its steerable beam, another satellite that can beam the cell will replace this satellite. This is to ensure that the end-to-end system is highly available.

When the cell and the tracking area are fixed and static, it remains relatively simple to manage the radio network. When the satellite's orbit can be accurately determined, the beam towards the UEs in the cell remains pretty stable and fixed. Any adjustments required can be, however, made by beam steering. The network design and engineering criteria remain standard, except that some timers need to be extended to deal with much higher round trip delays. Also, some new RAT values have been introduced by 3GPP. Below is a summary of the extended timer values (in Table 8.1) and new RAT values that the compatible satellite device and the 5G network functions need to consider and act accordingly.

Modified Timers:

Table 8.1 Increase of NAS/core timers due to satellite link.

	Timer #	RTTs propagation delay	(Minimum) Suggested Timer value increase
5GMM UE Side Timers	T3510	5 RTT	5 WCRTT
	T3517	3 RTT	3 WCRTT
5GMM AMF Side Timers	No need for changes. AMF determines appropriate values		
	T3580	4 RTT	4 WCRTT
5GSM UE Side Timers	T3581	1 RTT	1 WCRTT
	T3582	1 RTT	1 WCRTT
5GSM SMF Side Timer	No need for changes. SMF determines appropriate values		

New RAT values:

- "NR (LEO)".
- "NR (MEO)".
- "NR (GEO)".
- "NR (OTHERSAT)"

When the satellite beam covers multiple countries or the coverage is moving/dynamic, these simple choices of fixed cell id and TA cannot be contemplated anymore. In those circumstances, virtual cells and virtual TAs may be introduced near the international borders.

As per classical radio network design, mobile operators divide coverage areas into fixed cells within well-defined geographic regions. But with this new concept of virtual cells, these cells are preferred to be square-shaped, and they do not relate to actual RF coverage from any satellites or existing cells for terrestrial NR access. The cells may be defined concerning an array of grid points, as shown below. A serving cell for any UE would then be explained by the cell associated with the grid point, which is closest to the current UE location. The definition of the cells may be closely aligned to the national borders of a country.

The virtual cells do not relate to the actual radio coverage of any satellites. Instead, an array of grid points are defined for rectangular or square cells (refer to Figure 8.8). If a UE is in the vicinity of a grid point, it will be served by the cell associated with the grid point. A cluster of these virtual cells can comprise a virtual TA, and these would be generally defined near the international borders. Grid point locations would be propagated by satellites along with the linked virtual cell IDs, TAIs, PLMN IDs (MCC, MNC), etc.

Below is a typical representation of what these virtual cell arrangements in proximity to the international borders may look like. This may be a solution to circumvent issues regarding satellite coverage blackspots, overlapping coverage, or moving coverage, where the UE may prefer to be served by virtual or pseudo cells (or grid points) in those cases to maintain session continuity.

8.4.2 Interference

The existence of terrestrial and satellite 5G networks could invoke channel interference issues [3]. Broadly, there can be four categories (Figures 8.9, 8.10, 8.11, 8.12) of interference though it is not exhaustive.

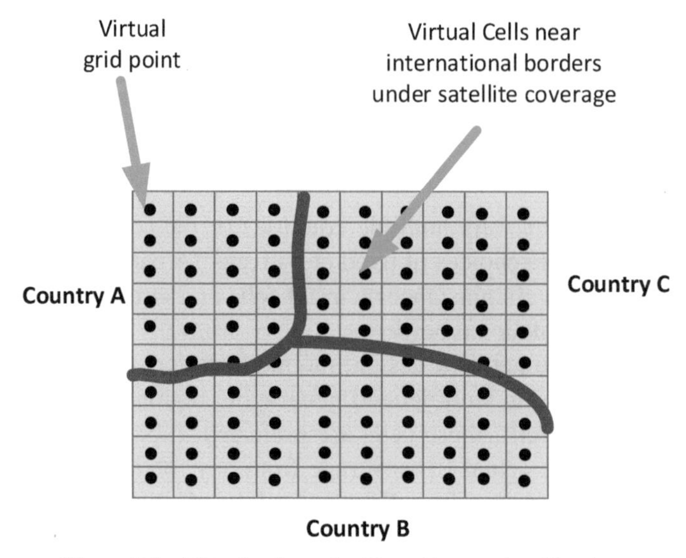

Figure 8.8 Virtual cells and grid near international borders.

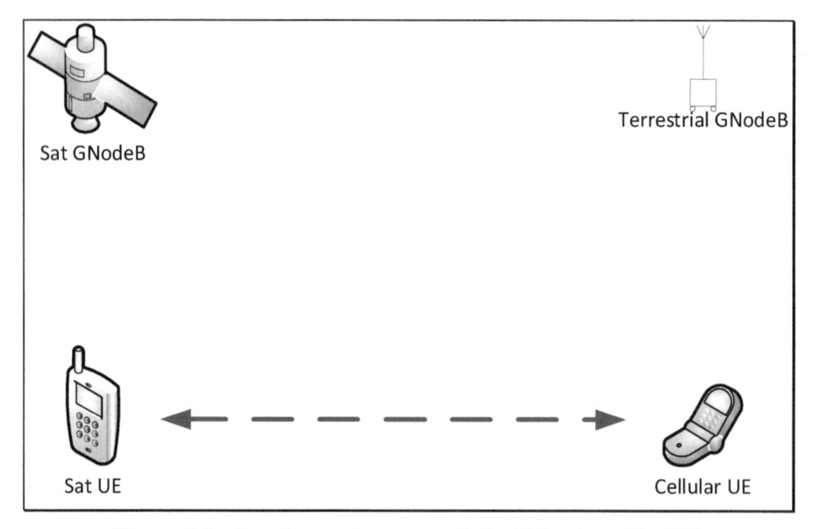

Figure 8.9 Interference between cellular UE and satellite UE.

Depending on the frequency bands, 5G satellite UE may use the same frequency band for uplink communication as that being used by the cellular UE for the downlink communication. This causes interference at the cellular UE. The satellite UEs operating in the S-band primarily house omnidirectional

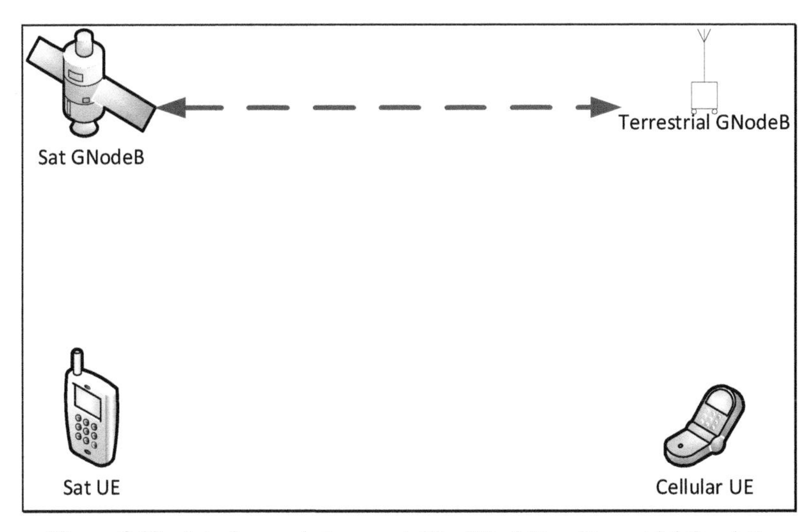

Figure 8.10 Interference between satellite GNodeB and terrestrial GnodeB.

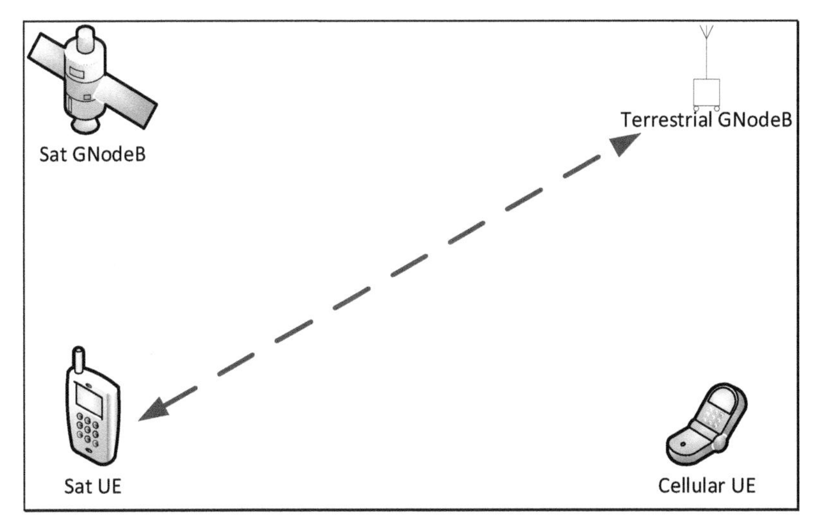

Figure 8.11 Interference between satellite UE and terrestrial GNodeB.

antennas that radiate the signal in all directions, causing heavy interference to the nearby cellular UEs and the terrestrial GNodeBs. However, satellite UEs operating in the Ka-band have VSAT antennas that are more focused towards the satellites mitigating the risks of interference.

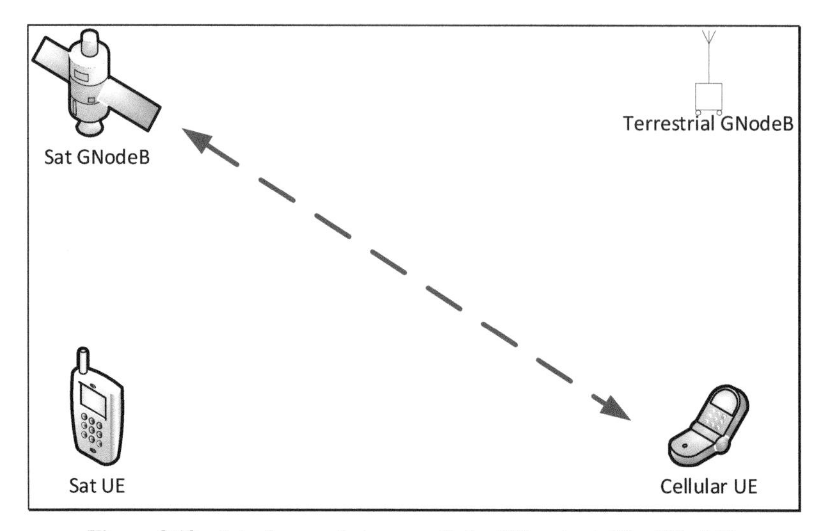

Figure 8.12 Interference between cellular UE and satellite GNodeB.

With the beam forming and steering technologies in the modern high throughput satellites and the 5G sat devices, the UL /DL beams are now more pointed, increasing the SNR, reducing channel interference, and improving the overall signal quality.

As a regulatory requirement, the Isotropic radiated power (especially in the C band) of the terrestrial and satellite RAN must be defined to realise a hassle-free co-existence and interop between the two systems.

8.4.3 Latency

Radio signal nearly travels at the speed of light. Therefore, for a propagation delay of 1 ms between the terrestrial terminal (UE or VSAT) and the satellite, the satellite's distance from the earth needs to be around 300 KILOMETER. As LEO satellites are at a distance between 500 Kilometers to 1500 Kilometers from the earth, achieving the latency of 1 to 5 milli seconds may be nearly impossible. At the moment, an end to end latency of 20 to 25 milli second may be attainable with LEO satellites.

However, there are other sorts of high altitude aerial platforms (HAPS) based on drones or balloons at the height of, say, 80 to 100 Kilometers that may comply with the latency requirements of URLLC applications.

However, the scope of this chapter is not to elaborate on the HAPS systems.

8.4.4 Spectrum Availability

If morning shows the day, then it is most likely that the C band (frequency between 3.7 GHz to 4.2.Ghz) will be used for 5G over satellites in most countries. This is currently the case in the USA. The synergies are obvious.

The 3.7 GHz to 4.2. GHz is one of the most implemented bands for terrestrial 5G networks due to its propagation characteristics and block size. Incidentally, this coincides with the C band of the satellite operators. Hence, there is an endeavour in certain countries (mainly the US) to release a portion of the C bands for terrestrial or satellite mobile communication. Subsequently, a couple of satellite companies strive to enter the 5G communication space by sending Satellite GNodeBs or Satellite GNodeBs - DU or Satellite repeaters (acting as bent/relay between the satellite-enabled UE and the terrestrial GNodeBs).

8.5 Examples of 5G Use Cases

8.5.1 URLLC and EMBB: An Example

Would you like to drive a car where the 'in-car road safety' features are based on the C-V2X application that communicates with the RSU or connected car application via LEO 5G satellites? I would probably not dare to.

But wait. Is it even possible to realise this setup in the first place? The tolerance level for V2X applications is in the range of 1 ms to 5 ms. As deliberated in section 4.2, it may not be viable to achieve this latency on the satellite leg even with DSSN type networks. Moreover, the mobility and velocity of the automobiles can fling a challenge to the '5G-on-satellite' providers to render consistent and reliable coverage actuating continuous beam steering and coping with the high number of handovers.

However, there may be other applications where mobility conditions are moderate and which may be complacent to relatively higher latency. Let us consider the case of a remote mining unit [Figure 8.13]. The pilot miner at the control centre with an Augmented Reality /Virtual Reality gear controls the remote mining equipment/vehicle in an isolated mine connected by satellite-enabled 5G IoT devices. Latency requirements may be in the order of 20ms. However, unlike CV2X applications, these may demand high throughput. A high-resolution AR / VR remote gear rendering "retinal" 360° video streaming experience would require around 600 MBPS.

As this is a typical use case for real-time tactile communications that aim to support the high intuitiveness and reciprocity required between the control

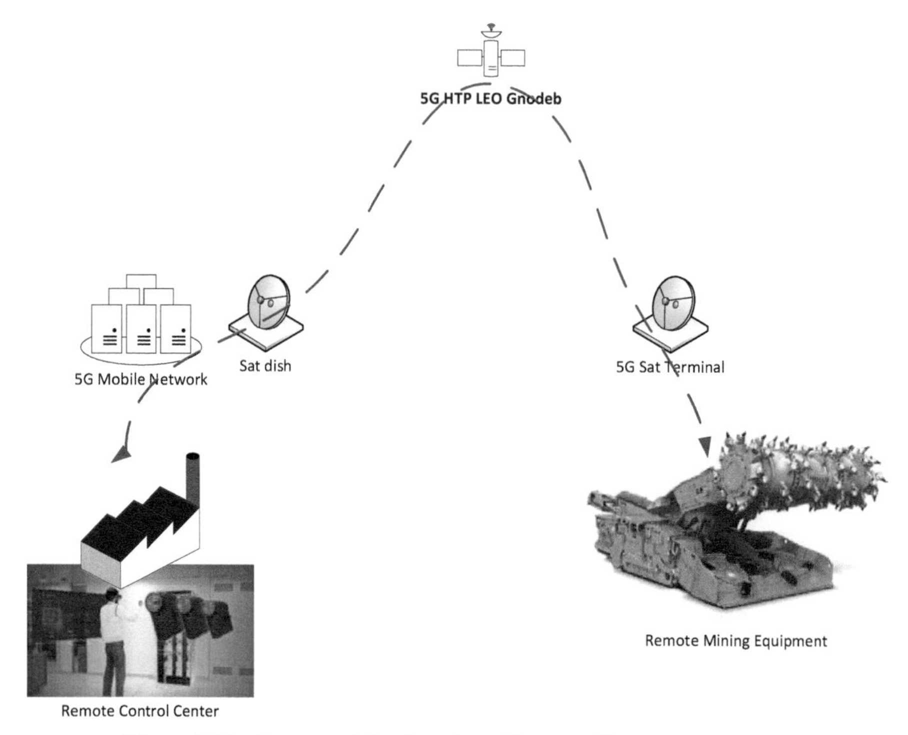

Figure 8.13 Remote mining based on 5G on satellite communication.

centre and the remote equipment, the combined throughput (considering the bandwidth needed for AR/VR) would be around two GBPS per user. Thanks to the advancements in satellite technology and antenna design, the LEO GNodeB satellites can support this bandwidth and latency on the N3 interface towards the terrestrial UPF (slice), supporting the QFIs for E-MBB and URLLC.

In the above examples, we use satellite-enabled 5G devices where the control and the user plane are conveyed over the satellite. We have some dependencies on satellite capabilities and constraints on bandwidth and latency per user. Alternatively, if there are remote local breakout points for the data traffic, then the satellite link may only be used for the control plane traffic for mobility and session management.

8.5.2 m-MTC: An Example

In the previous section, we have just touched upon the use case of a peer to peer communication where low latency and high bandwidth are required for both the forward and the return channels between the UEs / terminals and the network. However, there are multicast applications where reachability and the downlink bandwidth are key criteria, and latency may not be a prime factor. Let us take the example of FOTA (Firmware Over the Air) or SOTA (Software Over the Air) to keep the car software up-to-date continuously (Figure 8.14).

In the 'as is' setup, the car manufacturer would need to tie up with a mobile service provider that can render worldwide coverage via roaming. The E-SIM profile of this mobile operator embedded in the IoT of the car that would be subsequently used for 'over the air' upgrades is generally termed the bootstrap profile. For IoT use cases, NB-IoT and LTE-M accesses are most widely used nowadays. And shortly, this could be 5G.

The present end-to-end system used for any over the air upgrade for automobiles heavily depends on the terrestrial roaming footprint of the bootstrap operator. There can be reliability issues, primarily due to coverage glitches

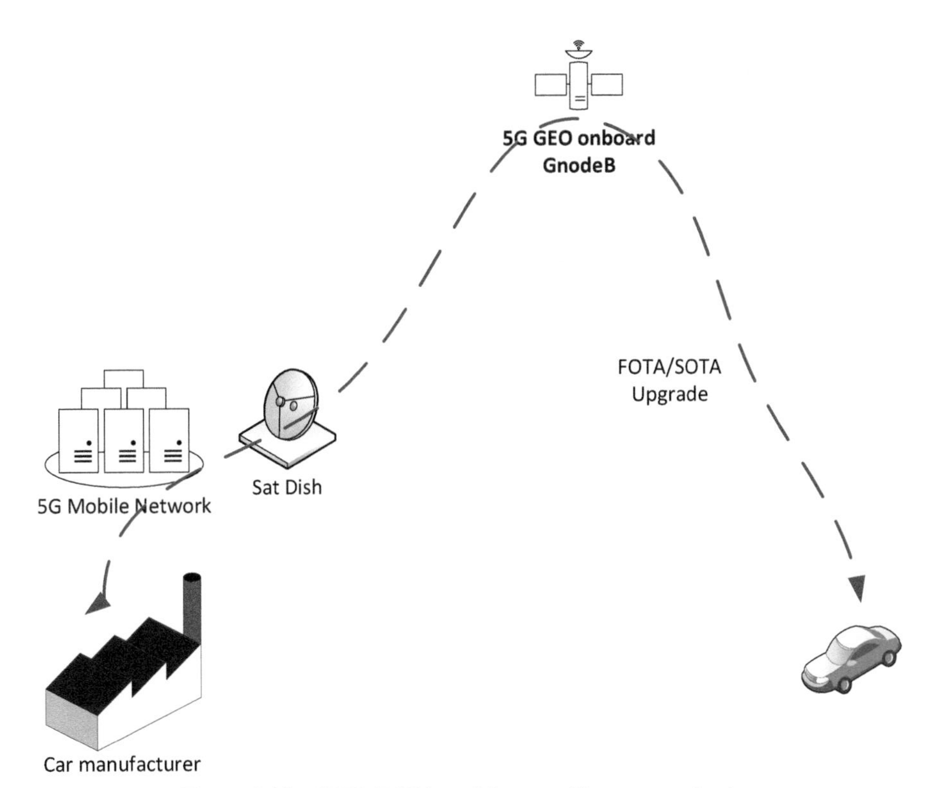

Figure 8.14 SOTA/FOTA on 5G on satellite communication.

in the international borders. The overall operational cost can be high due to the involvement of the roaming leg. Rendering global coverage for the FOTA/SOTA upgrade can be a challenge.

GEO satellites have a global view and can render coverage spanning continents. GEO satellites with onboard GNodeB can beam the software/firmware upgrades from the sky. The car manufacturer or bootstrap operator can tie up with a 5G satellite provider launching (or has launched) GEO satellites at more than 36000 Kilometers and have the required spectrum to operate in the target countries. The end to end latency can shoot up to 300 milli second, but that would still be within the allowable limit for these applications.

8.5.3 Mission-Critical Services/Disaster Management

Iridium has been at the forefront of disaster management and mission-critical communication systems using satellite phones and has been used many times during hurricanes, floods, earthquakes, oil spills, etc. The Iridium satellite phones have an SOS butter (as in the picture in Figure 8.15) which can send messages / and call emergency centres in times of catastrophes.

However, in recent times, thanks to the advancements in onboard electronics and software, it may be possible shortly to place a GNodeB and slice or a '5G in a box' sort of network within the satellites themselves. As it will be a self-contained system, the N6 interface will trombone via an inbuilt mission critical service (MCS) micro application hosted in the same onboard slice to achieve peer to peer communication.

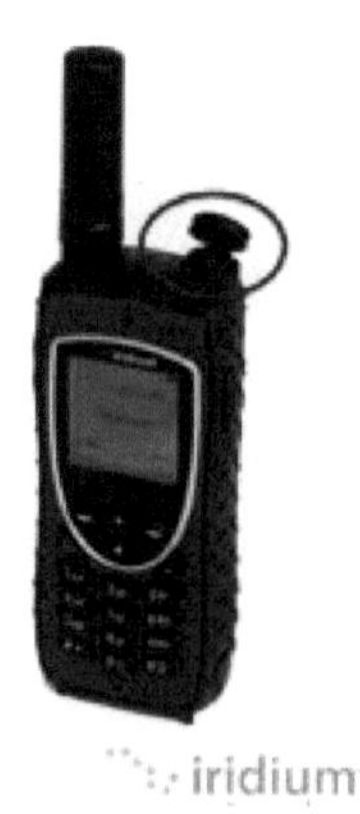

Figure 8.15 Iridium satellite device with SOS button.

These applications can be placed in MEO satellites covering wide areas where user density may not matter as in most mission-critical use cases. However, beam steering and spotting may be utilised to attain high-density coverage in specific geographic locations in hurricanes or earthquakes where a large population can be impacted.

8.6 Regulatory Constraints

Often, we come across questions on the degree of disruptiveness of this technology. Can the global players be able to spawn up mobile access anywhere in the world and potentially cannibalise the local operators? No, it does not seem as painless as it sounds, not only due to technological hurdles but also because of regulatory restrictions [1]. Some of the main regulatory challenges have been identified in this section, though there is no solution for all of them.

Generally, the local telecom regulations to the country of operations should apply here. Some typical directives and by-laws that can pose to be problematic for an international mobile service provider rendering service in another country or continent via satellite links may be the following:

- Spectrum sourced in the country for cellular operations should not be used to beam radio signals to another country for rendering mobile services. The beam from MEO and GEO satellites may encompass more than one country, the diagram below. It may prove quite challenging to restrain the beam from crossing the national peripheries (refer to Figure 8.16.)
- Some countries mandate that the user data (user traffic, generated CDRs, BI data) should never leave the country's boundaries. In the case of satellite-based systems, the core network (including the slices) may be placed outside the boundaries of the operating country, which may be a problem for data retention within the country. More specifically, the user-specific data in the UDM, PCF and OCS, semi-permanent mobility data of the user in AMF, session data in SMF and UPF may be residing outside the country where the satellite establishes the network coverage.
- Emergency calls should be routed to the PSAP of the specific country where the user resides, i.e., the country where the 5G satellite establishes the radio network coverage. However, the call will be routed to the core network situated in the central location (maybe in another country) on a public/private or hybrid cloud. In such a case, the core network should have an existing network integration with the PSAP of the country where the user is located.

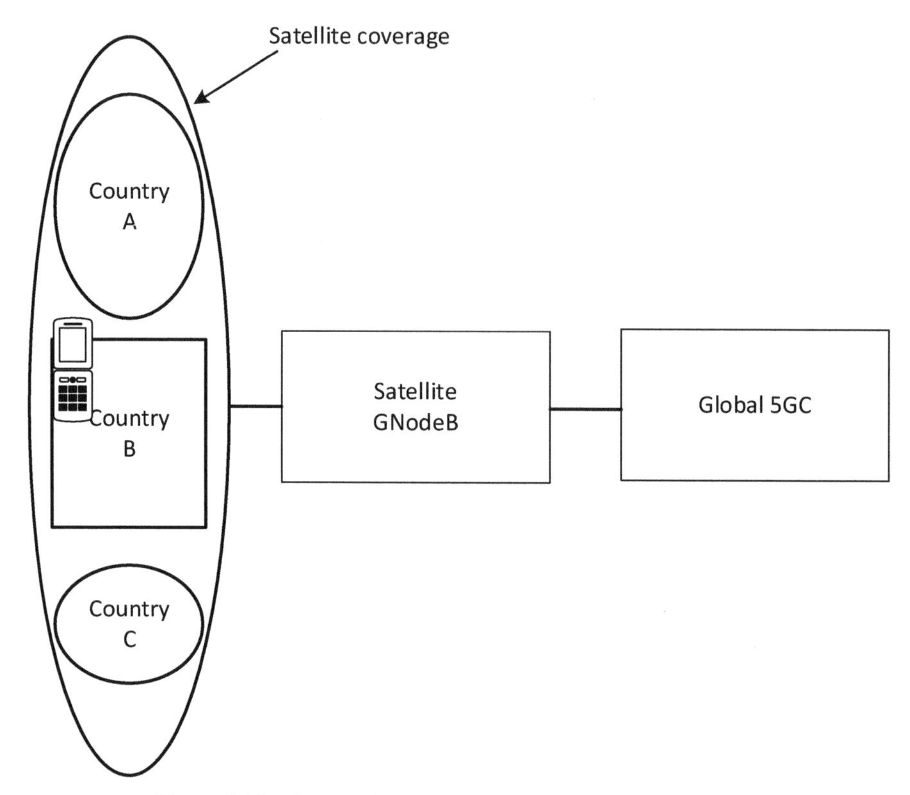

Figure 8.16 Super national mobile coverage by satellite beams.

However, this is somewhat similar to the scenario of a non-UE detected emergency call on a VoLTE network in a roaming scenario when the call is routed to the HPMN. To terminate the call to the target PSAP, the HPMN either needs to be able to resolve the PSAP number (of the VPMN network) and route it over there or needs to have an integration with PSAP. A similar arrangement may be considered for emergency calls from UE via 5G / 6G over satellite links.

- Lawful interception (LI) is a mandatory requirement for almost all countries. To adhere to laws about jurisdiction, the Intercept Access Points (or tap points) need to be in the operating region.

This invokes a few architectural and regulatory questions.

1: The satellite bearing 5G network components is in space and technically not within the periphery of the country defined by geographical borders and air space (class A to class E). There may be no network equipment in the country where the satellite beams are focused on rendering coverage. How can we realise "taps" within the country if the regulatory mandates LI tapping points within the country?

2: How easy or difficult would it be to establish a Lawful Interception Interface (H1/H2/H3) between the Core network at the central location (probably in another country) and the Law Enforcement Agency in the country of operations.

Incidentally, any public cloud-based Mobile Core Network will face the same issue. The physical location of the cloud hosting the network may be distant from the country where the network operates.

However, LI solutions are evolving to support cloud-native architecture. ATIS (Alliance for Telecommunications Industry Solution) has proposed a few alternatives to achieve LI for cloud-based XPAAS platforms [4]. In 3GPP, Service & System Aspects (SA) and Core Network & Terminals (CT), working groups are deliberating on solutions related to regulatory services such as lawful intercept, emergency calls, and public warning services.

We do not have all answers to all these issues currently and hence would need to keep an eye on the developments on this front.

8.6.1 Generic Issues

Power requirement: The satellites equipped with onboard 5G equipment, be it the gNodeB, a DU-gNodeB, or a bent, need to provide the RF signals to communicate with the terrestrial devices (User Devices). Also, the satellite has to cater for the power required by the onboard electro-mechanical systems required for beam steering and beam spot. A standard GNodeB requires between 12 Kilo Watt to 50 Kilo Watt of power depending on the make and specifications. Hence, it is a challenge to deliver this power from the solar cells in the satellite. With substantial solar cells, the weight of the satellite increases and keeping the satellites in the LEO or MEO space can be a challenge.

Maintenance: As the satellites are in space, maintenance can be an issue if the connection to the satellite is lost temporarily. In case of hardware issues for terrestrial GNodeB, the maintenance engineer may need to visit the data centre to change or fix the hardware. That would not be possible with Satellite 5G GNodeB. A stable 2-way IP connection is required between the terrestrial control centre and the satellites to maintain the Software or Firmware of the 5G components. So, activities like a software upgrade, implementation of new releases, security upgrades etc., may not be smooth and straightforward.

***Security*:** The satellites are not immune to security threats. Some of the common threats linked to 5G RAN on satellites are follows.

- Eavesdropping the SRI links: Hackers can listen to the control plane signalling and user plane data about end-users and derive sensitive user information.
- Modification of user data or control plane data and compromise data integrity.
- Channel Interference: Hackers can generate high power electromagnetic waves to interfere with the SRI links. This technique can also be used to launch denial of service attacks.
- Inject illegal signals via Uu or SRI interface from terrestrial sources or rogue satellites to consume the bandwidth and disrupt the services.
- Satellite hijacking: A malicious terrestrial unit or another rogue satellite can hijack the 5G RAN satellites and gain complete control.

8.7 Conclusions

Cellular communication piggybacking on satellite-based RAN may not be able to cater to and fuel all the use cases that we envisage for 6G. But it is imperative that many crucial scenarios can be supported. The technology is novel and is still evolving. This will be a key component of the 6G network. The satellites in the space bearing a part of the RAN will be powered by the sun, making the network environment friendly. The number of cell towers may eventually dwindle, reducing the terrestrial infrastructural cost and power requirements of a cell tower. As a collateral gain, we may substantially reduce the carbon footprint and conceive a greener, cleaner, and sustainable future network.

Acknowledgements

Thank you, Alain Brichard (Proximus Luxembourg), for providing critical and constructive comments and sharing niche ideas vital to shaping this chapter.

References

[1] [1]3GPP TR 23.737 V17.2.0 (2021-03)Study on architecture aspects for using satellite access in 5G : (Release 17)

[2] Anttonen, Antti; Ruuska, Pekka; Kiviranta, Markku, 3GPP nonterrestrial networks, VTT Technical Research Centre of Finland, 01/01/2019

[3] Interference analysis for terrestrial-satellite spectrum sharing, Document Number: H2020-EUK-815323/5G-ALLSTAR/D3.Project Name: 5G AgiLe and fLexible integration of SaTellite And cellulaR (5G-ALLSTAR)

[4] Cloud Services Impacts on Lawful Interception Study,ATIS-1000075

[5] Naveed UL Hassan, Senior Member, IEEE, Chongwen Huang, Member, IEEE, Chau Yuen, Senior Member, IEEE, Ayaz Ahmad, Senior Member, IEEE, and Yan Zhang, Fellow, IEEE, Dense Small Satellite Networks for Modern Terrestrial Communication Systems: Benefits, Infrastructure, and Technologies, IEEE Wireless Communications (Volume: 27, Issue: 5, October 2020)

Biography

Rajarshi presently works in Proximus in Luxembourg as a Telecom Engineering Specialist focusing on IMS/VoLTE/VOWIFI/Fixed Mobile Convergence, 5G Core Network design and implementation (NSA/SA), EPC, Roaming (VoLTE S8HR/5G), Enterprise UC and Voice core applications, 5G MEC research projects, Private 5G Networks. Previously, he worked in Brussels in BICS (Belgacom International Carrier Services) as a network architect for 11 years, delving into engineering, design, standardization, and technology strategy in 5G, Roaming Platforms MVNE/IoT/E-SIM (SM DP/SR), VoLTE.

Rajarshi has worked in many other organizations in R&D and engineering areas, for example, IBM as a technology consultant (mobile service delivery platform), Reliance Communications as a technical manager (wireless core network engineering), a French telecom startup as a telecom domain specialist, Upaid (mobile core network developments), Hutchison Telecom as telecom engineer (core network engineering & VAS), Indian Institute of Technology, Kharagpur as Research Associate. He had played instrumental roles in delivering state of the art solutions, technology consulting and guiding/mentoring a team of experts to realize niche products. He has over 23 years of industrial and research experience in wireless telecommunications. He has been a speaker in various international forums, like ITU, Computer Engineering World Congress, IEEE forums, Horizon 2020 (5G), etc., in Europe and USA. He obtained a PhD in Future Mobile Access Technology from Aalborg University Denmark in 2014. He has five granted patents (USA, Europe, Russia, India) in roaming, VoLTE, and 5G. He has 28 telecom research papers on GSM/LTE/IMS and 5G.

9

Virtualized, Open and Intelligent: The Evolution of the Radio Access Network

Atanas Vlahov[1], Dessislava Ekova[1], Vladimir Poulkov[1], and Todor Cooklev[2]

[1]Technical University of Sofia, Sofia, Bulgaria
[2]Purdue University Fort Wayne, Indiana, USA
E-mail: avlahov66@gmail.com; vkp@tu-sofia.bg; cooklevt@pfw.edu

Like any radio, the cellular base stations (BSs) have a radio frequency (RF) and a baseband (BB) part. For many years, the RF and BB parts of these significant components of the radio access networks (RANs) infrastructure have been tightly integrated. At these times, connecting the RRUs of one vendor to the BBU of another vendor usually could not be possible due to the propriety of the solutions. RANs have significantly evolved from analogue to digital signal processing units in the last few decades, where hardware components are being replaced with flexible and reusable software-defined functions allowing the RRU and BBU of cellular BSs to be an independent implementation of advanced access architectures. Following these driving forces, the BS architecture has evolved considerably towards virtualized, open and intelligent RAN architectures over the last few years. The main idea behind virtualizing and opening the RAN is to disaggregate the BS elements and develop open standards for the interfaces and interaction procedures between them. Currently, service providers worldwide are driving the adoption of "Open RAN" for 5G" and considering the implementation of intelligent functionalities for 6G. This chapter presents an overview of the evolution of RAN technologies and the latest trends toward virtualized, open and intelligent next generation RAN (NG-RAN).

9.1 Introduction

The cellular base stations (BSs) are major wireless infrastructure components. A BS, like any radio, has an RF part and a baseband (BB) part. The BB part is implemented in a combination of digital hardware and software. The BB functions include modulation/demodulation (bit-to-symbol mapping, IFFT/FFT), encoding and decoding; the MAC protocol, including radio scheduling; concatenation/segmentation of radio link control (RLC) protocol; and encryption/decryption procedures of packet data convergence protocol (PDCP), for the downlink and uplink directions. The radio frequency (RF) part is implemented using analogue RF hardware. However, the RF part is not entirely analogue because it needs to implement a digital interface. For more than 100 years, the BB and RF parts have been tightly integrated. This started to change around 2011 when distributed BSs appeared. Before that time, the way BSs were built was entirely proprietary; all that was expected of a BS was to implement the 3GPP standards for cellular communication. A distributed base station consists of a base band unit (BBU) and remote radio units (RRUs) also called Remote Radio Head (RRH), connected to the baseband unit typically via common public radio interface (CPRI).

The RRH contained only the radio functions and was located close to the antenna in the cell site tower, where the BBU contained all baseband processing functions. Each RRH and BBU pair is connected using a new network segment called the Fronthaul network. The Fronthaul network was most often a point to point connection, and the radio signals were transmitted using the common public radio interface (CPRI), open base station architecture initiative (OBSAI) or open radio interface (ORI) protocol.

The CPRI interface is the most common but is essentially proprietary. Connecting the RRUs of one vendor to the BBU of another vendor is impossible due to the propriety of the solutions. Only large equipment manufacturers of BS benefit from this network-building scheme since this scheme does not allow easy change of the equipment manufacturer at the operator's discretion. The losers are not only the operators who cannot flexibly build their network but also small and medium-sized enterprises (SMEs) that cannot develop a full range of BS equipment and its software but are ready to develop either software or separate elements of the microwave infrastructure.

Therefore, BSs architecture has evolved considerably over 5-10 years. We investigate the technology evolution of RAN. One of the main opportunities is virtualizing and opening the RAN, or the so-called "Open RAN". The main idea of Open RAN is to make the RF subsystem and BB subsystem of cellular

BSs independent. Virtualization technology has come to play a crucial role in the evolution of communication networks. Network function virtualization (NFV) refers to the software implementation of network functions in a way that is completely decoupled from network hardware. NFV is becoming possible for cellular networks. Eventually, communication networks and IT infrastructures will merge. This process will bring economic savings for mobile operators and new opportunities for technology companies.

The main trend is to centralize and virtualize higher-level radio access network (RAN) functions while keeping lower-level RAN functions in distributed units (near antennas). This leads to two new nodes, the so-called central unit (CU) and distributed unit (DU) by the 3GPP, and an interface referred to as the front-haul interface. A virtual BBU implements in software the higher-level network functions. The CU consists of multiple BBUs. Similarly, the centralized BB processing of radio signals in the cloud allows applying resource pooling and statistical multiplexing principles when allocating computing resources.

BBU-pool virtualization introduced the concept of shared processing, where it is possible to share the available processing resources amongst several sites and allocate extra processing efforts when needed in different areas. The network became adaptable to the non-uniform patterns of users' daily movements, such as going from a residential area to a business area in the morning and back again in the evening, referred to as the tidal effect. By sharing baseband resources between office and residential areas, a multiplexing gain on BBUs can be achieved.

Virtualizing network functions and running them on distant servers raise many issues in terms of performance. It is essential that the whole base-band processing of radio signals must meet strict latency requirements (namely, one millisecond in the downlink direction and two milliseconds in the uplink). Other requirements include Open RAN to support joint radio resource allocation for interference reduction and data rate improvements.

Several different organizations have started to work on these problems: ***the Open Networking Foundation, xRAN Forum, C-RAN Alliance, O-RAN Alliance, The O-RAN Software Community, Open RAN Policy Coalition, Telecom Infra Project, OpenRAN Group, Cisco's Open vRAN initiative, the OpenAirInterface Software Alliance*** and ***Facebook Connectivity.*** Although all of these groups claim that they are working on the same thing, which will make the RAN architecture more open using standardized interfaces and network elements, there are specific differences in a closer examination of the activities of these groups [1].

In March 2011, six companies that own and operate some of the largest networks in the world (Deutsche Telekom, Facebook, Google, Microsoft, Verizon, Yahoo) announced the formation of the ***open networking foundation (ONF)***, a nonprofit organization dedicated to promoting a new approach to networking called software-defined networking (SDN). Joining these six founding companies, members became many other companies, including major equipment vendors, networking and virtualization software suppliers, and chip technology providers. Today ONF is a leader in the implementation of open source solutions. One of ONF's projects has been related to software-defined RAN (SD-RAN) during the last few years. SD-RAN is ONF's 3GPP compliant software-defined RAN platform consistent with the O-RAN architecture built on recognized platforms, such as ONOS and Aether [2].

The ***xRAN Forum*** was formed in 2016 to standardize an open alternative to traditional hardware-oriented RAN. The group focused on three areas: separating the Control Plane RAN from User Plane, creating a modular evolved Node Base (eNB) software stack using Commercial-Off-The-Shelf (COTS) hardware, and publishing Open Interfaces (OIs). The group implied membership-primarily focused on telecom operators, including AT&T, Verizon, Deutsche Telekom, KDDI, NTT Docomo, SK Telecom, Telstra and Verizon. In February 2018, ***xRAN Forum* merged with the *C-RAN Alliance*** to form the ***O-RAN Alliance***.

The ***O-RAN Alliance*** is a worldwide organization with members operating in the Radio Access Network industry, including operators, vendors, research & academic institutions and SMEs. It is worth mentioning that the meaning of the abbreviation "O-RAN" in the name of this alliance can also be related to "Operator Defined Next Generation RAN", but whatsoever, its mission is to drive new levels of openness in the radio access network of next-generation wireless systems. The O-RAN Alliance consists of different Working Groups focused on various aspects of realising the future RAN networks. The primary focus of its working groups is realising a future open RAN solution built on a foundation of virtualized network elements, white-box hardware and standardized, interoperable interfaces that fully embrace O-RAN's core principles of intelligence and openness [3].

The ***O-RAN Software Community*** was jointly created by the O-RAN Alliance and the Linux Foundation in April 2019. Its mission is to provide a software solution to support the implementation of an open and intelligent Radio Access Network (RAN) compliant with the O-RAN Alliance specifications [4].

The ***Open RAN Policy Coalition,*** founded in May 2020, represents a group of companies aimed at promoting policies that will "*advance the adoption of open and interoperable solutions in the RAN as a means to create innovation, spur competition and expand the supply chain for advanced wireless technologies including 5G*". Coalition members represent a cross-section of the wireless communications industry, ranging from network operators to network solutions providers, systems integrators, cloud providers, edge device manufacturers, etc. They believe that by standardizing or "opening" the protocols and interfaces between the various subcomponents (*e.g.*, radios, hardware and software) in the RAN, networks can be deployed with a more modular design without being dependent on a single supplier. The Coalition promotes initiatives and policy priorities that: support new and existing innovative technology suppliers that are implementing these open interfaces, as well as small and large network operators; help create a competitive global ecosystem of diverse trusted suppliers and service providers; build and maintain U.S. and allies' technological leadership both in 5G and future wireless network development.

Telecom Infra Project (TIP) [5] was founded in 2016 by Facebook, Intel, Nokia, Deutsche Telekom and SK Telecom.Its task is to separate software and hardware, and the members of this project are more than 500 Internet companies, operators, suppliers and system integrators. In November 2017, Vodafone contributed its software-defined RAN project to the TIP. Vodafone and Intel created the ***OpenRAN Group*** [6], which will develop RAN technologies based on general purpose processing platforms (GPPP) and disaggregated software to establish a *vRAN* ecosystem based on 3GPP and disaggregated software. The OpenRAN Group focuses on introducing and describing how to build software and hardware, while the xRAN Group is more focused on developing specifications. However, both groups interact with each other, and many members of these groups participate in both projects. Many TIP member companies act only as observers and do not play an active role in developing the OpenRAN Group projects.

At the February Mobile World Congress 2018, Cisco announced a new open virtual radio access network (vRAN) initiative called ***Open vRAN*** [7]. The goal of Open vRAN is to build an open and modular RAN architecture based on GPPPs and disaggregated software that will support various use cases. The Cisco group has some of the same members as xRAN and OpenRAN, including Intel and Mavenir. Other vendors involved in Cisco's Open vRAN initiative include Altiostar, Aricent, Phazr, Red Hat, and Tech Mahindra.

The ***OpenAirInterface Software Alliance (OSA)*** [8] is a non-profit consortium fostering a community of industrial as well as research contributors for open source software and hardware development for the core network (evolved packet core; EPC), access network (AN) and User Equipment (UE) of 3GPP cellular networks. The Alliance sponsors the work of the Research Center "EURECOM" [9] to create an Open-Air Interface for the development of 5G Cellular Stack on COTS hardware. OSA has served as an innovation agent and catalyst for developing 5G Open Software in its four years of history. In March 2020, ***Facebook Connectivity*** [10] joined the OSA organization's Strategic Board. Facebook Connectivity works closely with partners, including network operators, equipment manufacturers and more, to develop programs and technologies that increase the availability, affordability and awareness of high-quality internet access.

At the end of this introduction, it is essential to note that open RAN groups, software-defined networking (SDN), network functions virtualization (NFV), and network virtualization (NV) have a vast open source community committed to growing projects that promote open standards. Groups like the Linux Foundation and its ***OpenDaylight Project*** and ***the Open Networking Foundation*** have been instrumental in creating open-source projects and products, such as OpenFlow and open source SDN Controllers. Several companies and groups contribute to such open-source projects, including the newly announced open platform for NFV project (OPNFV) [11]. It should be expected that these communities consider working toward opening the RAN and implementing intelligence in NG-RANs.

Further, this chapter is organized as Follows. The next section presents the evolution of the RAN architectures and their specific characteristics. Following the xRAN and O-RAN architectures are considered. Section 3 discusses some of the opportunities and challenges with the implementation of Open RAN. The last section concludes the chapter with some findings and comments.

9.2 Evolution of RAN

Radio access networks (RANs) have significantly evolved from analogue to digital signal processing units in the last few decades, where hardware components are often replaced with flexible and reusable software-defined functions. In a pure software-defined radio (SDR) system, the entire radio function runs on a general purpose processor (GPP). It only requires analogue-to-digital and digital-to-analogue conversions, power amplifiers, and antennas. In contrast, in typical cases, the system is based upon programmable dedicated hardware, e.g. application-specific integrated circuit (ASIC), application-specific instruction-set processor (ASIP), digital signal processor (DSP) or GPP and associated control software. Thus, the flexibility offered by a pure SDR improves service life-cycle and cross-platform portability at the cost of lower power.

9.2.1 Overview of RAN Architectures

We present a summarized overview of the development of different RAN architectures throughout the evolution of the mobile networks towards the cloud-RAN (C-RAN) architecture in this section. Different types of C-RAN, namely heterogeneous Cloud-RAN, virtualized Cloud-RAN, and Fog-RAN, are considered based on surveys [12] and other related publications. The overview illustrates how each of the practically implemented RAN architectures responds to new service requirements that cannot be satisfied by the existing legacy systems.

9.2.1.1 GSM/GPRS (RAN1/RAN2)

The architecture of the early RAN1 and RAN2 technologies is shown in Figure 9.1. The base station subsystem (BSS) consists of the base transceiver station (BTS), the base station controller (BSC), the *Air* interface, the *Abis* interface and the *A*-interface. With its radio front-end, BTS realizes the direct wireless connection to mobile stations (MSs). The BSC controls the mobility management and radio resource management of all BTSs and their connected MSs. A typical base station subsystem (BSS) includes tens of BSCs and hundreds of BTSs. The *Air*-interface connects the MSs to the BTS, and the *Abis*-interface connects BTSs to the BSC [13]. The **general**

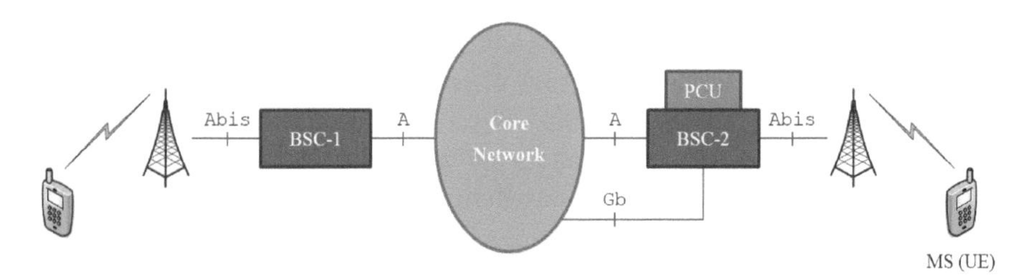

Figure 9.1 GSM/GPRS (RAN1/RAN2) architectures.

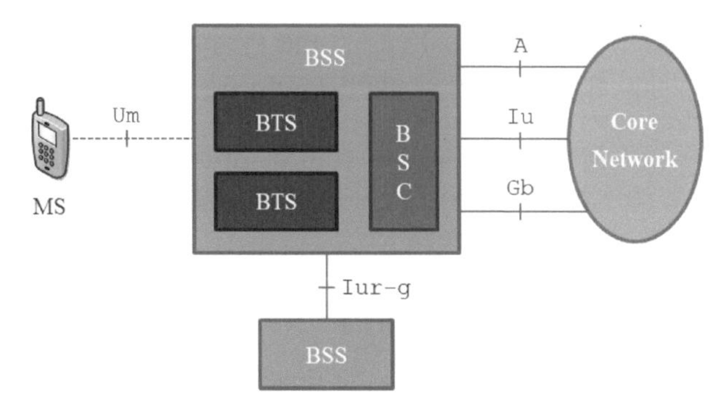

Figure 9.2 The GERAN architecture.

packet radio service (GPRS) technology uses packet-switching for bandwidth sharing among users, i.e. efficient resource utilization. Adding GPRS functionalities to the GSM network is practically an upgrade of the BSS architecture by adding the packet control unit (PCU), which could be positioned at various locations within the network, i.e., close to BTS BSC or CN [14].

9.2.1.2 GERAN

The **GSM/Enhanced data rate for GSM evolution (EDGE) RAN (GERAN)** introduces significant modifications to the radio protocols in GSM/GPRS for the goal of throughput improvement and end-user quality of service (QoS). GERAN specifies several new interfaces for connection with other architectures' core networks and RANs.

9.2.1.3 UTRAN

To support simultaneously more user links, thus achieving increased overall throughput, Wideband Code Division Multiple Access (WCDMA) was implemented in the *Uu* air interface of the **UMTS terrestrial radio access network (UTRAN).** UMTS consists of radio network subsystems (RNSs), each with one or more radio network controllers (RNCs) (responsible for the mobility and resource management) and several BSs, each named a Node B [15]. The RNC communicates with Node B and the CN over the *Iub* interface and two types of *Iu* interfaces for circuit-switching and packet-switching CNs (Figure 9.3) [16].

9.2.1.4 E-UTRAN

An improved version of UTRAN with reduced latency and increased efficiency is the Evolved UTRAN (E-UTRAN), standardized for the first time

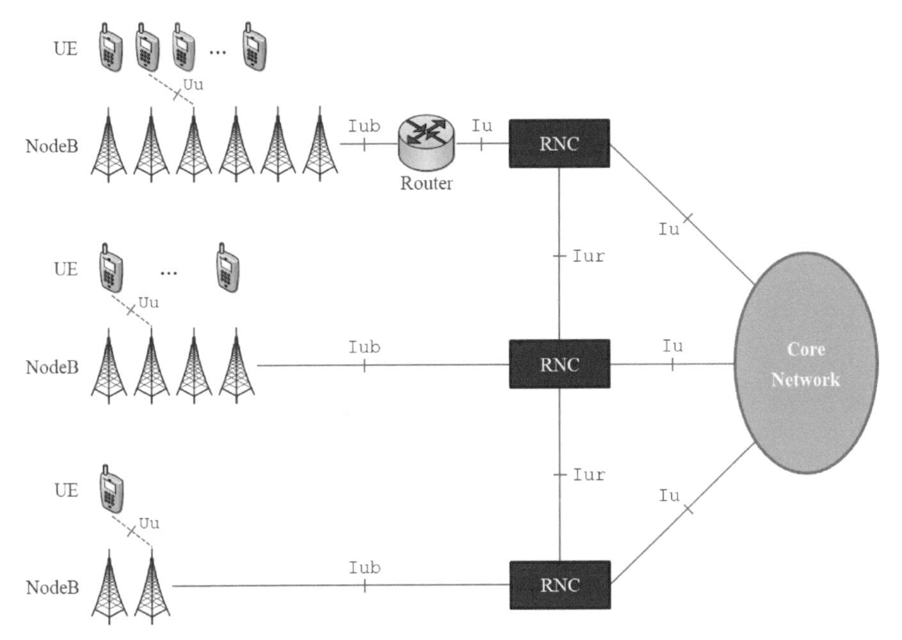

Figure 9.3 UTRAN architecture.

along with LTE in Rel. 8 and Rel. 9 [17]. Unlike former RANs, E-UTRAN integrates many functions such as RRM, header comparison, security, etc., into the eNodeB. Such an approach enables load sharing and increases network reliability, eliminating the risk of single-point failure for the EPC nodes. E-UTRAN does not have a centralized controller, but only BSs called an evolved Node B (eNodeB; eNB). The eNodeBs are interconnected to the evolved packet core (EPC) through the X2 and S1 interfaces, respectively (Figure 9-4). Additionally, every eNodeB is connected to the mobility management entity (MME) and the serving gateway (S-GW). User experience is improved by implementing OFDM (for downlink) and SC-FDM (for uplink), carrier aggregation, enhanced Inter-Cell interference coordination (eICIC), etc. All these are standardized in Rel. 10 [8]. Further improvements of the E-UTRAN were brought later in releases (Rel. 11 - Rel. 14) by supporting services such as narrow band internet of things (NB-IoT), mMTC and D2D communications [19].

9.2.1.5 D-RAN

In a traditional 3G/4G macro BS, the radio and baseband processing functions are integrated. A distributed unit (DU), responsible for functions such as

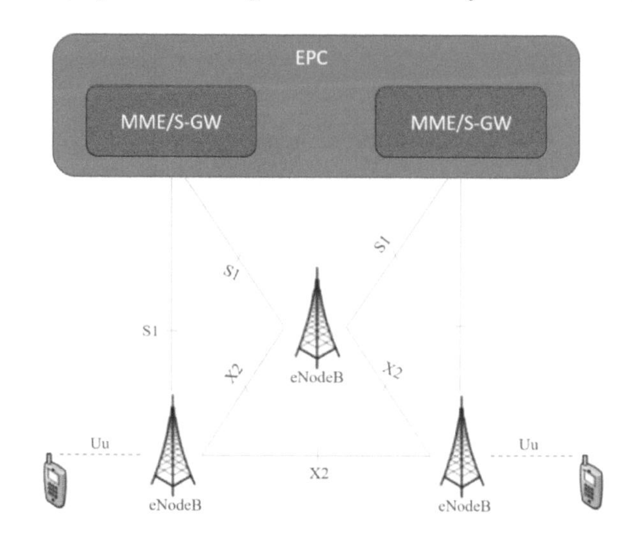

Figure 9.4 E-UTRAN architecture.

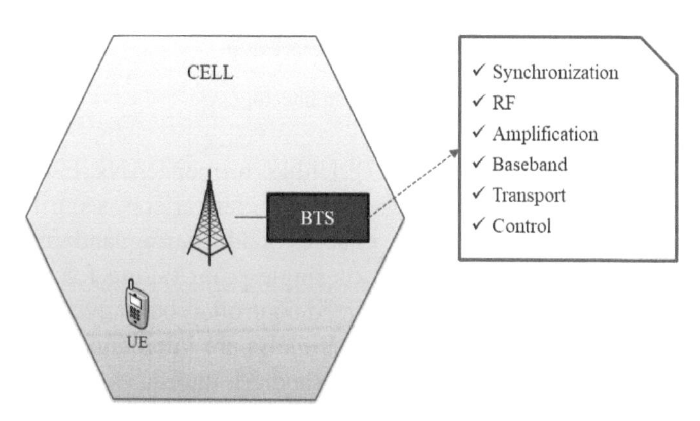

Figure 9.5 Traditional macro BS.

amplification, modulation and demodulation, frequency conversion, radiofrequency filtering, digital-to-analogue and analogue-to-digital conversion, etc., and a radio equipment controller (REC) responsible for baseband signal processing, controlling and managing of BSs, and interfacing with the RNC (Figure 9.5) are connected to the antenna located at the top of the tower.

In UTRAN and E-UTRAN, the radio and signal processing units, i.e. RRH (or RRU) and BBU (or DU) are separated. The BBU dynamically allocates network resources to its corresponding RRHs concerning the network requirements [20]. Such RAN architectures are also called **distributed RAN**

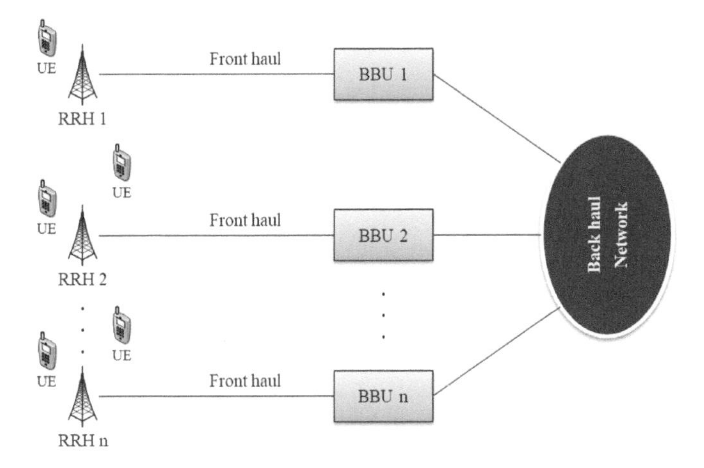

Figure 9.6 D-RAN architecture.

(D-RAN) (Figure 9.6). In D-RAN, each RRH is interconnected to its own dedicated BBU through a transport network called Fronthaul using Common Protocol Radio Interface (CPRI) to transmit In-phase and Quadrature (IQ) signals.

9.2.2 Cloud-Radio Access Networks

IBM initially proposed the Cloud RAN (C-RAN) under the name wireless network cloud (WNC) [21], later described in detail in a white paper of the China Mobile Research Institute [22]. C-RAN has two main goals – to optimize network performance and decrease the CAPEX and OPEX of mobile networks [23]. This is achieved by virtualising the baseband processing and centralising management and control. C-RAN allows the energy consumption to be decreased, the resource management to be improved, and the network management, maintenance and scalability to be optimized. At the same time, the second C-RAN aims to virtualise and centralise higher RAN functions (CU) in the network while keeping lower-RAN functions in DU, which enables flexible and scalable functional splits. In addition, the co-location of CUs with Edge Computing facilities allows for the implementation of services requiring more minor delays and low latency [24].

As depicted in Figure 9.7, the central concept behind C-RAN is to separate all BBUs from their corresponding RRHs and pool them into a centralized, cloudified, shared, and virtualized BBU pool. Every BBU pool can

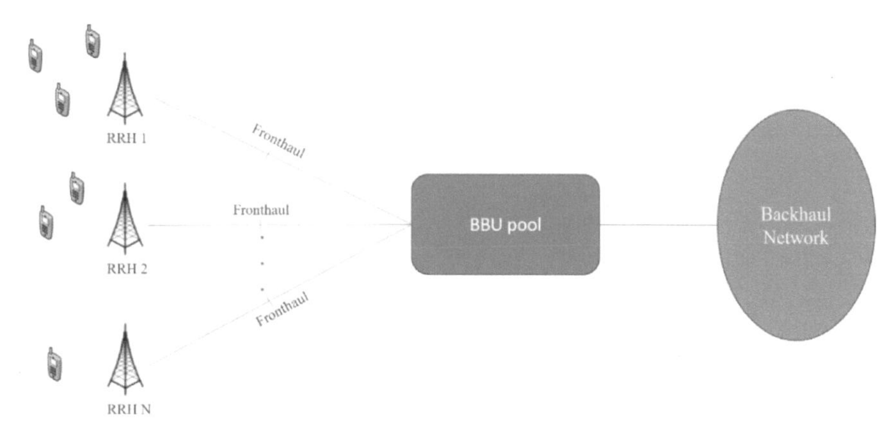

Figure 9.7 C-RAN architecture.

support up to tens of RRHs and connect to the core network via a backhaul. In such a way, cloud computing is embedded into the RAN architecture, thus allowing the implementation of:

a) *Commoditization and Softwarization*. This refers to the software implementation of network functions on general-purpose processors (GPPs) with no dependency on dedicated hardware.
b) *Virtualization and Cloudification.* This refers to the execution of network functions on top of virtualized (and shared) computing, storage, and networking resources controlled by a cloud operating system.

As shown in Figure 9.7, C-RAN aims to implement the whole base-band processing of radio signals in software meeting strict latency requirements (i.e. one millisecond in the downlink direction and 2 milliseconds in the up-link).

9.2.2.1 Fully, Partially and hybrid centralized C-RAN

Concerning splitting functions between RRH and BBU, C-RAN can be categorized into **Fully Centralized C-RAN** and **Partially Centralized C-RAN**. In **Fully Centralized C-RAN,** all the physical, MAC and network layers' functionalities are moved to the BBU. All functions related to Layers 1,2,3, such as sampling, modulation, resource block mapping, antenna mapping, transport-media access control, radio resource control, etc., are located in the BBU (as shown in Figure 9.8). Some of the advances achieved through this complete centralization are easy operation and maintenance, easy expansion of network coverage, easy upgrading of network capacity, the efficiency of

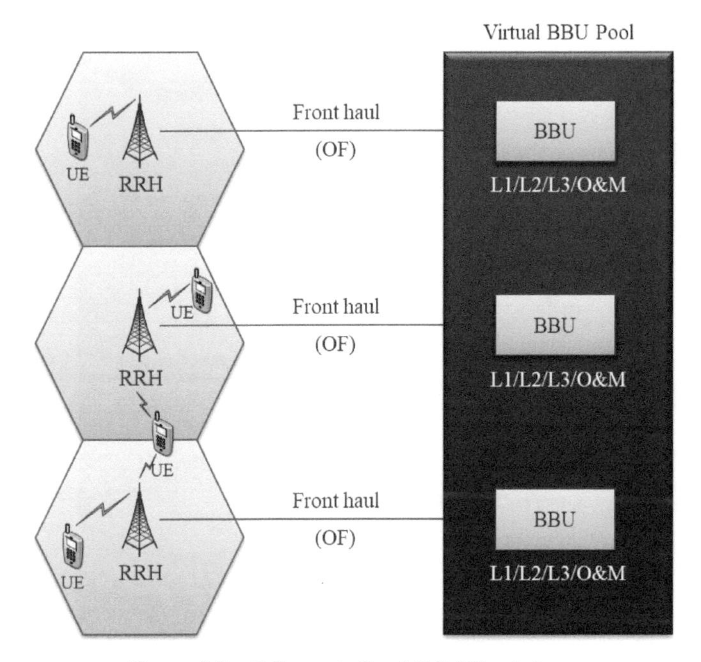

Figure 9.8 Fully centralized C-RAN solutions.

network resource utilization, and support for multi-cell joint operation and signal processing. Such architecture faces challenges related to the high bandwidth requirements for transmission baseband I/Q signals between RRH and BBU.

These drawbacks are avoided in the **partially Centralized C-RAN**, where the radio and baseband processing functions are integrated into the RRH, whereas all functions related to high layers are integrated into the BBU (as shown in Figure 9.9). In this architecture, the physical layer functions are done at RRHs while MAC and network layers' functions are performed at the BBUs, thus requiring a lower transmission rate and channel bandwidth between RRH and BBU. The drawbacks of such decentralization are lower flexibility in network upgrades, complex physical and MAC communication and the limited possibilities for efficient resource management between different RRUs.

One structure that can be the more flexible in resource sharing is the **Hybrid Centralized C-RAN**, where parts of the physical layer functions are done in RRHs while others are performed in BBU.

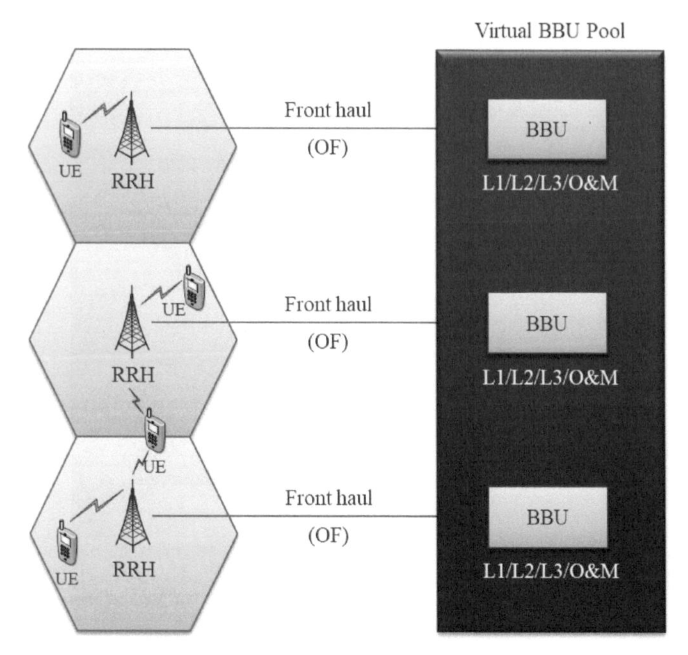

Figure 9.9 Partially centralized C-RAN.

The deployment of either one of the a.m. structures depends on the network requirements and OAM strategies of the network operator. If the latter is interested in improving the network's capacity, it will be easy to deploy new RRHs and connect them to the BBU pool, but if there is increasing traffic, this will require upgrading the hardware in the BBU pool to enhance processing capabilities.

9.2.2.2 Heterogeneous C-RAN

In a heterogeneous cellular network (HetNet), the macrocells provide wide coverage over large geographic areas. In contrast, smaller cells (Femto-, pico-, micro-) are deployed in small areas to improve overall network capacity and coverage, decrease network cost, increase spectral efficiency, etc., by moving the computational and communication resources close to the end-users. The so-called Heterogeneous-CRAN or H-CRAN practically aims to combine the advantages of HetNets and C-RAN to improve spectral and energy efficiencies and enhance the overall data rates in the network. The H-CRAN architecture consists of two cellular layouts, the macro BSs cellular layout and the small BSs or RRHs cellular layout [25, 26].

Such architecture decouples both control and user planes to enhance the functionalities and performance of C-RAN architecture in which control plane functions are only implemented in the macro BSs. In H-CRAN, as in the traditional C-RAN, the RF and simple signal processing functionalities are executed in the RRHs and functionalities that are related to upper layers are implemented in the BBU pool. But there are many differences in interfaces, such as decreased requirements òî the Fronthaul and resource management.

9.2.3 FOG Radio Access Networks

Cloud computing allows for the storage of all the volume and variety of data that devices and networks generate. But current cloud computing techniques also have many limitations related to increased end-to-end delay and latency, traffic congestion, data processing time and communication costs. This is also the reason for one of the main issues with Cloud-RAN: the required bandwidth to transmit radio signals between the BBU pool and each of the DUs. A new model for storing, processing, managing, and analyzing data at the edge of the network was recently proposed, called Fog Computing [27]. The term "Fog Computing" was initiated by Cisco [28], which means "fog is a cloud close to the ground". Fog Computing allows the implementation of a new RAN architecture called ***Fog RAN (F-RAN)*** [29], in which processing, storage, communication, control and management functions are implemented at the edge of the cellular network, thus aiming to utilize the advantages of both Fog Computing and C-RAN to cope with increasing traffic demands and better QoS provision to the users [30].

9.2.4 Software Defined Radio Access Network

SDN offers a logically centralized and flexible control model and the possibility of resource sharing that could be implemented in RAN. A centralized C-RAN solution and open platforms based on software defined radio (SDR) can upgrade the RAN architecture. Thus, a software-defined radio access network (SD-RAN) architecture with its benefits is considered an effective approach to meet the diverse QoS requirements in transforming next-generation RAN architectures and the evolution of the air interface technologies. Figure 9.10 shows one of the relatively early proposed SD-RAN architectures [31]. The reason to describe this architecture here is that the SD-RAN framework could be considered a C-RAN with complementary

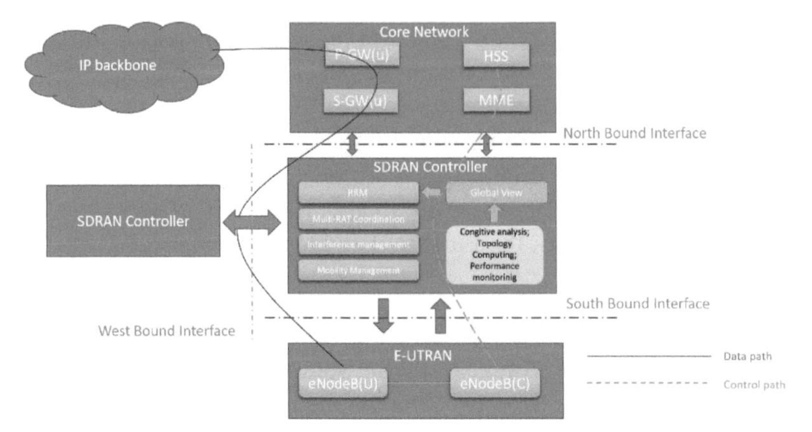

Figure 9.10 SD-RAN as per [31].

technologies such as cognitive radio, SDN, NFV, SON, and Big Data analytics. Compared to traditional RAN architectures, the main differences are separating the data and control paths, the SD-RAN centralized controller and flow-based policy control.

Figure 9.10 separates the control path (green dash line) and data path (red full line). On the RAN side, the eNodeB could be logically split into two parts: eNodeB(U) and eNode(C), where the former implements the radio transmission with configurations interpreted by eNode(C). The eNode(C) reports network status (network load, interference, etc.) to the central controller and could implement control functions that are not suitable for virtualization or centralization. On the CN side, the P-GW and S-GW could also be virtualized into P-GW(U) and S-GW(U), which the CN controller manages. In this architecture, the controllers could be logically divided into RAN controllers and CN controllers, depending on the control entities' location. Figure 9.10 shows that the SD-RAN controller obtains the local view from eNode(C) and constructs the global resource and network topology view from the collected local views and UE profiles. There are four interfaces defined for SD-RAN controller:

- *Northbound interface to external API.* It is controlled by operators allowing them to dynamically change the share of resources and policy at any time or location.
- *Northbound internal interface to CN controller.* Information between the SD-RAN controller and CN controller is exchanged via the interface, enabling the cooperation and coordination of RAN and CN functions.

- *Southbound interface to RAN entity (eNodeB).* The SD-RAN controller uses this interface to enforce different policies according to requests from the virtual operators and realize effective virtualization of the RAN and resources, enabling quick configuration of network parameters.
- *East-West bound interface to other SD-RAN controllers.* Considering the scalability and capability of the SD-RAN controller, the east-west bound interfaces to provide policies exchanged among inter-domain adjacent SD-RAN controllers.

Comparing C-RAN with SD-RAN, it could be seen that both have centralized resource management. But in C-RAN, baseband data are collected by Fronthaul, while the SD-RAN controller implements centralized control functions by gathering signalling and information rather than the data. The models and approaches to implementing the SD-RAN architecture are similar to C-RAN. An evolutionary model allows for a gradual deployment based on existing networks and a clean slate model in which the centralized control functions are constructed directly independent of the legacy network. There are three possible ways of implementation: centralized, distributed and hybrid, each of them with its Pros and Cons similar to those in C-RAN.

9.2.5 Virtualized Radio Access Networks

Virtualization technology facilitates the logical isolation of resources, such as network, computing or storage resources, while the physical resources are shared in a dynamic and scalable way. Network virtualization enables a flexible control mechanism, efficient resources, low cost, and diverse applications and is realized through multiple nodes and links deployed on one physical machine [32]. The virtualization of a cellular network includes various scopes of virtualization: *infrastructure virtualization, spectrum virtualization, air-interface virtualization, virtualization of multiple radio access technologies, and virtualization of computing resources.* Virtualization brings advantages such as efficient resource utilization, improved network performance, simplified migration to new technologies, reduced CAPEX/OPEX, increased revenue and new business opportunities.

Several virtualization approaches can be considered as an enhancement to the RAN virtualization: Network Function Virtualization (NFV), Software-Defined Networking (SDN) and Self Organizing Networks (SON). NFV refers to implementing telecommunication network functions by software running on GPPs. SDN is another technology that can be utilized to implement wireless virtualization in RAN. Such technology, essentially evolving in networking, separates the control plane from the data and centralizes the control. Network switches are considered forwarding devices monitored by a centralized entity, and they are the BBUs in one pool. Usually, SDN and NFV are linked together and considered complementary. Self-Organizing Networking (SON) allows the configuration, O&M, optimization and healing of communication networks to become simpler and faster through autonomous network functionalities. SON is not a replacement but rather a supplement to provide additional adaptability for NFV and SDN. A RAN architecture incorporating SON functionalities can provide abstraction and network control functions via APIs; therefore, it can realize SON operation via an OI to support devices from different vendors [33]. Implementing SON functionalities can lead to reductions in CAPEX and OPEX, increased QoS and QoE for users and provision of efficient resource utilization through adapting the network to the environments and other operational conditions.

9.2.5.1 Virtualized C-RAN

With the deployment of C-RAN, network-related resources are deployed close to the end-user, making it easier and allowing flexibility to implement all core network-related functions and applications in a cloud data centre. MNOs are linking the data centre to the mobile edge computing (MEC) to lower the latency and improve the performance, i.e. massive number of UEs, lower latency services, and higher capacity demands of users and vertical industries are the requirements mobile networks are expected to fulfil. An excellent approach to meet these requirements would be to deploy NFV, SDN and SON in the C-RAN to virtualize all functions and resources in the RAN architecture. Such virtualization of the access network forms a new type of C-RAN, called ***Virtualized-CRAN or V-CRAN.***

V-CRAN has to be implemented considering some specific characteristics related to the wireless access network. The UEs' statistical distribution, mobility, and varying channel conditions require different virtualization mechanisms than the core. In addition, from the viewpoint of the MNOs, there are some additional challenges, like a fair distribution of the wireless resources, proper interference management and other technical, O&M and related managerial problems.

As a concept, the virtualization of the computing resources of a BS in V-CRAN, named **virtualized base station (V-BS),** has also been proposed. The V-BS shares radio equipment at the hardware level and runs multiple protocol stacks of a BS in software. The hardware virtualization solution has already been standardized, and traditional mobile operators have been using this scheme to decrease OPEX and increase energy efficiency [34].

9.2.5.2 Virtual RAN

vRAN differs from traditional RAN in that it decouples the RRU from BBU on a GPPP. Virtualization of the RAN on GPPPs offers many benefits for RAN deployments, such as more flexibility, faster upgrade cycles, resource pooling gains, and centralized scheduling. This is a process in which the proprietary baseband unit hardware, which sits below the antenna and the radio, is replaced with a general-purpose computer.

In vRAN, the BBU is virtualized, thus becoming a vBBU and connected to the RUU via cost-effective "non-ideal" transport links such as copper and cable. The realization of vRAN on a GPP and non-ideal transport allows reduction of manufacturing costs for technology providers and helps MNOs decrease the Time-to-market (TTM), while through decoupling, MNOs can optimize each component of the RAN and improve QoS, network reliability and performance. Through vBBUs, more effective resource management can be applied to improve radio performance and increase spectral efficiency.

9.3 xRAN and O-RAN Architectures

As mentioned in the introduction, xRAN Forum's goal was to standardise traditional hardware-oriented RAN, focusing on three areas: separating the Control Plane RAN from User Plane, creating a modular evolved Node B (eNodeB) software stack using COTS hardware, and OIs. The xRAN Forum was formed to develop, standardize and promote an open alternative to the traditionally closed, hardware-based RAN architecture. xRAN fundamentally advances RAN architecture in three areas: decouples the RAN control plane from the user plane, builds a modular eNB software stack that operates on COTS hardware and publishes open north- and south-bound interfaces the industry. Standard specifications for Fronthaul interfaces with a view to the next generation RAN were released by the xRAN Forum in April 2018. The specification ensures compatibility between BBU and RRU, even from different manufacturers. Then, with the integration of the xRAN Forum into the O-RAN Alliance in March 2019, these specifications continued as O-RAN Fronthaul specifications, the latter considered the first standard to enable interoperability between different vendors. To allow possible relaxation of the stringent Fronthaul requirements, in these specifications, the functional splits between the BBU and RRU are defined [35, 36].

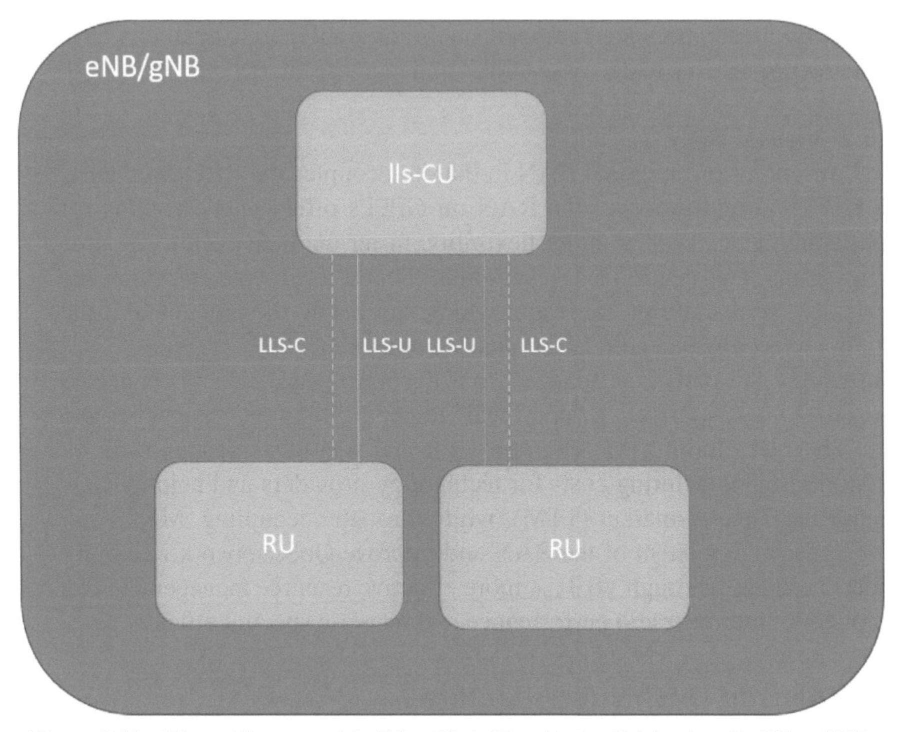

Figure 9.11 The architecture of the BSs eNb / gNb with the division into lls-CU and RU.

9.3.1 The xRAN Base Station Architecture

Figure 9.11 shows the architecture of the base stations eNb and gNb divided into two types of nodes: lls-CU (lower layer split central unit) and RU. The presented LLS-C (low-layer split control-plane) and LLS-U (low-layer split user-plane) interfaces are responsible for transmitting control-level (C-plane) and data-level (U-plane) messages in the LLS interface. In this architecture, the lls-CU and RU nodes can be defined as follows:

- lls-CU - a logical node that includes eNb/gNb functions up to the Split point 7-2x, except for functions allocated exclusively in RU . lls-CU controls the operation of the radio modules.
- The radio module (RU) is a logical node that includes a subset of the eNb/gNb functions, after the Split point 7-2x.

The Fronthaul interface is the interface between the lls-CU and RU nodes.

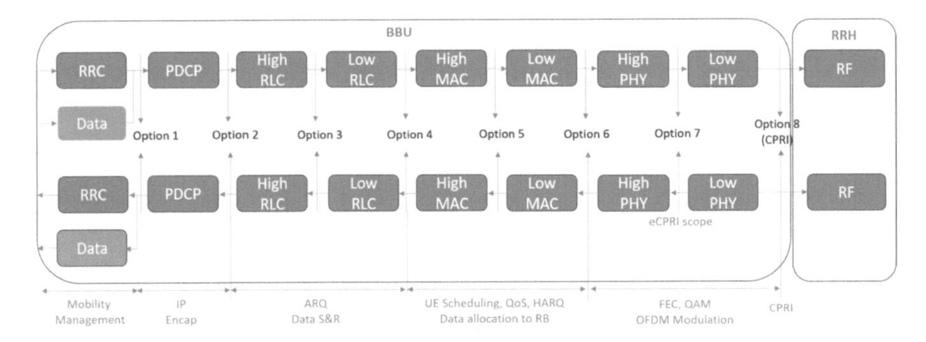

Figure 9.12 Split options for LTE eNB.

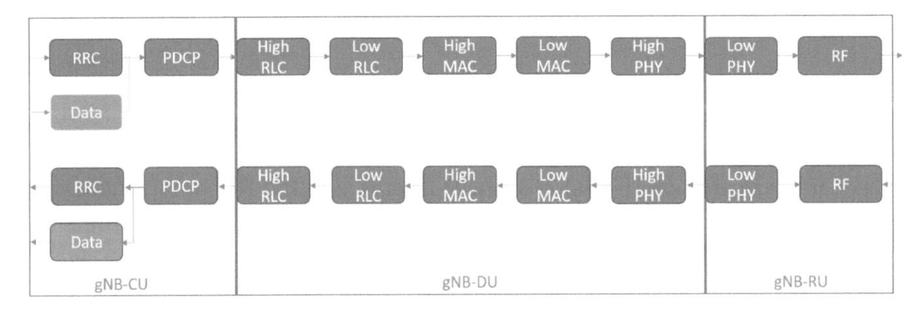

Figure 9.13 Split options for 5G gNB.

9.3.1.1 Functional separation of the Fronthaul interface

The possible functional separations of the interface between the CU and RU are shown in Figures 9.12 and 9.13.

When analyzing the options for functional separation of the Fronthaul interface, two opposite points arise:

- On the one hand, the task is to simplify the RU as much as possible because the size, weight, and power are essential parameters of the radio modules, and the more complex the RU, the more challenging and more energy-intensive the RU is.
- On the other hand, arranging an interface with RU at a higher transport-level reduces the bandwidth requirements of the interface, but the higher the level of the exchange interface between lls-CU and RU, the more complex the RU unit is.

This trade-off and the full processing chain are illustrated in Figure 9.14.

As part of the search for an alternative for both tasks, the xRAN group chose one split point – "split point 7-2x but allowed variations in which the

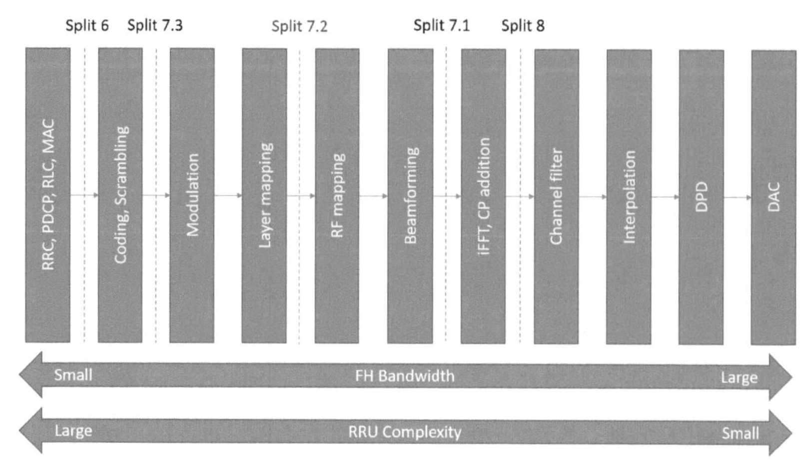

Figure 9.14 xRAN functional split 7.2 showing full processing chain.

precoding function should be located either "above" the Fronthaul interface in the lls-CU node or "below" the Fronthaul interface in the RU node. Radio modules in which precoding is not performed (simpler radio modules) are called Category A radio modules. Radio modules in which precoding is performed are called Category B radio modules. Figure 9.15 shows this dual concept of radio modules.

9.3.2 The O-RAN Architecture

As discussed in the introduction, the O-RAN Alliance is an industry consortium that promotes the definition of an open standard for the vRAN, but with two main goals in focus. The first is the definition of open architecture, enabled by well-defined interfaces between the different elements of the RAN. The second is the integration of machine learning (ML) and artificial intelligence (AI) techniques in the RAN, thanks to intelligent controllers deployed at the edge [35]. All O-RAN components must expose the same APIs, allowing O-RAN-based 5G deployments to integrate elements from multiple vendors and make it possible to utilize COTS hardware.

Figure 9.16 illustrates the high-level architecture and the interfaces of O-RAN.

O-RAN is composed of a non-real-time, and a near-real-time RAN intelligent controller (RIC) and by the eNBs and gNBs. Service management and orchestration (SMO) operates the non-real-time RIC, which performs control decisions higher than one-second granularity. The near-real-time RIC

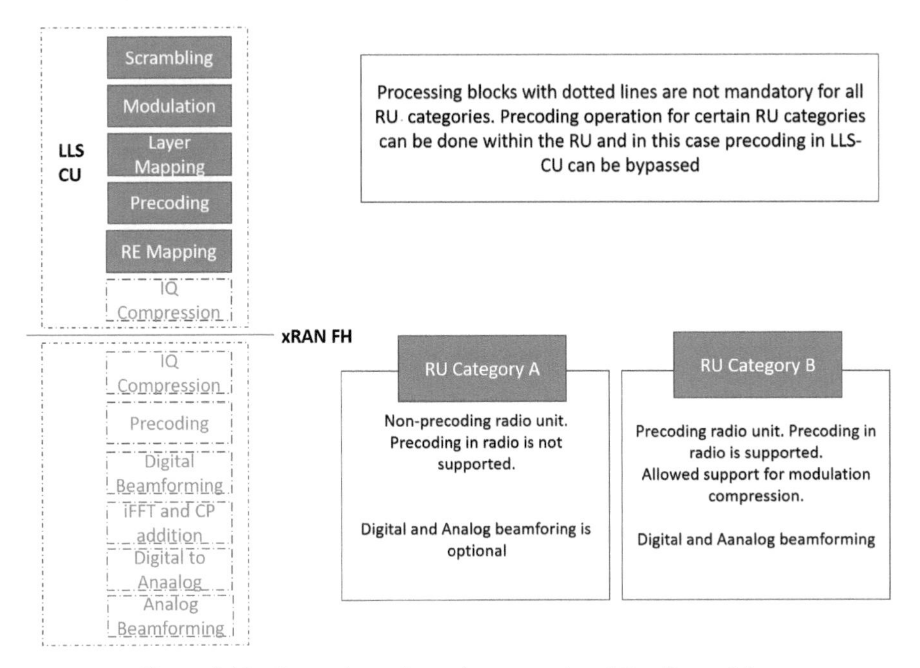

Figure 9.15 Separation point and category A and B radio modules.

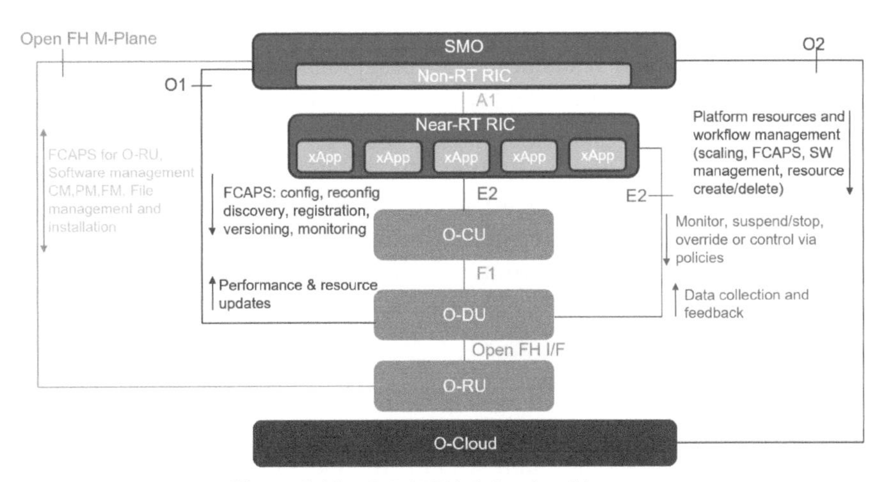

Figure 9.16 O-RAN high-level architecture.

performs a control loop with a much tighter timing requirement (as short as 10 ms), relying on different start, stop, override, or control primitives in the RAN, e.g., for radio resource management. Different processes could be controlled through the near-real-time RIC, such as handovers, allocation of resources, load balancing, traffic steering, etc. The near-real-time RIC can also leverage ML algorithms trained in the non-real-time RIC. The remaining components of the ORAN architecture on the bottom of Figure 9.16 concern the CU/DU/RU into which 5G gNBs are split and the 4G eNBs. The CU is further divided into a control plane CU and a user plane CU. Among the different options investigated by the 3GPP, O-RAN has selected split 7-2x for the DU/RU split, in which coding, modulation and mapping to resource elements are performed in the DU, and the inverse FFT, the cyclic prefix addition and digital to analogue conversion are carried out in the RU.

9.4 Open RAN Opportunities and Challenges

The Open RAN concept will be a new wave in communications and open up new doors for disrupting innovations and businesses. Open RAN is a driver for developing new technical solutions and business models that will reduce costs, increase business efficiency, and enable more significant innovation. By disaggregating hardware and software components and leveraging open interfaces, this technology has the potential to enrich the mobile ecosystem with new technological solutions and business models. Open RAN technologies promise that more competition will be introduced to the market to encourage diversity and resilience. The disaggregation of hardware and software will lead to decreased deployment costs. To successfully implement Open RAN solutions and technologies, expertise, specific skills and capabilities for developing and integrating new technologies will be needed, enabling innovation and opening new opportunities.The move to vRAN and Open RAN represents the latest step in transforming networks that have been happening for decades, migrating from primarily voice networks to today's data-only networks [38, 39].

9.4.1 Use Case Opportunities

Open RAN's flexible and open nature will foster opportunities to realise a rich set of new use cases. Some examples already under consideration in the research and development community are mentioned below.

Data analytics. The Open RAN could be a suitable platform for applying different types of Data Analytics. Having access to the other open interfaces will be straightforward to collect different data types. This will allow the implementation of various data analytics approaches even at the baseband level (IQ data), such as RF data analytics [36].

Context-based dynamic handover management for V2X. The main challenge in this use case is supporting frequent handover requests in high-speed heterogeneous environments. To address this challenge, Open RANs can exploit past navigation and radio statistics using Near-Real-Time RAN Intelligent Controller's ML models to customize handover sequences on a User Equipment level.

UAV applications. This use case is related to developing new UAV related applications such as a dynamic adjustment of radio resource allocation policies to optimize the service experience of the users and the mobility performance of UAVs jointly.

QoE optimization. The main objective of such a use case is to exploit ML models implementing traffic classifiers, QoE predictors and QoS decision-making engines to optimize the QoE of specific applications.

QoS based resource optimization. This use case is related to the ability to enforce fine-grained policies to drive QoS optimization in the RAN.

Traffic steering. The main goal of this use case is to exploit ML models to enable intelligent and proactive traffic steering control at the UE granularity and in the presence of multi-frequency cells.

Massive MIMO beamforming optimization. The ability to configure massive MIMO parameters will be paramount in future RANs.

RAN sharing. RAN sharing support among different operators is one of the cornerstones of virtualized RANs.

Intelligent Processing Algorithms. Open RAN is designed to make the radio access network more cost-effective and flexible. Open RAN has the potential to drastically reduce the volume of physical infrastructure that MNOs need to deploy and operate. And this is the case even if the centralized BBUs (cBBUs) are not virtualized and running on COTS platforms. Consolidating baseband processing in a much smaller number of sites enables much less hardware to be used, as processing capacity can be intelligently adjusted to network load, and there is no need to maintain masses of physical equipment at each BS, which is standing idle for much of the time. Operators are expected to focus first on using the open RAN to deliver better 4G and 5G connectivity at a significantly lower cost, most notably by consolidating and rationalizing the sheer volume of physical equipment they need to deploy and maintain.

Edge computing-based services. The Open RAN has considerable potential to support future edge compute-enabled use cases by distributing, virtualizing and, in the case of an open RAN, properly disaggregating RAN functionality: providing more choice and flexibility about the RAN components that can be deployed (and where to deploy them) to support service innovation.

Specific use cases. The new Open RAN architecture, software-defined, unbundled, programmable, and flexible, can meet enhanced mobile broadband and ultra-low latency requirements. It supports 5G network slicing, enabling the creation of multiple virtual networks from a single shared infrastructure to support specific use cases for dynamic and large-scale networks.

The Open RAN architecture gives a service creation environment that realizes the more advanced use cases in specific sectors and supports the modernization and digital transformation of many industries, including connected robotics, industry 4.0, smart agriculture, and smart cities.

9.4.2 Implementation of Artificial Intelligence and Machine Learning

The integration of Artificial Intelligence/Machine Learning (AI/ML) is a cornerstone in the design of the O-RAN architecture. The O- RAN architecture aims to extend the SDN concept of decoupling the control plane from the user plane in the RANs, by fostering embedded intelligence. Extending the CP/UP split, this approach further enhances the traditional RRM functions with embedded intelligence by introducing a RAN intelligent controller (RIC) or the so-called Software Defined, AI-enabled RAN Intelligent Controller. All this allows more advanced control functionalities to deliver increased efficiency and better radio resource management.

Moreover, the AI/ML integration represents a unique business opportunity. The goal is to exploit AI/ML models to carry out tasks that have traditionally been done quasi-statically by human operators in the past or are very complex tasks that never transferred from academia into business. These include tasks such as (but in any case, not limited to), for instance, zero-touch and automated resource control tasks, anomaly detection or traffic classification, etc.

It is worth also mentioning that the O-RAN Alliance leverages deep learning techniques to embed intelligence in the RAN architecture to automate operational network functions, optimize network-wide efficiency and reduce OPEX. This AI-powered radio control will help drive more new use cases for mobile RAN from private networks to ultra-dense urban deployments and low-cost coverage, taking a step further to launch 5G commercially around the world.

9.4.3 Implementation Challenges

Fronthaul design plays a crucial role in the overall system performance of a virtualized wireless network such as the open RAN. In RAN virtualization, the main resource bottleneck is in the Fronthaul link capacity. Housing together several tens of virtual base stations will necessitate a Fronthaul capacity that spans hundreds of Gbps for future systems, even in the order of Tbps. In addition, low-latency Fronthaul is of utmost importance for the proper operation of the RAN. The end-to-end latency between the RRH and the BBUs should be approximately an order of magnitude lower than the latency requirement of any system algorithm or service.

A significant issue in deploying an open RAN solution is deciding where to "split" the RAN architecture. A split with more centralized processing of the current baseband protocol stack offers better performance (improved inter-site coordination and pooling gains for the centralized baseband processing). In contrast, the requirements on Fronthaul (the transport connecting the remote radio unit (RRU) and the virtualized baseband unit (vBBU)) become tougher to meet as the split gets lower in the baseband protocol stack. A functional split determines the number of functions left locally at the antenna site and the number of functions centralized at a high processing powered data centre. The Fronthaul network is affected by choice of the functional split, both in terms of bitrates and latency, and as the different functional splits provide different advantages and disadvantages.

The contributions for RAN open source software are still minimal. There is a lack of robust, deployable, and well-documented software. As of now, most of the frameworks and libraries cannot be used in existing networks, as their open-source component is either incomplete, requires additional integration and development for actual deployment, or lacks robustness. To reach the quality of commercial solutions, the open-source community should deliver well documented, easy-to-deploy, and robust software, specifying all the necessary dependencies and the additional software components that guarantee the correct and efficient system functioning. Also, a very stringent requirement is the development of secure open-source software. Openness facilitates useful scrutiny of the code, and implementing security "by design" will be a challenge. The exposure of APIs to third party vendors (e.g., for the

RIC apps), for instance, could introduce new vulnerabilities in the network in case the APIs are not properly securely designed and contain weaknesses that attackers can exploit.

9.5 Conclusions

Open RAN groups have a vast open source community committed to growing projects that promote open standards. Several different organizations currently work on the topic of Open RAN: the Open Networking Foundation, xRAN Forum, C-RAN Alliance, O-RAN Alliance, The O-RAN Software Community, Open RAN Policy Coalition, Telecom Infra Project, OpenRAN Group, Cisco's Open vRAN initiative, the OpenAirInterface Software Alliance and Facebook Connectivity. Although all of these groups claim that they are working on the same thing, which will make the RAN architecture more open using standardized interfaces and network elements, there are specific differences in a closer examination of the activities of these groups.

From the overview of RAN architectures discussed in Section 1, an important thing should be noted related to C-RAN. What could be seen is that there are different interpretations when discussing C-RAN. In most cases, the issues of Centralized RAN, Centralized BBU, Cloud RAN and virtualized Cloud RAN are considered all under the general topic of C-RAN. From a business perspective, they are similar. All these approaches to C-RAN are intended to reduce CAPEX and OPEX for MNOs by centralizing and then virtualizing processing functions traditionally performed by dedicated networking equipment located at the antenna sites (cell sites). However, as seen from the overview and analysis, they are quite different from a technical point of view.

Considering the overview and analysis of the basic functionalities of the different RAN technologies, especially those related to open source and virtualization, it is also important to stress the issue of the difference between Open RAN, C-RAN and vRAN. Generally, Open RAN can be considered an upgrade to vRAN or an evolution process, starting with the C-RAN with the concentration and consolidation of the baseband functionality across a smaller number of sites across the telco's network and cloud. Followed by the vRAN, the baseband unit is virtualized to run as software on generic hardware platforms. Finally, in the Open RAN, the legacy, proprietary interfaces between the baseband unit (BBU) at the foot of the cell tower and the remote radio unit (RU) at the top of the building are replaced with open standard interfaces.

vRAN is not Open RAN as it is not entirely open; it still contains proprietary interfaces and purpose-built hardware. With vRAN, the proprietary hardware remains, but the BBU gets replaced by a COTS server rather than proprietary hardware. The software that runs on the BBU is virtualized to run on any COTS server. The proprietary interfaces remain as they are. The Open RAN vision is that the RAN is open within all aspects, with the interfaces and operating software separating the RAN control plane from the user plane, building a modular BS software stack that operates on commercial-off-the-shelf (COTS) hardware with open north- and south-bound interfaces. This software-enabled Open RAN network architecture enables "white box" RAN hardware – meaning that baseband units, radio units and remote radio heads can be assembled from any vendor and managed by Open RAN software to form a truly interoperable and open network.

One of the O-RAN Alliance principles states that the O-RAN architecture and interface specifications shall be consistent with the 3GPP architecture and interface specifications to the extent possible. But it is essential to note that the O-RAN Reference Architecture is designed to enable next-generation RAN infrastructures. The architecture is based on well-defined, standardized interfaces to allow an open, interoperable supply chain ecosystem in full support of and complementary to standards promoted by 3GPP and other industry standards organizations. Empowered by principles of intelligence and openness, the O-RAN architecture is the foundation for building the virtualized RAN on open hardware, with embedded AI-powered radio control that operators around the globe have envisioned.

In conclusion, virtualized and Open RAN is a revolutionary, though complicated, technology with all the disruptive characteristics. The traditional monolithic RAN solution approaches no longer meet today's quickly evolving network requirements. The O-RAN vision will "revolutionize not only the modus operandi and business of telecom operators, but also the world of researchers and practitioners that will be able to run a modern, open-source, full-edged RAN control infrastructure in their lab and investigate, test and eventually deploy all sorts of algorithms (e.g., AI-inspired) for cellular networks at scale" [37].

Acknowledgements

This research was supported by Open code systems Ltd, Sofia, Bulgaria. https://opencode.com/.

References

[1] https://itechinfo.ru/itech-articles

[2] https://www.opennetworking.org/sd-ran/

[3] https://www.businesswire.com/news/home/20180227005673/en/xRAN-Forum-Merges-C-RAN-Alliance-Form-ORAN

[4] www.o-ran-sc.org.

[5] https://telecominfraproject.com/openran/

[6] https://www.sdxcentral.com/articles/news/xran-open-vran-and-openran-whats-the-difference/2018/04/

[7] https://www.sdxcentral.com/articles/news/cisco-launches-open-vran-initiative/2018/02/

[8] https://www.openairinterface.org/?page_id=466

[9] http://www.eurecom.fr/fr

[10] https://connectivity.fb.com/

[11] https://www.sdxcentral.com/open-source/

[12] M. A. Habibi, M. Nasimi, B. Han and H. D. Schotten, "A Comprehensive Survey of RAN Architectures Toward 5G Mobile Communication System," *in IEEE Access*, vol. 7, pp. 70371-70421, 2019.

[13] F. Hillebrand, GSM UMTS: The Creation Global Mobile Communication. New York, NY, USA: Wiley, 2002.

[14] G. Heine and H. Sagkob, GPRS: Gateway to Third-Generation Mobile Networks. London, U.K.: Artech House, 2003.

[15] Technical Specifications and Technical Reports for UTRAN-Based 3GPP System, 3GPP TR 21.10. 2003.

[16] H. Holma, A. Toskala, WCDMA for UMTS: HSPA Evolution and LTE. New York, NY, USA: Wiley, 2007.

[17] Evolved Universal Terrestrial Radio Access (E-UTRA) and Evolved Universal Terrestrial Radio Access Network (E-UTRAN); Overall Description, document TS 36.300, 3GPP, Release 9, 2006.

[18] Evolved Universal Terrestrial Radio Access (E-UTRA) and Evolved Universal Terrestrial Radio Access Network (E-UTRAN); Overall Description, document TS 36.300, 3GPP, Release 10, 2009.

[19] C. Cox, An Introduction to LTE: LTE, LTE-Advanced, SAE and 4G Mobile Communications. Hoboken, NJ, USA: Wiley, 2014.

[20] E. Dahlman, 3G Evolution: HSPA LTE for Mobile Broadband. Amsterdam, The Netherlands: Elsevier, 2010.

[21] Y. Lin, L. Shao, Z. Zhu, Q.Wang, and R. K. Sabhikhi, "Wireless network cloud: Architecture and system requirements," *IBM J. Res. Develop.*, vol. 54, no. 1, pp. 4:1_4:12, Jan./Feb. 2010.

[22] C-RAN the Road Towards Green RAN-White Paper, China Mobile Res. Inst., China Mobile, Beijing, China, Oct. 2011.

[23] J. Wu, Z. Zhang, Y. Hong, and Y. Wen, "Cloud radio access network (CRAN): A primer," *IEEE Netw.*, vol. 29, no. 1, pp. 35_41, Jan. 2015.

[24] Study on new radio access technology Radio access architecture and interfaces, 3GPP, 3rd Generation Partnership Project, 3 2017, v14.0.

[25] M. Peng, Y. Li, J. Jiang, J. Li, and C. Wang, "Heterogeneous cloud radio access networks: A new perspective for enhancing spectral and energy ef_ciencies," *IEEE Wireless Commun.*, vol. 21, no. 6, pp. 126_135, Dec. 2014.

[26] Y. Li, T. Jaing, K. Luo, and S. Mao, "Green heterogeneous cloud radio access networks: Potential techniques, performance trade-offs, and challenges," *IEEE Commun. Mag.*, vol. 55, no. 11, pp. 33_39, Nov. 2017.

[27] C. Mouradian, D. Naboulsi, S. Yangui, R. H. Glitho, M. J. Morrow, and P. A. Polakos, "A comprehensive survey on fog computing: State-of-the art and research challenges," *IEEE Commun. Surveys Tuts.*, vol. 20, no. 1,pp. 416_464, 1st Quart., 2018.

[28] M. Mukherjee, L. Shu, and D. Wang, "Survey of fog computing: Fundamental, network applications, and research challenges," *IEEE Commun. Surveys Tuts.*, vol. 20, no. 3, pp. 1826_1857, 3rd Quart., 2018.

[29] R. K. Naha, S. Garg, D. Georgakopoulos, P. P. Jayaraman, L. Gao, Y. Xiang, and R. Ranjan, "Fog computing: Survey of trends, architectures, requirements, and research directions," *IEEE Access*, vol. 6, pp. 47980_48009, 2018.

[30] K. Liang, L. Zhao, X. Zhao, Y. Wang, and S. Ou, "Joint resource allocation and coordinated computation offloading for fog radio access networks," *China Commun.*, vol. 13, no. 2, pp. 131_139, 2016.

[31] Xu, F., Yao, H., Zhao, C. et al. Towards next generation software-defined radio access network–architecture, deployment, and use case. J Wireless Com Network 2016. https://doi.org/10.1186/s13638-016-0762-6

[32] M. Hoffmann, M. Staufer, "Network Virtualization for Future Mobile Networks: General Architecture and Applications," *Communications Workshops (ICC), 2011 IEEE Int. Conf. on Comm.*, Kyoto, 2011, pp. 1-5.

[33] P. Semov, H. Al-Shatri, K. Tonchev, V. Poulkov, A. Klein, "Implementation of Machine Learning for Autonomic Capabilities in Self- Organizing Heterogeneous Networks", *Springer, Wireless Personal Communications*, vol. 92, no. 1, pp. 149-168, 2017, DOI: 10.1007/s11277-016 -3843-2.

[34] X. Wang, et.al. “Virtualized cloud radio access network for 5G transport,” *IEEE Commun. Mag.*, vol. 55, no. 9, pp. 202_209, Sep. 2017. Specification (V1.0),” Tech. Rep., Aug. 2017.

[35] M. Polese, et.al. Machine learning at the edge: A data-driven architecture with applications to 5G cellular networks, submitted to IEEE Transactions on Mobile Computing (8,2019). https://arxiv.org/pdf/1808.07647.pdf

[36] Cooklev, T., Poulkov, V., Bennett, D., Tonchev, K. Enabling RF data analytics services and applications via cloudification. (2018) *IEEE Aerospace and Electronic Systems Magazine*, 33 (5-6), pp. 44-55. doi: 10.1109/MAES.2018.170108

[37] https://www.businesswire.com/news/home/20200605005027/en/Mavenir-Goodman-Networks-Partner-Deliver-OpenRAN-Solutions

Biographies

Professor Vladimir Poulkov has received his M.Sc. and PhD degrees from the Technical University of Sofia (TUS), Sofia, Bulgaria. He has over 30 years of teaching, research, and industrial experience in telecommunications managing numerous industrial, engineering, R&D and educational projects. He has been Vice Chairman of the General Assembly of the European Telecommunications Standardization Institute. Currently, he is Head of the "Teleinfrastructure" R&D laboratory at TUS; Chairman of the "Cluster for Digital Transformation and Innovation", Bulgaria; Chairman of the Board of the "Research and Development and Innovation Consortium" at Sofia Tech Park, Bulgaria. He is a Fellow of the European Alliance for Innovation and a Senior IEEE Member.

Todor Cooklev received a PhD degree in electrical engineering from the Tokyo Institute of Technology, Japan, in 1995. Dr Cooklev is Harris Professor of Wireless Communication and Applied Research at Purdue University Fort Wayne, Indiana, USA.

His current research interests include most aspects of modern wireless systems and standards. Dr Cooklev is the author of more than 100 publications and 30 patents issued in the United States. Among his honours, he received the Best Paper Award at the 1994 Asia-Pacific Conference on Circuits and Systems, the NATO Science Fellowship Award in 1995, the 3Com Inventor Award in 1999, and the Wireless Educator of the Year Award in 2006, an award from the IEEE Standards Association in 2012. He was inducted into the Purdue Innovators Hall of Fame in 2019.

Dr. Cooklev is currently a member of the Board of Governors of IEEE Standards Association and the editorial board of the IEEE Communications Standards Magazine.

Desislava Ekova is a PhD student at the Faculty of Telecommunications of the Technical University of Sofia, Bulgaria. She received her BSc and MSc degrees in Telecommunications from the Technical University of Sofia, Bulgaria, in 2013 and 2015, respectively, graduating with both degrees with the highest performance records. She has been a member of the Math Team of the Technical University of Sofia and received awards from various mathematical competitions. His current research interests in telecommunications are related to resource management, crowd management, user Localization, and Open Radio Access Networks.

Atanas Vlahov has received his BSc and MSc degrees in Telecommunications from the Technical University of Sofia (TUS), Bulgaria, in 2019 and 2021, respectively. Currently, he is a PhD student in the Faculty of Telecommunication of TUS. His current research interests in telecommunications are related to Open Radio Access networks and Software Defined Networks.

10

Aerial Infrastructure Sharing in 6G

Navin Kumar

ECE Dept., Amrita School of Engineering Bengaluru, Amrita Vishwa Vidyapeetham
E-mail: navinkumar@ieee.org

In Terrestrial networks, infrastructure sharing and challenges like coverage area, remote access, disastrous prone locations, quality of service (QoS), etc. are critical and cost-effective solutions for network providers. The non-terrestrial networks can eliminate these issues but are not efficient in terms of cost and deployment ease. However, an attempt is being made from the 4G network to integrate terrestrial and non-terrestrial networks and provide ubiquitous coverage. But non-terrestrial networks are primarily defined as satellite-based, either in low or higher orbit. However, recent development in an unmanned aerial vehicle (UAV) has opened several opportunities for onboard cell station band options and thus, allows defining various layers of space-based platforms such as low altitude platforms (LAP), high altitude platforms (HAP, etc. Such hierarchy has become essential for the next generation (6G) cellular communication. This chapter briefly introduced infrastructure sharing and different kinds of sharing, including aerial, and discussed various platforms supported by 5G and 6G communication.

10.1 Introduction

Cellular mobile communication has evolved, and we see the next-generation networks almost every decade [1–2]. Significant development took place in technology, architecture, applications and services. Up to fourth generation (4G) networks, the main focus has been on enhancing link capacity at the same time ensuring good coverage in the two-dimensional (2D) plane [3].

Also, from the 4G-LTE all-IP architecture, networks provide communication coverage and support for cloud integration. The cloud services are efficient in data retrieving/storage and support services to mobile Internet users. Fifth-generation (5G) network requirements are multi-dimensional [4–5] and augment severe constraints. For example, mission-critical services require high reliability and low latency communications (URLLC), a massive number of devices (mMTC), a more extensive range, and higher operational costs (OPEX) in the communication infrastructure. In new radio (NR) Rel.15, 3GPP has standardized radio technologies where the main focus is on enhanced mobile broadband (eMBB), URLLC and massive connectivity. 5G is expected to provide technical support for future industries such as industry 4.0 [6–7] and new features to society integrated with artificial intelligence (AI) and the Internet of Things (IoT). It also offers further upgradation of multimedia communication services with the support of high speed and high capacity.

Furthermore, 5G allows new opportunities for sharing the primary infrastructure among remote and self-contained networks using network slicing [8–9]. Additionally, several new emerging 5G services include ubiquitous coverage/capacity, service for new use cases, varieties of application scenarios, and traffic conditions. This would be a real challenge for the one-network like LTE-A or 5G to accommodate them.

While the rollout of 5G is still underway, the International Telecommunication Union (ITU) has recently announced the 5G detailed specifications [10], the researchers across the world have started working on the next generation of wireless networks. Next-generation communication sector systems (6G) [11–12] aim to achieve even higher spectral and energy efficiency than 5G. Additionally, it will support the extensive growth in the number of IoT devices. The 6G will provide even greater bandwidth and low latency than the 5G network. The 6G network will use high frequencies to improve Internet speed and connectivity. It is believed that 6G will offer fast Internet speed up to 1-Terabit per second.

5G has an enormous impact on various industries, and with 6G, it is expected to have a more significant impact on numerous aspects of the market. Multiple industries, academia and research bodies have already started the research activities for the 6G networks taking existing technologies as the foundation [11]. So, 5G is under laying the groundwork, and it will work as the stepping stone for 6G technology. It is believed that 6G technology will provide massive improvements in imaging, AI and machine learning (ML), presence technology, etc. [13]. Significant advancement is expected in the health care sector after the introduction of 6G as it will help remove time and space barriers. Currently, there are many network issues, but 6G will solve the problem of mobile networks in the closed areas with the help of distributed antenna systems or femto cells [14–15] or many others. Some of the 5G technologies that will be further developed for 6G are Millimeter-wave technology [16] for using much higher frequencies, multiple-input and multiple-output (MIMO) for multiplying the radio link's capacity dense network for the effective use of the spectrum. The improvement of these technologies will not be enough. So, many new technologies will have to be introduced for the development of the 6G network. Some of the key performance indicators (KPIs) in this context are ubiquitous connectivity, scalability, and affordability. Enabling 3D coverage would require the placement of network elements up in the sky and space.

The sixth-generation will use satellite, aerial, and terrestrial platforms to increase radio access capability and expose the support of on-demand edge cloud services in three-dimensional space. It might require mobile edge computing (MEC) functionalities on aerial platforms and low-orbit satellites [17]. This will encompass the MEC support to devices and network elements in the sky, resulting in space-borne MEC. It will enable intelligent, personalized, and scattered on-demand services. End users will have the experience of being surrounded by a distributed computer, satisfying their requests with approximately zero latency. The 6G wireless communication network is expected to incorporate terrestrial, aerial, and nautical communications into a robust network. Such a network would be more reliable, faster, and support a massive number of devices with ultra-low latency requests [18]. Satellite communications can be used in conjunction with 5G and beyond in several combinations and deployment scenarios. This includes collaborative access networks, shared access and also satellite backhaul.

This work highlights the shared infrastructure for 5G and beyond 5G. We discuss the architecture that can support ubiquitous connectivity and integration techniques of terrestrial and non-Terrestrial networks. A brief discussion on non-terrestrial communication is also explained. Different approaches to improve the coverage of shared infrastructure are discussed. Some of the results on the performance of such networks are also presented.

10.2 Infrastructure Sharing in Mobile Communication

Infrastructure sharing (IS) or mobile infrastructure sharing [19] is a technique where two or more operators join together to share various portions of their network infrastructure for their service provisioning. Infrastructure sharing across the telecommunication sector is a requirement and a trend. It allows mobile service providers to share both electronic and non-electronic infrastructure. Where electronic infrastructure includes spectrum, antenna, radio nodes, transmission networks etc. (active sharing), nonelectronic sharing includes building premises, sites and masts (Passive sharing). Technology is the main criteria for sharing. The infrastructure sharing opportunities not only allows the firms to cut cost but also to lower their investments. Also, through infrastructure sharing, developing economies tend to foster investment concentration toward accessible and affordable mobile services. Figure 10.1 shows an overview of types of infrastructure sharing based on units that can be shared. In passive infrastructure sharing, non-electronic infrastructure at a cell site, for example, power supply and management system, physical elements like backhaul transport networks, etc., are shared. This type is further classified into i) Site sharing – in this, physical sites (location) of base stations are shared, and; ii) Shared backhaul – In this, the transport networks from base stations to radio controller are shared. Passive infrastructure sharing can be implemented per sites. This allows operators to share sites and keep their planned competitiveness. The process is also simple as network equipment remains separated in this type. However, the possibility of cost-saving is limited in this kind of sharing.

When the electronic infrastructure of the network is shared, it is known as active infrastructure sharing. This includes radio access and core networks (servers and core network functionalities). This type is further classified into i) Multi-operator radio access network (MORAN) – In this, radio access

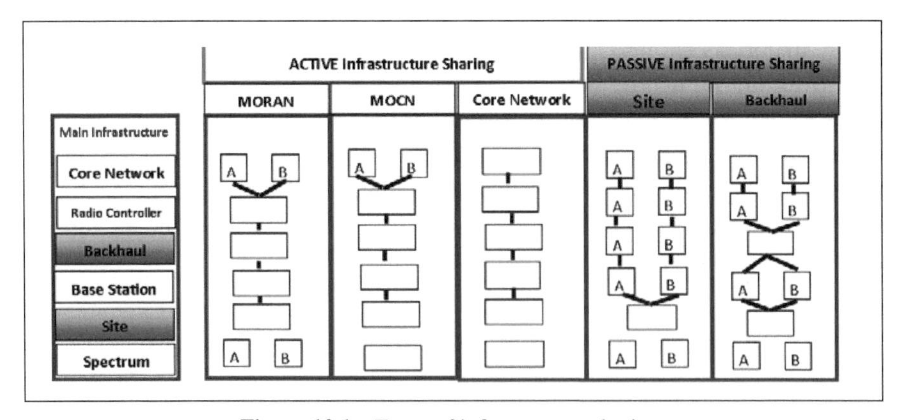

Figure 10.1 Types of infrastructure sharing.

networks (RANs) are shared, and each sharing operator uses a set of the different spectrum; ii) Multi-operator core network (MOCN), where RANs and spectrum are shared, and in the core network sharing, servers and core network functionalities are shared. In this kind, the operation of network equipment requires to be shared and therefore, the complexity of sharing increases as compared to the site sharing. The core network allows significant cost-saving, but operational complexity is higher and challenging to maintain strategic differentiation.

10.3 Benefits and Issues with Infrastructure Sharing

The above discussion clearly shows that cost reduction remains the main driver for network sharing. To find possible sharing options, operators require considering five dimensions: scope of technology, geographical coverage, architectural capacity, potential partners and sourcing. Eventually, network sharing is driven by the necessity to maximise enterprise value. The main benefit of network sharing is a net drop in network stocktickerCAPEX and OPEX. Depending on the sharing options, it can vary from 10- 40% of the in-scope costs. Another issue is network densification in indoor environments, which has led to challenges in acquiring sites for radio access networks (namely, base stations), such as the availability of limited or restricted space where indoor base stations can be installed. Exceptionally few choices might be available considering the coverage demand to be satisfied. Additionally, having more than one mobile operator further complicates the problem.

10.4 Constraints in Sharing

10.4.1 Technical Constraints in Passive Sharing

For site sharing, the operators must ensure:

- the suitable sites to share (in respect of electromagnetic compatibility, area, optimizing legacy networks),
- fixing of equipment on the shared site (accessibility and safety),
- operation and maintenance of equipment (on-site monitoring, debugging).

10.4.2 Technical Constraints in Active Sharing

Active sharing can be:

10.4.3 Antennas Sharing

- the need for common choices (radio planning, diversity communication, antenna architecture),
- antenna for the number of frequency bands,
- coupling loss consideration.

10.4.4 NodeB Sharing

- the use of NodeB containing at least two different carriers,
- a limited number of operators (typically 3 or 4),
- quality of service may be because of power-sharing,
- maintenance and operation of shared assets.

10.4.5 RNC Sharing

It has a similar constraint as NodeB. Additionally,

- managing split of the RNC functionalities (access configuration, quality of services),
- interoperability of equipment from different manufacturers (software and hardware configuration).

10.4.6 Sharing Core Network

- the design choice of equipment (Mobile switching centre-MSC, RNC, NodeB, Serving GPRS Support Node-SGSN) to handle the traffic related to the delivery of services of each operator,
- a design suite from core network management and quality of service,
- require to support intelligent network protocols to ensure continuity of user service of each operator during roaming on the shared network.

10.4.7 Enablers for Infrastructure Sharing, Especially in 5G and Beyond

Software-defined networking (SDN) and network function virtualisation (NFV) offer an operator to use commodity hardware in place of physical equipment, and all occurrences of network units are virtualised and exist as logical units (i.e. have various logical occurrences of a network unit such as S-GW on a single physical node). Also, SDN and NFV allow network slicing, which permits the "slicing" of one physical network into many virtual networks enjoying a different QoS and topology [20]. This will enable operators to deploy virtual networks with diverse necessities on physical infrastructure.

SDN and NFV will reduce the hardware dependencies to a great extent and enable numerous services through slicing. Also, moving from 2D to 3D range requires placing network elements up in the sky and space, a non-terrestrial network. The 6G wireless network is likely to integrate the terrestrial, non-terrestrial (aerial and maritime) communications into a robust network that meets the specification of the current 5G network and offers superior performance.

10.5 Non-terrestrial Infrastructure Sharing with Terrestrial Infrastructure

A non-terrestrial network position the elements of the network infrastructure overhead in the form of satellites, HAPS (High Altitude Platform Stations) or Drones (Unmanned Aircraft Systems).

10.5.1 Non-terrestrial Network Architecture

Figure 10.2 illustrates the high-level architecture of a non-terrestrial network [21]. It comprises a satellite gateway that connects terrestrial and non-terrestrial elements, a feeder link, optional inter satellite link, and a service link. The extended architecture proposed by 3GPP with two options:

- Transparent payload – The satellite acts as an analogue radio frequency repeater for both the feeder and service links. In the 5G network, the satellite (UAS) repeats the 5G NR-Uu radio interface from the feeder link to the service link.
- Regenerative payload – The satellite regenerates the signal from the earth with the NR-Uu interface operating on the service link and N2 and N3 operating over a satellite radio interface on the feeder link, i.e. the gNB or 5G base stations are now in the space.

Although this is expensive, it offers several unique advantages, including:

- Extending coverage – In addition to providing connectivity to those rural communities with currently no or little access to the Internet, industries operating in remote areas such as mining, oil and gas exploration, not to mention research stations, could also benefit from a significant increase in both coverage and capacity.

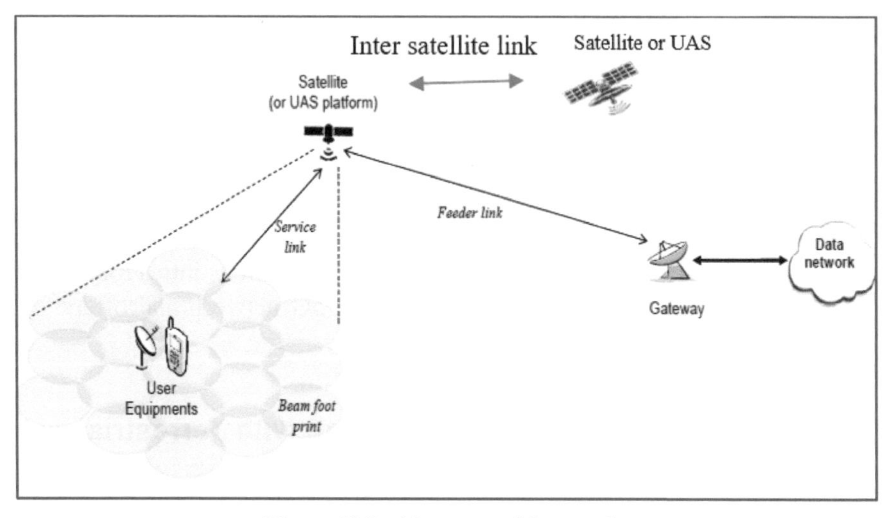

Figure 10.2 Non-terrestrial network.

Figure 10.3 3Typical beneficial scenario of non-terrestrial networks.

- Mission Critical Services – Natural disasters such as earthquakes or tsunamis often render traditional network infrastructure inoperable at the most critical time (Figure 10.3). Restoring the terrestrial network fast is too tricky. Thus the immediate use of satellites and drones to support essential communications could provide necessary respite and save numerous lives.
- Providing Reliability and Resilience –Instances when huge capacity may be required, such as supporters leaving a sports stadium wanting to watch highlights on their smartphones, large gatherings during festivals, etc.

10.5.2 Non-terrestrial Network Challenges

There are several technical and nontechnical challenges in deploying non-terrestrial networks. Some of them are:

- Network deployment is costly.
- Latency - The distance between the satellites or UAS with the mobile/device will directly impact the transmission delay or latency. Some 5G applications require latencies of 1ms, which will not be achievable using satellites but may potentially be supported by using drones at very low altitudes.
- Coordinated Handover – Satellites operating in a low earth orbit (LEO) or medium earth orbit (MEO) will be moving relative to the device on the ground. This would require coordinated handovers to be supported since one satellite moves out of coverage and another move in.
- Doppler Effect - Satellites travelling faster may require Doppler effects to be considered, especially on the inter-satellite links.
- Regulatory Authority - Since beam footprints (as shown in Figure 10.2) is very large, greater than 1,000km in diameter. It introduces many issues; for example, a "cell" may span the border between two countries, so how will it be possible to identify which country the device is located in and, therefore, which regulatory authority must be complied with.
- Closer to Earth UAS Life - If usage of UAS closer to the earth is considered, this particular device has many unique problems, such as the longevity of UAS, given the weight of the radio equipment and fuel? Will these platforms operate autonomously, or will they need to be piloted?

10.6 Terrestrial and Non-terrestrial Networks Deployment Scenario

Deployment of the 5G network would not provide complete terrestrial coverage. Connecting the remaining unconnected population is not only complex but costly. However, non-terrestrial networks such as satellite communication combined with terrestrial networks might cover a larger footprint. Satellite communications can be used in conjunction with 5G in various combinations and deployment scenarios, including combined access networks, shared access, and satellite backhaul.

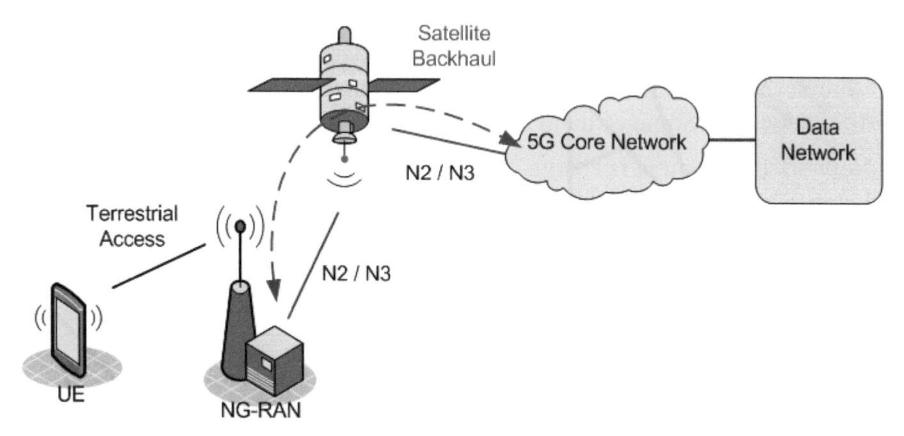

Figure 10.4 Satellite as backhaul in shared access.

10.7 Satellite Backhaul

Backhaul over satellite can deliver services to remote sites. A satellite backhaul is employed between the core and terrestrial access networks and transports 5G N1/N2/N3 reference points, as shown in Figure 10.4. This method can offer high bit rate connectivity to remote locations across a large coverage area with the ability to multicast content, e.g. video, HD (High Definition)/UHD (Ultra High Definition) TV, and non-video data. This could also provide flexibility for local storage and consumption. Similarly, the IoT (Internet of Things) scenario can provide an efficient backhaul of aggregated IoT traffic from multiple sites.

10.8 Cooperation Among Terrestrial and Non-terrestrial Networks

Integrating satellites with 5G and beyond 5G (B5G) infrastructure would improve the quality of experience (QoE) [22, 23]. By routing and offloading traffic intelligently, satellites save important spectrums and enhance the resilience of each network. LEO satellites or high-altitude platforms (HAP) or low altitude platforms (LAP) would play an essential role in extending cellular 5G networks coverage to space, sea and remote areas uncovered by small cell networks. The satellite can support a seamless extension of 5G services from the city to aeroplanes and vehicles in remote locations to a cruise liner. Similarly, IoT sensors and machine-2-machine (M2M) connections on farmhouses and remote locations like mines can also be covered by 5G satellites.

Similarly, HAP stations [13] which are unmanned aircraft normally placed at an altitude of over 20km and have very long-duration flights (approximately in years). HAPSs provide wide area coverage, ease of deployment, small delays and less attenuation.

Unmanned aerial vehicles (UAVs) provide radio on-demand coverage [13]. UAVs and HAPs have received significant attention in data traffic management, network coverage improvement, improving the QoS, exploiting network access [24], etc. Figure 10.5 shows the overall structure of integrated terrestrial-non terrestrial networks. It can be seen that various platforms in space integrated with the terrestrial networks would improve ubiquitous connection.

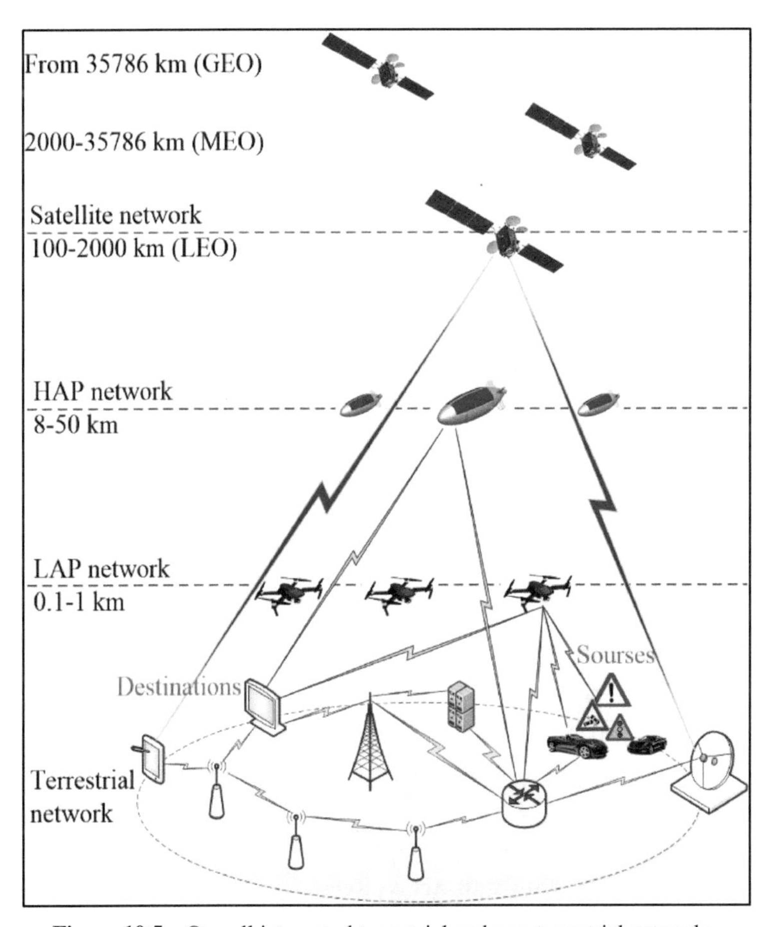

Figure 10.5 Overall integrated terrestrial and non-terrestrial networks.

10.9 Low Altitude Platform (LAP) and Sharing

LAP has become significantly important recently. In LTE-A, aerial base stations (AeNB) using LAP are implemented in the ABSOLUTE project [25]. The primary purpose of this design is to support wireless coverage and capacity for public safety applications during and after large-scale unexpected and temporary events. 3GPP TR22.829 V17.1.0 [26] focuses on a wide range of applications and scenarios using low-altitude UAVs in various commercial and government sectors. Numerous use cases are envisioned with LAP.

10.10 Radio Access Node On-board UAV

LAP can also offer UAV-assisted communications, as shown in Figure 10.6. Here, LAP (e.g., flying base stations (a) or mobile relays (b)) can provide/enhance communication services to ground users in high traffic demand

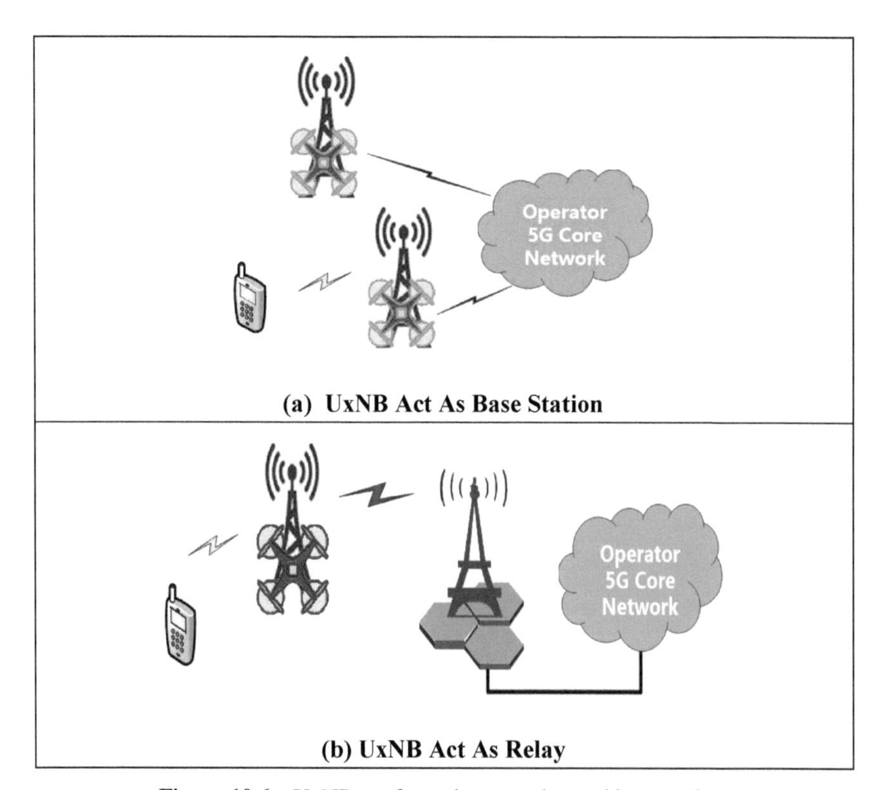

Figure 10.6 UxNB configurations as relay and base station.

and overloaded situations (Figure 10.7). Additionally, UAV infrastructure can solve multiple problems of existing networks. As shown in Figure 10.8, UAVs can relay traffic from users if the base station is malfunctioning, a link fails, an uncovered area, etc.

UAVs find many commercial applications such as disaster monitoring, border surveillance, emergency service and so on because of ease of deployment, less maintenance cost, etc. Using radio access node on-board UAV (i.e.

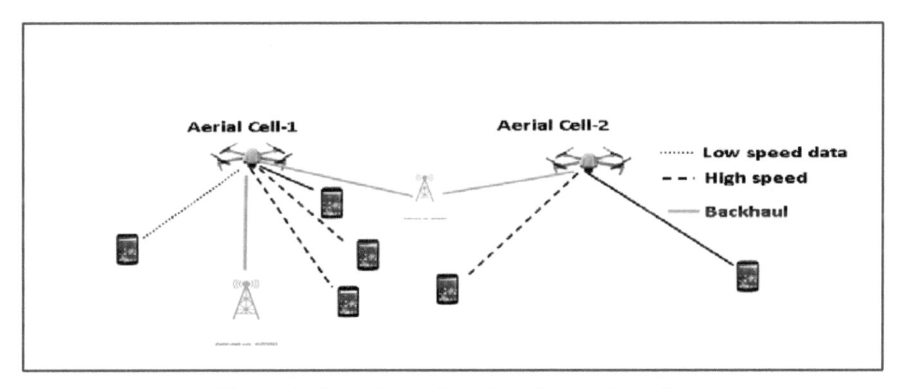

Figure 10.7 Enhanced services from aerial cells.

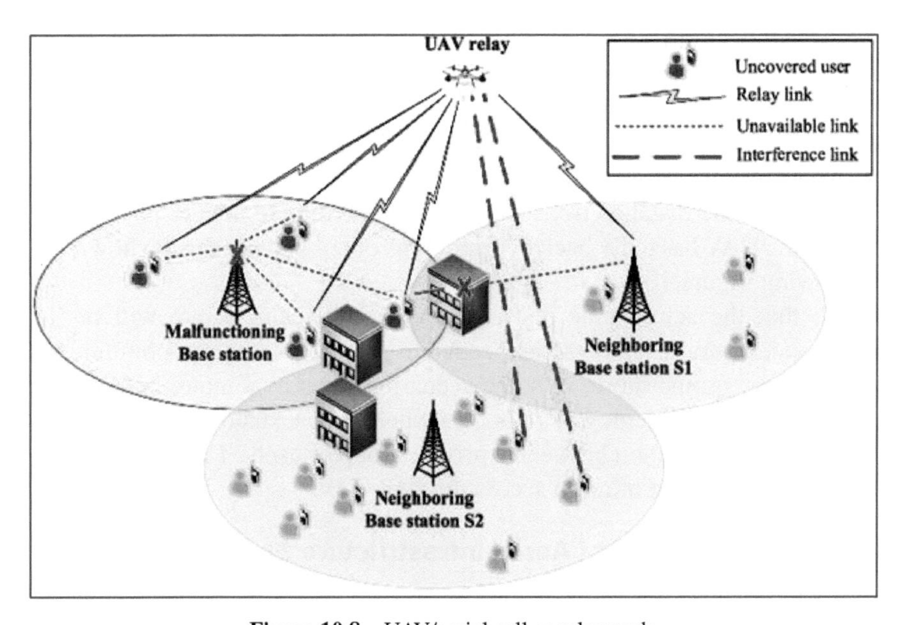

Figure 10.8 UAV/aerial cell as relay node.

UAV eNB/gNB/ng-eNB, UxNB) enhances coverage in several scenarios, e.g. emergencies and temporary coverage for mobile users and hot-spot events like a festival or large gatherings. UAS and high altitude pseudo satellite (HAPS) as base stations are described in TR 38.811. The altitude of the UAS is suggested to be between 8km and 50km. However, as UAV has a lower altitude (usually around 100m) and has an on-board base station (i.e. UxNB), this platform is more flexible than that of UAS as far as the coverage and fast deployment are concerned. The UxNB act as either base station or relay, as shown in Figures 10.6(a) and (b).

In some instances, say bandwidth-limited resources, a base station may need to simultaneously support data transmission for eMBB users on the ground and UAVs in the air. For example, during a live broadcast, a UAV over 100m height requires to transmit the captured video or pictures to the base station in real-time. This would require a high transmission rate and large bandwidth. At the same time, quality of service (QoS) for ground users (e.g., eMBB users) is also expected. Of course, the interference between both kinds of users should be minimized.

10.11 Radio Access Through UAV

UAVs have limitations in flying time. When a UAV is working as a flying RAN, the weight of the UAV will further increase due to the payload of RAN equipment. This will further reduce the flight time of the UAV. Additionally, the efficiency of a UAV as a mobile RAN platform will further be dependent on deployment scenarios. For example, a flying RAN is typically needed in the area where ground-based RAN equipment cannot be installed. Hence, the UAV has to fly some distance from the origin to the remote area to provide service, and the UAV has to fly back before it runs out of power. This is shown in the following Figure 10.9.

We see that the actual time that the flying RAN can operate will be smaller. Hence, several UAVs may be needed to provide continuous communication services. Furthermore, in some scenarios, such as remote, isolated locations, a UAV can be deployed in an area where no backhaul connectivity is needed for communication between a private group of users. This is known as the remote deployment of radio access through UAV.

10.11.1 The architecture of Aerial Infrastructure Sharing

The assets in aerial infrastructure are: i) Spectrum; ii) LAP; ii) Aerial cell; and; iv) Backhaul. There could be different combinations of sharing these

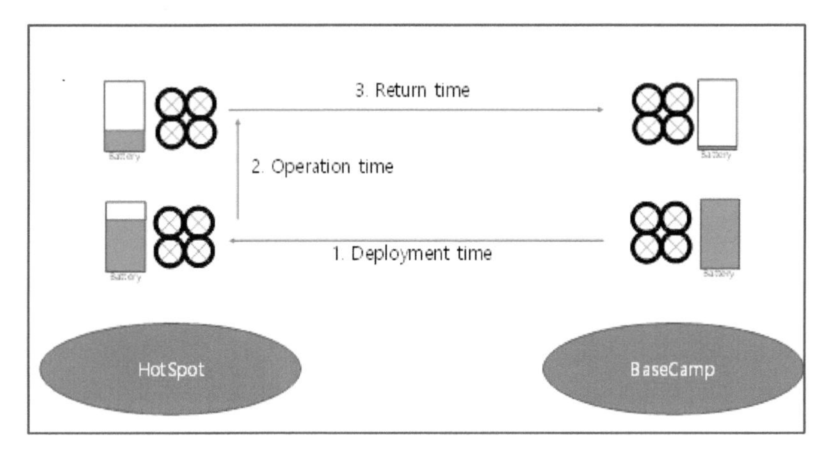

Figure 10.9 UAV operational time analysis

Table 10.1 Possible architecture for aerial infrastructure sharing.

Aerial Assets	Architecture for sharing			
	Shared LAP	Shared Backhaul	Aerial MORAN	Aerial MOCN
Aerial Cell	No	No	Yes	Yes
LAP	Yes	Yes	Yes	Yes
Backhaul	No	Yes	Yes	Yes
Spectrum	No	No	No	Yes

assets. For example, the aerial cell accommodated on the LAP can be shared between the operators. Sharing can also be done for the network nodes in the backhaul (CN). Table-10.1 presents the architecture for aerial infrastructure sharing [27]. It shows what can be shared and what can't be shared. Of course, there might be a possibility for more sharing options. Since sharing of aerial assets is available across operators, provided an agreement is in place, asset utilization will improve. Therefore, it is possible to compare different architectures, and analysis can identify the best architecture.

Nevertheless, there would always be a trade-off. Consider the architecture, aerial MOCN where all four assets are shared, as shown in Figure 10.10. The UEs from an operator are simultaneously connected using either carrier aggregation (CA) or dual connectivity (DC) [27] to the respective operator's terrestrial cell and the shared aerial cell (LAP). Various performance studies would be carried out to understand the pros and cons. It is also possible to use machine learning or deep learning to understand the best architecture and utilization of shared resources.

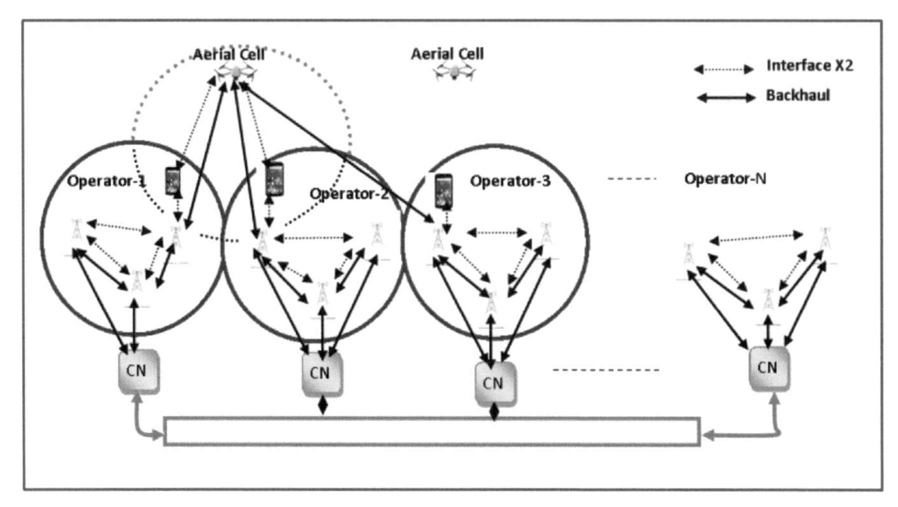

Figure 10.10 Shared architecture with LAP: an example.

10.12 Conclusions

In this chapter, we introduced the concept of infrastructure sharing. We also presented the advantages, disadvantages and technical enablers for the sharing. We progressed through 4G cellular networks through 5G, highlighting the ITU's standardisation efforts. We also briefly discussed the challenges in designing such integrated networks. However, it became evident through the end that various platform in space is necessary for ubiquitous coverage in 5G and beyond. Some of the other aerial infrastructure sharing architecture are highlighted. However, this would enable researchers to investigate these architectures' suitability further.

References

[1] Guangyi Liu, Dajie Jiang, "5G: Vision and Requirements for Mobile Communication System towards Year 2020", *Chinese Journal of Engineering*, vol. 2016, Article ID 5974586, 8 pages, 2016. https://doi.org/10.1155/2016/5974586

[2] Alsharif, M.H., Nordin, R. Evolution towards fifth generation (5G) wireless networks: Current trends and challenges in the deployment of millimetre wave, massive MIMO, and small cells. *Telecommun Syst* 64, 617–637 (2017). https://doi.org/10.1007/s11235-016-0195-x

[3] A. Roessler, J. Schlienz, S. Merkel, M. Kottkamp, "LTE- Advanced (3GPP Rel.12) Technology Introduction", *Rohde-Schwarz white paper*, June 2014.

[4] Edward J. Oughton, Zoraida Frias, Sietse van der Gaast, Rudolf van der Berg, Assessing the capacity, coverage and cost of 5G infrastructure strategies: Analysis of the Netherlands, Telematics and Informatics, Volume 37, 2019, Pages 50-69. https://doi.org/10.1016/j.tele.2019.01.003.

[5] S. E. Elayoubi *et al.*, "5G service requirements and operational use cases: Analysis and METIS II vision," *2016 European Conference on Networks and Communications (EuCNC)*, 2016, pp. 158-162, doi: 10.1109/EuCNC.2016.7561024

[6] Steve Doyle (GSMA), "5G for Industry 4.0 operational technology networks", GSMA, March 2021.

[7] Ondrej Burkacky et, al, "The 5G era New horizons for advanced electronics and industrial companies", : McKinsey analysis, January 2020.

[8] Kevin Sparks et. al, "5G Network Slicing" White paper, FCC Technological Advisory Council 5G IoT Working Group, 2018.

[9] Lingli Deng, Hui Deng and Andy Mayer, "Harmonizing Open Source and Standards: A Case for 5G Slicing", *ONAP*, March 2020.

[10] Recommendation ITU-R M.2150-0, "Detailed specifications of the terrestrial radio interfaces of International Mobile Telecommunications-2020 (IMT-2020)", February, 2021.

[11] F. Tariq, M. R. A. Khandaker, K. -K. Wong, M. A. Imran, M. Bennis and M. Debbah, "A Speculative Study on 6G," in *IEEE Wireless Communications*, vol. 27, no. 4, pp. 118-125, August 2020, doi: 10.1109/MWC.001.1900488.

[12] Mohammed H. Alsharif et al, "Sixth Generation (6G) Wireless Networks: Vision, Research Activities, Challenges and Potential Solutions", *MDPI journal*, December 2020, doi:10.3390/sym12040676

[13] You, X., Wang, CX., Huang, J. *et al.* Towards 6G wireless communication networks: vision, enabling technologies, and new paradigm shifts. *Sci. China Inf. Sci.* 64, 110301 (2021). https://doi.org/10.1007/s11432-020-2955-6

[14] Online: https://www.waveform.com/pages/das-distributed-antenna-systems Accessed on, March 2021.

[15] Z. Fan, F. Cao and Y. Sun, "A distributed antenna system for in-building femtocell deployment," *2011 IFIP Wireless Days (WD)*, 2011, pp. 1-5, doi: 10.1109/WD.2011.6098163.

[16] W. Hong *et al.*, "The Role of Millimeter-Wave Technologies in 5G/6G Wireless Communications," in *IEEE Journal of Microwaves*, vol. 1, no. 1, pp. 101-122, winter 2021, doi: 10.1109/JMW.2020.3035541.

[17] Calvanese Strinati, E, Barbarossa, S, Choi, T, et al. 6G in the sky: On-demand intelligence at the edge of 3D networks (Invited paper). *ETRI Journal*. 2020; 42: 643– 657. https://doi.org/10.4218/etrij.2020-0205

[18] Ian F. Akyildiz et al, "6G and Beyond: The Future of Wireless Communications Systems", *IEEE Access*, July 2020.

[19] GSMA, Mobile infrastructure Sharing, GSMA report, September 2012.

[20] Long, Q., Chen, Y., Zhang, H. *et al.* Software Defined 5G and 6G Networks: a Survey. *Mobile Netw Appl* (2019). https://doi.org/10.1007/s11036-019-01397-2

[21] F. Rinaldi *et al.*, "Non-terrestrial Networks in 5G & Beyond: A Survey," in *IEEE Access*, vol. 8, pp. 165178-165200, 2020, doi: 10.1109/ACCESS.2020.3022981.

[22] J. Kim *et al.*, "5G-ALLSTAR: An Integrated Satellite-Cellular System for 5G and Beyond," *2020 IEEE Wireless Communications and Networking Conference Workshops (WCNCW)*, 2020, pp. 1-6, doi: 10.1109/WCNCW48565.2020.9124751

[23] Konstantinos Liolis et. Al. ,"Use cases and scenarios of 5G integrated satellite-terrestrial networks for enhanced mobile broadband: The SaT5G approach", International Journal of Sat. Communications Networks, John Wiley & Sons, Ltd. 2018.

[24] Bithas, Petros S et al. "A Survey on Machine-Learning Techniques for UAV-Based Communications." *Sensors (Basel, Switzerland)* vol. 19,23 5170. 26 Nov. 2019, doi: 10.3390/s19235170

[25] UoY, David Grace et al. "Integrated Project ABSOLUTE-Aerial Base Stations with Opportunistic Links for Unexpected & Temporary Events." (2014).

[26] 3GPP Technical Report TR 22.829 V17.1.0, "Technical Specification Group Services and System Aspects; Enhancement for Unmanned Aerial Vehicles; September 2019.

[27] T. Vrind, S. Rao, L. Pathak and D. Das, "Deep Learning-based LAP Deployment and Aerial Infrastructure Sharing in 6G," *2020 IEEE International Conference on Electronics, Computing and Communication Technologies (CONECCT)*, 2020, pp. 1-5, doi: 10.1109/CONECCT50063.2020.9198319

Biography

Navin Kumar obtained his PhD (2007-2011) in Telecommunication Engineering from the University of Minho, Aveiro and Porto of Portugal, Europe. He did his M.E in Digital system engineering from Motilal Nehru regional engineering college (now National Institute of Technology) Allahabad, UP –India, in the year 2000 and BE in Electronics and Telecommunication engineering from IETE New Delhi – India, in 1996. He has over 24 years of working experience in Government, Industry and academia in the IT & Telecommunication area. Currently, he is working as Assoc. Professor and chairman ECE Dept. in Amrita School of Engineering Bangalore campus of Amrita Vishwa Vidyapeetham (University). Navin has over 90 publications in peer-reviewed international journals and IEEE conference proceedings. In addition, he has also authored and edited books and book chapters. Navin has been awarded the Fraunhofer Challenge award in the academic year 2010-2011 for the best PhD thesis work. He had also received a research grant from the foundation of science and technology (FCT) Govt. of Portugal for his PhD research work. He is the recipient of the Gowri Memorial award, India, in the year 2009, for the best journal paper. Many of his papers were adjudged as best paper awards at International IEEE conferences outside India. He is a Sr Member of IEEE, AIENG (HK), a Life Member of IETE and a Fellow of and Charter Engineer Institution of Engineers IE-India). His research area includes 5G (mmWave and Massive MIMO), Visible Light Communication, Optical Wireless Communication, IoT and Intelligent Transportation Systems. He is also a part of IEEE 5G and Future Network Initiative.

11

Radio Frequency Spectrum for 5G and Beyond Applications – ITU's Perspective

P S M Tripathi[1], Ambuj Kumar[2] and Ashok Chandra[3]

[1]Ministry of Communications, Government of India, New Delhi, India
[2]Department of Business Development and Technology, Aarhus University, Aarhus, Denmark
[3](Former) Wireless Adviser to the Government of India, New Delhi, India
E-mail: psmtripathi@gmail.com; ambuj@btech.au.dk; drashokchandra@gmail.com

The radio spectrum is a natural fuel that gives life and drives 41 radio-communication services. As defined by the International Telecommunication Union (ITU), these services include inter-alia Fixed (FX) mobile, broadcasting, aeronautical, navigation, satellite etc. Today, the radio spectrum is earmarked/assigned in the frequency range VLF to mm bands for the existing wireless applications. The International Mobile Telecommunications (IMT) is the most widely used application by the public under the 'Mobile (MO) Service' category. Its journey started in 1979 as 1 G followed by 2 G in 1991, 3 G in 2000 and so on. The frequency bands assigned for these applications, including 4 G/LTE, are UHF bands up to 2 GHz. The unprecedented growth of mobile traffic and high-capacity data transfer led to the need for an additional spectrum. After detailed studies, the frequency bands are identified by the World Radio Conferences (WRCs), held normally during 3-4 years. The WRC held in 2019 identified some frequency bands in the mm range 24.25 -86 GHz for 5 G applications. For 6 G applications, one has to look in TeraHz (THz), i.e., beyond 100 GHz, may be up to 275 GHz and until 1000 GHz. Half of the portion is earmarked for Fixed and MO services. Because of the very short wavelength, thorough propagation studies must be made scientifically before establishing IMT networks. This paper explores deep insight into various issues in this regard.

11.1 Introduction

The radio frequency spectrum is a unique, non-depletable, but limited natural resource. It is the essential raw material for radiocommunication services, i.e. wireless services. As per the Radiocommunication Sector of the International Telecom Union (ITU) [1], 41 different radio communication services require a spectrum. These services include aeronautical, maritime, radio-navigation, radiolocation, radio astronomy, meteorological, broadcasting, satellite broadcasting, cellular mobile, fixed-satellite, mobile-satellite, space services, etc., which require radio spectrum for their existence. Mobile service is one of the radiocommunication services, and mobile cellular telecom is a part of mobile service.

The journey of mobile cellular telecom was started in 1979 in Japan by Nippon Telegraph and Telephone (NTT) and Advanced Mobile Phone System (AMPS) in the United States for analogue voice communication only. This was termed 1st generation (1G) of mobile communication and introduced in 450MHz/800/900 MHz [2].

Following the success of 1G, digital communication was launched with second-generation (2G) in 1991 across western Europe with GSM (Global System for Mobile Communication) based on the TDMA (Time Division Multiple Access) and FDMA (Frequency Division Multiple Access) technologies in 800/900 MHz/1800 MHz band. CDMA (Code Division Multiple Access) was another 2G system developed in the USA [3–5]. Data services in the form of Short Message Service (SMS) and digital voice were the main attributes of the 2G system. The 2G system was upgraded with General Packet Radio Service (GPRS), which became known as 2.5G and provided data speeds of around 115 Kbps. Enhanced Data Rates further upgraded this for GSM Evolution (EDGE), which became known as 2.75G with a speed of over 384 Kbps.

The 3rd generation (3G) [3–6] was introduced in 2000 with the objective "anywhere, anytime" under the ITU-R umbrella IMT-2000 (International Mobile Telecommunication). Compared to 2G, 3G had four times the data speed with a maximum of up to 2 Mbps on average and a maximum of up to 10 Mbps. Mobile Internet access, Video streaming, video conferences, mobile TV, Global Positioning System (GPS) etc., were the main attributes of 3G. Emails also became another standard form of communication over mobile devices. ITU-R approved a family of five 3G standards, which were part of the 3G framework known as IMT-2000:

- *Three standards based on CDMA, namely CDMA2000, WCDMA, and TD-SCDMA.*
- *Two standards are based on TDMA, namely, FDMA/TDMA and TDMA-SC (EDGE).*

CDMA based technology was more popular compared to TDMA based technologies. Out of three CDMA technologies, WCDMA was more popular and used in most countries. The 3G networks operate in the 800 MHz, 900 MHz, 1,700 MHz, 1,900 MHz and 2,100 MHz bands. The 3G networks led to massive growth in the use of mobile data through USB dongles and smartphones [5].

Due to the convergence of services and the advent of smartphones, data usage increased continuously, and the existing 3G network could not handle the data growth rate. Therefore, in November 2008, ITU-R specified a new standard for the 4th generation, named the International Mobile Telecommunications Advanced (IMT-Advanced) [3, 4, 7]. Generation is usually defined by data speed rate. Each generation has almost ten times that of the previous generation. Thus, the required speed of 4G may be about ten times higher than 3G. Accordingly, ITU-R set peak speed requirements for 4G service at 100 Mbps for high mobility communication (such as from trains and cars) and 1 Gbit/s for low mobility communication (such as pedestrians and stationary users) [7]. A new radio interface and core network LTE (Long Term Evolution) was introduced in 2010 by 3GPPP in its Release 8 [8] to meet the enhanced data demand, which is termed as 4th generation mobile communications. LTE, an IP (Internet Protocol) network, is based on the Orthogonal Frequency Division Multiplexing (OFDM) and MIMO (Multiple Input Multiple Output) technologies [9]. LTE was developed on a new platform based on the OFDM Technology contrary to earlier technologies based on TDM/FDM/CDM technology. The LTE has peak data rates of up to 100 Mbit/s for high mobility and up to 1 Gbit/s for low mobility and latency less than 10 ms with scalable channel bandwidths of 5–20 MHz, maximum up to 40 MHz. 4G is also an all-IP (internet protocol)-based standard for both voice and data, different from 3G, which only uses IP for data. To provide higher-speed services and the already allocated spectrum bands, new frequency spectrum bands 700 MHz, 2300 MHz, 2500 MHz, 3400 MHz, and 4400 MHz bands have been identified for IMT services.

The 5th generation mobile communication (5G) system was introduced in 3GPP Release 15 [10] in 2018, which enables wireless services and applications at a data rate of more than one terabit per second (Tbit/s) with coverage extending from a city to a country to the continents and the world. 5G network covers a wide range of applications, mainly enhanced Mobile Broadband (eMBB), Ultra-reliable and Low Latency Communications (URLLC) and massive Machine Type Communications (mMTC). The eMBB is an extension of the 4G/LTE network, and URLCC and mMTC are new dimensions added to the 5G network. 5G is now in the deployment phase. About 157 mobile operators have launched commercial 5G services in 62 countries [11]. Ericsson, in its report, anticipates that 5G will cover up to 65 percent of the world's population with more than 2.6 billion subscriptions globally by the year 2025, and nearly half of the world's mobile data traffic will be over 5G networks by 2025 [12]. To meet the 5G requirements, additional spectrum bands 24.25-27.5 GHz, 37-43.5 GHz, 45.5-47 GHz, 47.2-48.2 and 66-71 GHz have been identified for the 5G network[13]. The generations in the pictorial form are given in Figure 11.1 below:

Cellular mobile telecom is one of the fastest-growing sectors in the world. Mobile Telecom has dramatically accelerated the growth of a country's economic and social sectors and contributed significantly to the country's economic growth. Presently, the growth of any other industry has a direct or indirect link with telecom, which leads to extraordinary demands for mobile data. In its mobility report, Ericsson estimated that mobile traffic will grow by 27 percent annually between 2019 and 2025 and will reach 160 exabytes per

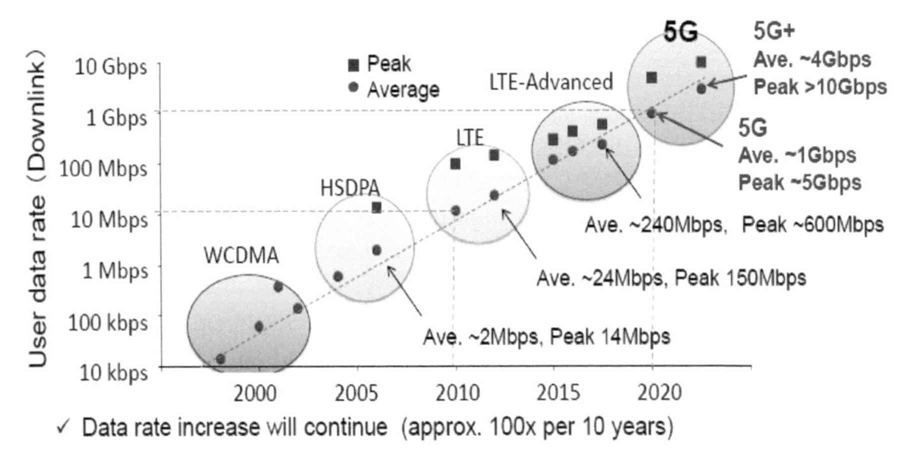

Figure 11.1 Growth of mobile technology [14].

month in 2025 [12]. There will be 8.9 billion mobile subscriptions by the end of 2025, out of which around 90 percent will be for mobile broadband [12]. In its Report ITU-R M.2370 [15], ITU estimated global IMT traffic estimates for the years 2020 to 2030 from several sources and anticipated that global IMT traffic would grow in the range of 10-100 times over this period. The overall mobile data traffic volume will exceed 5 ZBytes per month by 2030. The number of mobile subscriptions will reach 17.1 billion compared to 5.32 billion in 2010. Growth in mobile communications can be attributed to the evolution in technology and the availability of the radio spectrum.

New technology has been developed and commercially exploited almost every ten years. Due to strong growth, demand for radio spectrum in the mobile telecom sector is much higher than any other radiocommunication services due to the convergence of services and the need for a high data rate. The demand for spectrum increases with the intersection of more services and the desire for high data speed. We could see that the first generation was started with a meagre spectrum in 450/900 MHz band with voice service only, and demand increases with every next generation due to the desire for high data speed and convergence of services. Right now, we are in the 5th generation, for which several frequency bands up to 100 GHz have been identified while squeezing the allocation of spectrum for other radiocommunication services.

The radio spectrum is a scarce natural resource, and manned spectrum bands have already been allocated to 41 radiocommunication services. So what we could say is that no new spectrum is available for any service. Therefore, making available an additional spectrum for mobile telecom, i.e. IMT services, is a bit complex and is operated at an international level through the Radiocommunication Sector of ITU. This paper provides insight into the spectrum allocation process for IMT services starting from the beginning of mobile telecom services. The structure of the chapter is as follows: functioning of ITU-R is explained in section 2, section 3 describes the process for identification of spectrum for IMT services, the requirement of spectrum beyond 5G is given in section 4, and finally, the conclusion is in section 5.

11.2 The Function of the Radiocommunication Sector of ITU

Radio Spectrum has some unique properties. It does not recognize political boundaries; no two separate communications can be transmitted on the same frequency in a given geographical area, and being a limited resource, it has to be shared amongst 41 different radiocommunication services. Further, each spectrum band has its propagation characteristics, making it suitable for some specific radiocommunication services. Therefore, regulation of the radio spectrum is inherently international due to its peculiar properties.

The history of radio regulation started on 24 May 1844, when Samuel Morse sent his first public message over a telegraph line between Washington and Baltimore. Initially, it was limited to country boundaries, and it crossed the boundaries of countries in a short period. At that time, there was no management system for the radio spectrum. Due to this, a wireless user could choose any frequency for their transmission [16]. As applications and services grew, interference became an increasing problem. It was felt to evolve a mechanism for regulating the radio spectrum. With this objective, the first International Telegraph Convention was convened on 17th May 1865 and participated by 20 countries. This marked today's International Telecom Union (ITU) [17]. The formal name of ITU was adopted in 1934 and became part of the United Nations on a provisional basis in 1947. The ITU is a UN specialized agency for information and communication technology (ICT), providing an international forum for over 190 Member States and more than 700 Sector Members and Associates from industry, international and regional organizations [18]. ITU works towards its commitment, i.e. "ITU is committed to connecting all the world's people – wherever they live and whatever their means" [19].

ITU works through its three Sectors; the Radiocommunication Sector (ITU-R), the Telecommunication Standardization Sector (ITU-T) and the Telecommunication Development Sector (ITU-D).

ITU's work related to radiocommunications is focused on the ITU-R Sector, which works toward a worldwide consensus on using space and terrestrial radiocommunication services [18]. ITU-R plays a vital role in the management of the radio spectrum and satellite orbits, finite natural resources that are increasingly in demand from a large number of services such as fixed, mobile, broadcasting, amateur, space research, meteorology, and global positioning systems, monitoring and communication services that ensure the safety of life on land, at sea and in the skies.

The mission of ITU-R is to ensure the rational, equitable, efficient and economical use of the radio-frequency spectrum by all radiocommunication services, including those using satellite orbits, and to carry out studies and approve Recommendations on radiocommunication matters [20]

ITU-R works through Study Groups & its working parties, Conference Preparatory Meetings (CPM), and World Radiocommunication Conferences (WRC). The functioning of ITU-R is a closed-loop system built around

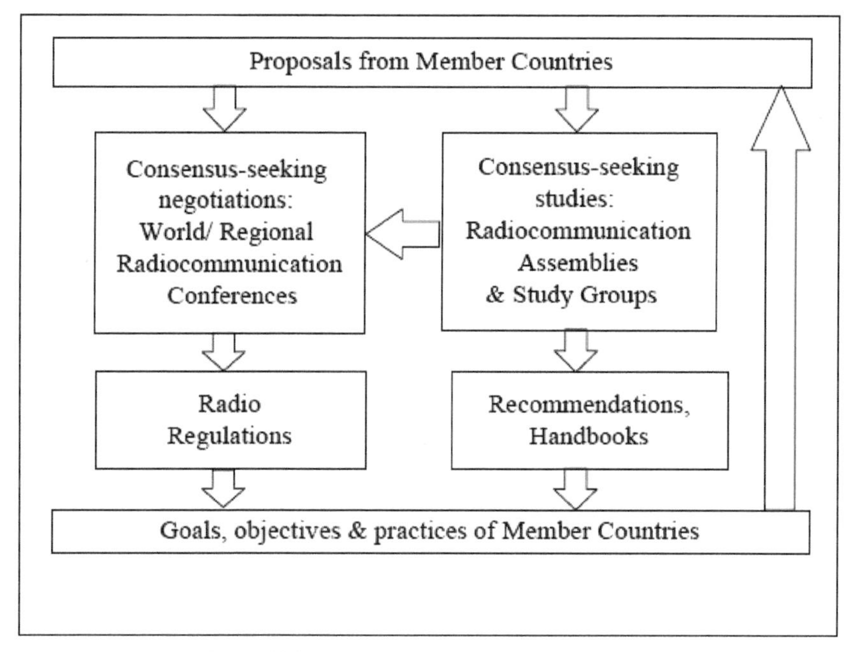

Figure 11.2 Spectrum management in ITU [21].

consensus-seeking studies and negotiations in the study groups and WRC, as shown in Figure 11.2

World Radiocommunication Conference (WRC) is the supreme body in radio spectrum management, usually held every three to four years. The last WRC was held in 2019 at sharm el-sheikh, Egypt, and the next one is scheduled for 2023. The primary job of WRC is to review and revise the Radio Regulations (RR), the international treaty governing the use of the radio-frequency spectrum and the geostationary-satellite and non-geostationary-satellite orbits [22].

The Study Groups (SGs) study a range of questions relating to radio-communication issues, focusing on RF spectrum in terrestrial and space communications, radio systems characteristics and performance, spectrum monitoring and emergency radiocommunications for the public protection and disaster relief. Study Groups accomplish their work in cooperation with other international radiocommunication organizations. Currently, 06 Study Groups are working: SG 1 Spectrum management, SG 3 Radiowave propagation, SG 4 Satellite services, SG 5 Terrestrial services, SG 6 Broadcasting service, and SG 7 Science services. These SGs work through various Working

Parties (WPs) and Task Groups (TGs) constituted within the SG to study the Questions assigned to the respective Study Groups [18]-Study Groups 2020]. The Working Party 5D of Study Group 5 studies all the aspects of IMT services.

Based on studies conducted by study groups between two WRCs, on regulatory, technical and operational aspects of new and existing radiocommunication services and systems, radiofrequency spectrum allocations are modified. WRCs also address any radiocommunication matter of worldwide character and determine agenda items for future WRCs. The outcome of WRC is Final Acts and subsequently revised Radio Regulations.

The Radio Regulations [23], an international treaty for administrations to govern the use of radio-frequency spectrum and satellite orbits at the global level. RR provides fundamental principles and terminologies for managing the radio spectrum. The RR contains definitions and spectrum allocation for all 41 radiocommunication services, categorized based on operational, administrative, and technical requirements. Based on the RR provisions, administrations have to develop a frequency allocation plan for the usage of spectrum in their country [23].

11.3 Radio Spectrum Identified for IMT Services

In the mid-1980, work began by ITU to define the next "generation" of mobile radio standards/mobile networks on a global basis led to the allocations of the new globally available frequency bands. The World Administrative Radiocommunications Conference (WARC)-1992 [24], 230 MHz new spectrum was identified in the frequency bands 1885-2025 MHz and 2110-2200 MHz in addition to previously identified 800, 900 MHz frequency Bands on a worldwide basis for the future public land mobile telecommunication systems (FPLMTS). The spectrum estimation was based on a model where voice services were considered the primary source of traffic, and only low data rate services were also considered. WARC-92 had also established the framework for the development of 3G with the titled International Mobile Telecommunications-2000 (**IMT-2000**) [6], as an advanced mobile communication applications concept intended to provide telecommunication services worldwide scale regardless of location, network or terminal used.

Spectrum so far identified for IMT services was insufficient to meet the requirements due to the rapid growth of mobile data services and wireless internet access. It was felt that an additional spectrum would be needed for IMT 2000. Accordingly, an agenda item no. 1.6 for WRC 2000 [25] was set to consider the additional demand for a globally harmonized spectrum for IMT services. The agenda item no. 1.6 as enumerated below:

Agenda Item No. 1.6: Issues related to IMT-2000 [25];

1.6.1. *review of the spectrum and regulatory issues for advanced mobile applications in the context of IMT-2000, noting that there is an urgent need to provide more spectrum for the terrestrial component of such applications and that priority should be given to terrestrial mobile spectrum needs and adjustments to the table of frequency allocations as necessary;*

1.6.2. *identification of a global radio control channel to facilitate multimode terminal operation and worldwide roaming of IMT-2000;*

As a result, at the WRC-2000, all the major existing cellular bands were added, increasing the potential IMT-2000 spectrum availability by approximately three times. It was also decided that three common bands, one below 1GHz, another at 1.8 GHz and the third one in the 2.5 GHz range, be made available on a global basis for countries for the implementation of the terrestrial component of IMT-2000 for a high degree of flexibility to allow operators to evolve towards IMT-2000 according to market and other national considerations. The three bands identified for use by IMT-2000 are 806-960 MHz, 1710-1885 MHz and 2500-2690 MHz [25].

To increase the spectrum availability for IMT services, an Agenda Item No. 1.4 was set up for WRC-07 [26].

Agenda Item No. 1.4

1.4 to consider frequency-related matters for the future development of IMT 2000 and systems beyond IMT 2000, taking into account the results of ITU R studies in accordance with Resolution 228 (Rev.WRC 03).

As a result, WRC-07 identified the additional frequency bands viz. (i) 450–470 MHz, (ii) 790–960 MHz, (iii) 1710–2025 MHz, (iv) 2110–2200 MHz, (v) 2300-2400 MHz, and (vi) 2500-2690 MHz, for IMT application for all the three regions. In addition, frequency bands 698-790 for Region 2 and 3400-3600 MHz for Region 1 and 3 were also identified, making a total availability of 1177 MHz spectrum for IMT services [26].

WRC-2000 and WRC-07 also provided the framework for 4G by opening up the 1.8 GHz and 2.6 GHz bands [27].

A Joint Task Group (JTG) 4-5-6-7 was established in WRC-12 [28] for the following two agendas of WRC-15 [29]:

- Agenda Item 1.1: *Additional spectrum allocations to the mobile service on a primary basis and identification of additional frequency bands for International Mobile Telecommunications (IMT) and related regulatory provisions; and*
- Agenda Item 1.2: U*se the frequency band 694-790 MHz for the mobile service, except aeronautical mobile, in Region 1.*

The Joint Task Group 4-5-6-7 recommended more than 20 additional frequency bands for IMT applications, considered during WRC-15 [26]. The WRC-15 considered the recommendations of JTG. After detailed deliberations, WRC-15 identified frequency bands 470-698 MHz, 1427-1518 MHz, 1885-2025 MHz, 2110-2200 MHz and 3300-3400 MHz or portions thereof for IMT applications [29]. WRC-15 had also requested ITU-R to study the potential use of additional spectrum above 6 GHz for IMT, and the results of those studies would be considered at the next WRC. WRC-15 passed Resolution 809 to recommend holding a world radiocommunication conference in 2019 (i.e. WRC-19) [30] and the proposed agenda items. WRC-19 agenda item 1.13 given additional spectrum to be required for 5G.

WRC-19 Agenda Item 1.13

To consider the identification of frequency bands for the future development of International Mobile Telecommunications (IMT), including possible additional allocations to the mobile service on a primary basis, in accordance with Resolution 238 (WRC-15)

Resolution 238 (WRC-15) [31]

- to conduct appropriate studies to determine the spectrum needs for IMT terrestrial components in the frequency range 24.25–86 GHz, taking into account:
    - technical and operational characteristics of terrestrial IMT systems;
    - deployment scenarios envisaged for IMT-2020 systems;
    - needs of developing countries; and
    - the timeframe in which spectrum would be needed.
- to conduct and complete in time for WRC - 19 the appropriate sharing and compatibility studies, taking into account the protection of services to which the band is allocated on a primary basis, for the frequency bands:

- o 24.25-27.5 GHz, 37-40.5 GHz, 42.5-43.5 GHz, 45.5-47 GHz, 47.2-50.2 GHz, 50.4-52.6 GHz, 66-76 GHz and 81-86 GHz, which have allocations to the mobile service on a primary basis; and
- o 31.8-33.4 GHz, 40.5-42.5 GHz and 47-47.2 GHz, which may require additional allocations to the mobile service on a primary basis

To conduct the study for Agenda Item No. 1.3, a Task Group 5/1 was constituted [32]. The Terms of Reference (ToR) of TG 5/1 was to estimate spectrum needs for the terrestrial component of IMT in the frequency range between 24.25 GHz and 86 GHz taking into account results submitted by various administrations on sharing and compatibility studies in the identified frequency bands. The frequency bands under consideration were: Item A (24.25-27.5 GHz), Item B (31.8-33.4 GHz), Item C (37-40.5 GHz), Item D (40.5-42.5 GHz), Item E (42.5-43.5 GHz), Item F (45.5-47 GHz), Item G (47-47.2 GHz), Item H (47.2-50.2 GHz), Item I (50.4-52.6 GHz), Item J (66-71 GHz), Item K (71-76 GHz), and Item L (81-86 GHz) [32].

The finding of TG5/1 were [32]:

- Frequency bands 45.5-47 GHz and 47-47.2 GHz were not proposed for IMT because no studies were carried out.
- Frequency bands 24.25-27.5 GHz [IMT – 26 GHz], 37-43.5 GHz, 45.5-50.2 GHz and 50.4-52.6 GHz [IMT 40/50 GHZ] and frequency bands 71-76 and 81-86 GHz [IMT 70/80 GHz] and 66 – 71 GHz [IMT 66/71 GHz] proposed for IMT.

Based on the TG5/1 report, WRC-19 identified additional radio-frequency bands 24.25-27.5 GHz, 37-43.5 GHz, 45.5-47 GHz, 47.2-48.2 and 66-71 GHz for International Mobile Telecommunications (IMT), to facilitate the development of fifth-generation (5G) mobile networks. While identifying these frequency bands, WRC-19 also took measures to ensure appropriate protection of the Earth Exploration Satellite Services, including meteorological and other passive services in adjacent bands.

17.25 GHz of the spectrum has been identified for IMT by the WRC-19, compared to 1.9 GHz of bandwidth available before WRC-19. Out of this 17.25 GHz, 14.75 GHz of the spectrum has been harmonized worldwide, reaching 85% global harmonization. Most of the administrations are ready to adopt/ have adopted part of the 24-27 GHz band for IMT services. A list of frequency bands identified in various WRCs is given in Table 11.1.

Cumulative allocation of spectrum for IMT services in different WRCs has been given in Figure 11.3. Initial allocation of 230 MHz was made in

Table 11.1 List of frequency bands identified in various WRCs.

IMT Frequency Band	Identified in WARC/WRC
450-470 MHz	WRC-2007
470-608 MHz	WRC-2015
694-960 MHz	WRC-2000, 2007
1427-1518 MHz	WRC-2015
1710-2200 MHz	WARC-1992, WRC-2000
2300-2400 MHz	WRC-2007
2500-2690 MHz	WRC-2000
3300-3400 MHz	WRC-2015
3400-3600 MHz	WRC-2007
4800-4990 MHz	WRC-2015
24.25-27.5 GHz, 37-43.5 GHz, 45.5-47 GHz, 47.2-48.2 and 66-71 GHz	WRC-19

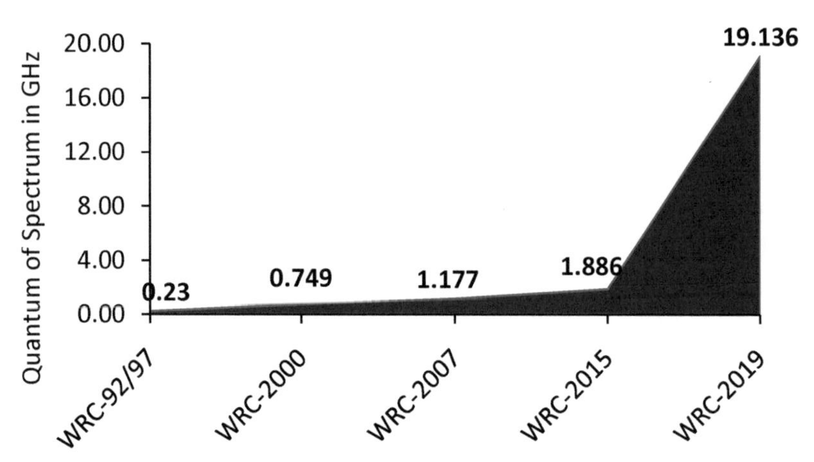

Figure 11.3 Cumulative allocation of spectrum for IMT in different WRCs.

WRC-92/97, which has now been enhanced to 19.136 GHz in WRC-19 to support the 5G network.

Figures 11.4 depicted the percentage-wise spectrum allocation for IMT services in various WRCs. Out of the total allocation of 19.136 GHz spectrum for IMT services, about 90% of the spectrum has been identified by WRC-19, and less than 10% spectrum was identified in earlier WRCs. This is clearly showing the growing importance of mobile communication.

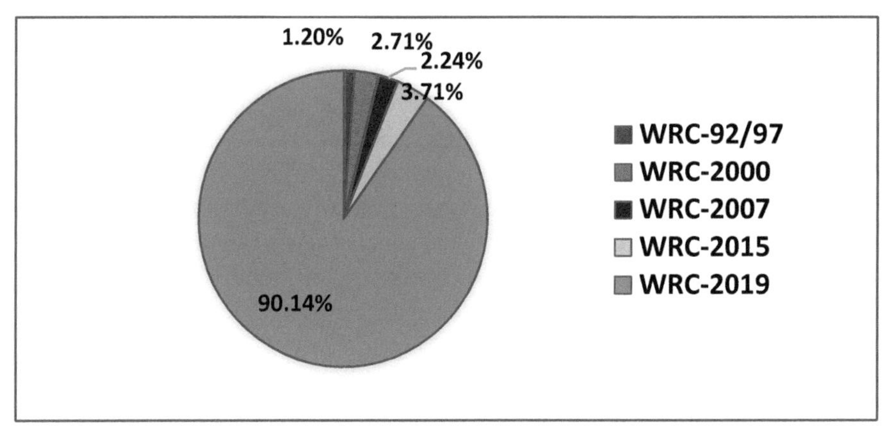

Figure 11.4 Percentage-wise distribution of spectrum allocation in WRCs.

WRC-19 has also sent an Agenda Item No. 1.2 for WRC-23 [33] to enhance the allocation for IMT services in all three regions. This agenda item considers the identification of additional frequency bands 3 300 - 3 400 MHz for Region 2 and amends footnote in Region 1, 3 600 - 3 800 MHz for Region 2, 6 425 - 7 025 MHz for Region 1, 7 025 - 7 125 MHz for all the three Regions, i.e. globally and 10.0 - 10.5 GHz for Region 2 for IMT including possible allocations to the mobile service on a primary basis. WRC-23 will consider the additional allocation of 700 MHz for Region 1, 900 MHz for Region 2 and 100 MHz for region 3 in mid-band for IMT services.

The spectrum so far identified for IMT services can be divided into three different categories:

Coverage layer – spectrum below 2 GHz (e.g. 700 MHz, 800 MHz, 900 MHz, 1800 MHz) is considered a coverage layer due to its comprehensive coverage and better penetration. Approximately 700 MHz spectrum below 2 GHz (about 35%) has been identified for IMT services. Further allocation below 2 GHz seems to be a difficult task.

Coverage and capacity layer –Spectrum bands in the 2 – 6 GHz band(2100 MHz, 2300 MHz, 2600 MHz, 3300 MHz, 3400 MHz and 4400 MHz) fall under this category as frequency bands in this category could provide a better balance between capacity and coverage. Approximately 1000 MHz spectrum between 2-6 GHz has been identified for IMT services, which is almost 25% of the total available spectrum. Further allocation in this range is a difficult task.

Capacity layer –Spectrum bands above 6 GHz are considered capacity layer due to their vast data-carrying capacity. The coverage area decreases as we move to the higher side of the spectrum. Therefore, identified spectrum bands beyond 6GHz are best suited for establishing a hot spot. 17.25 GHz spectrum has been identified for IMT services in frequency bands between 6-100 GHz, which is about 18% of the total spectrum available between 6-100 GHz. This band is lightly occupied with other services, which will go beyond 100 GHz. Therefore, there is ample opportunity to identify new frequency bands for IMT services in this category. In the future, most news will be coming in the capacity layer only.

11.4 Radio Spectrum for Beyond 5G

Networks beyond 5G are envisaged to provide unprecedented performance excellence, which requires extreme data rates with agility, reliability, zero response time and artificial intelligence. In future, new vertical applications with more stringent specifications will emerge. For example, Virtual Reality (VR) coupled with Augmented Reality (VR) over wireless will be a key application in future, which require a minimum of 10 Gbps data rate [34]. Future several IoE (Internet of Everything) applications in the integrated form will be embedded into cities, vehicles, homes, industries, and other environments. Hence, a high data rate with reliable connectivity will be required to support these applications [35]. To meet all these challenges, there has been an increasing demand for higher data rates at least 100 times beyond current networks; lower latency of around one millisecond, reduced energy consumption, improved reliability and security, and higher scalability [36]. We could say that 5G networks cannot deliver a completely automated and intelligent network coupled with enhanced data demand.

The 6G network will become more intelligent, with learning mechanisms to modify itself based on users' experience; situation awareness will lead to decision making and networking [37]. 6G will provide more advanced and enhanced capabilities compared to 5G. It is envisioned that 6G will require to deliver approximately 1 Tb/s per-user bit rate in many cases with mobility 1000Km/hour, provide simultaneous wireless connectivity 1000 times higher than 5G and latency less than 1-ms. The most exciting feature of 6G is the inclusion of fully supported Artificial Intelligence (AI) for driving autonomous systems [35]. The drivers of 6G will be a convergence of past generations and emerging technologies that include new services and the recent revolution in wireless devices.

To meet the data rate envisaged for 6G, one has to look at Terahertz Wireless Communication. So far, a spectrum below 100 GHz (up to 71 GHz) has been identified for IMT services. We have to look at radio spectrum beyond 100 GHz, i.e. THz.

The terahertz (THz) band is the part of the electromagnetic waves in the frequency between 100 GHz–30 THz, i.e. between microwaves and infrared waves. It is also called the sub-millimetre band. The THz band offers a much larger bandwidth, ranging from tens of GHz to several THz and is suitable for Nano cells [38]. The THz communication system is highly directional, more energy-efficient, less latency, less susceptible to free space diffraction and able to address the capacity limitations of current wireless systems. It can be used for access as well as a backhaul network [39].

According to Friis Law, free space propagation loss is directly proportional to the frequency square. Therefore, high propagation loss drastically reduces the coverage with the THz communication system. Besides free space loss, atmospheric absorptions (e.g., oxygen and water molecule absorptions) result in additional path loss. However, atmospheric absorption has not any fixed pattern. In general, atmospheric absorption increases with frequency but exhibits high absorption at some frequency. The absorption peaks create spectral windows, which drastically change with the distance variation, as shown in Figure 11.5. It can be seen that molecular absorption increases with

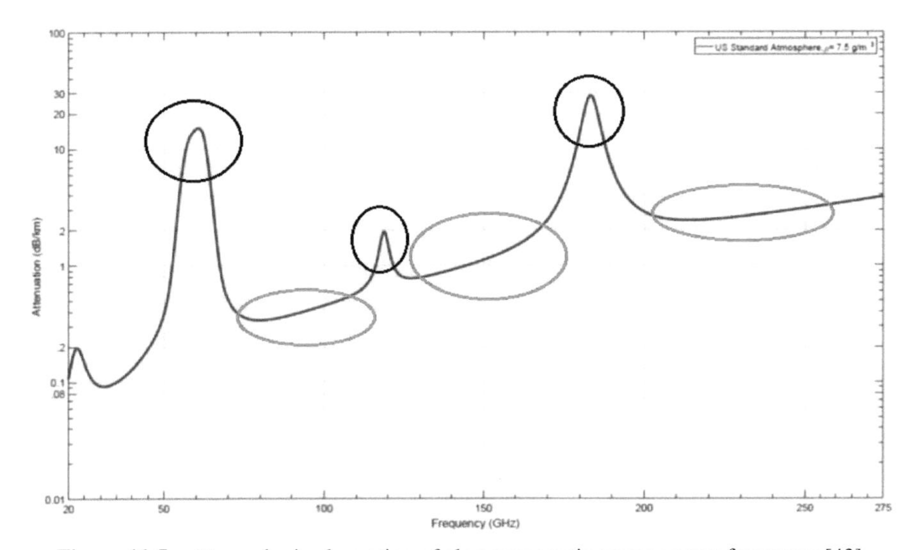

Figure 11.5 Atmospheric absorption of electromagnetic waves versus frequency [43].

frequency but shows a peak around 60 GHz, 120 GHz, and 183 GHz. The black circles represent the absorption peak. The spectrum bands defined by the green circle, i.e. 100-115 GHz, 125-170 GHz and 200-250 GHz, are the spectrum windows where molecular absorption is comparatively low. Spectrum bands in these spectrum windows could be potential candidate bands for mobile communication in future [40].

Standard bodies and organizations such as 3GPP, 5G Channel Model (5GCM), Mobile and wireless communications Enablers for the Twenty-twenty Information Society (METIS), and Millimetre-Wave Based Mobile Radio Access Network for Fifth Generation Integrated Communications (mmMAGIC) are working on THz communications [41].

The IEEE constituted the IEEE 802.15.3d task force in 2017 for global Wi-Fi use at frequencies from 252–325 GHz, the first wireless communications standard for the 250–350 GHz frequency range, with a data rate of 100 Gbps and channel bandwidths from 2 to70 GHz [42]. The Federal Communication Commission (FCC) has made provision for new equipment and applications for spectrum between 95 GHz and 3 THz and made available 21.2 GHz of spectrum for unlicensed use in the frequency bands 116-123 GHz, 174.8-182 GHz, 185-190 GHz and 244-246 GHz. FCC has also allowed experimental activities in the spectrum up to 3 THz [43]. The European Commission established the Radio Spectrum Policy Group to address THz-band communication management matters. Japan's Ministry of Internal Affairs and Communications (MIC) officially allocated an 18 GHz wideband from 116 GHz to 134 GHz for broadcasting service [44]. Besides these, several research institutes across the globe are working on THz communication [45].

As per Radio Regulation, radio spectrum allocation beyond 100 GHz is allocated mainly to Radio Astronomy and Satellite-based services such as Earth Exploration Satellite, Space Research, Amateur Satellite, Radio Navigation Satellite and Mobile Satellite etc. on a primary basis. Besides Satellite and Radio Astronomy, the allocation has also been made for Fixed and Mobile services as Primary services in 12 frequency bands with about 100 GHz spectrum between 100-275 GHz, which accounts for more than 55% of the total spectrum available between 100-275 GHz [23]. The percentage of allocation to different services in the frequency band 100-275 GHz is given in Table 11.2.

As of now, no frequency band beyond 100 GHz has been identified for IMT services. There is a possibility that IMT applications could come in those 12 frequency bands in which Mobile services are available as primary service in future.

Table 11.2 Percentage-wise allocations.

S. No.	Service	Allocation (in %)
1	Fixed	55.54
2	Mobile	55.54
3	Earth Exploration Satellite	33.60
4	Space Research	40.43
5	Inter Satellite	20.13
6	Fixed Satellite	20.00
7	Mobile Satellite	19.26
8	Amateur	2.29
9	Amateur Satellite	2.29
10	Radio Astronomy	61.60
11	Radiolocation	12.29
12	Radio-navigation	17.26
13	Radio-navigation Satellite	17.26

There is no allocation beyond 275 GHz. WRC-19 introduced a footnote 5.564A for allocating frequency band 275-450 GHz for land mobile and fixed services, allowing the coexistence of fixed and land mobile services with active and passive services in the frequency band 275-1000 GHz. Specifically, the radio astronomy service occupies 275-450 GHz, while the earth exploration-satellite and space research services operate in 296-306 GHz, 313-318 GHz and 333-356 GHz bands. The footnote also allows the frequency bands 275-296 GHz, 306-313 GHz, 318-333 GHz, and 356-450 GHz to implement land mobile and fixed service applications without protection Earth exploration-satellite service (passive) applications [23].

WRC-15 has framed a study question, QUESTION ITU-R, 256-1/5 on Technical and operational characteristics of the land mobile service in the frequency range 275-1 000 GHz under which sharing studies between the land mobile and passive services, as well as the land mobile and other active services, are being undertaken during the two study cycles, i.e. 2015 to 2019 and 2019 to 2023 and response, will be discussed during the WRC-23. These studies' results shall be in ITU Recommendations, Reports or Handbooks [46].

11.5 Conclusions

Spectrum requirements for mobile telecom services increase with the change of generation. ITU-R, through WRCs, identify frequency bands as per requirements of next-generation mobile communication. So far, more than

19 GHz spectrum has been recognised for IMT services, which is sufficient to run the 5G network. Beyond 5G network envisaged with data speed more than 1Tb/s and latency less than 1ns. The frequency bands up to 100 GHz identified do not have such a high data-carrying capacity. As per Radio Regulation, about 55% of the spectrum between 100 GHz – 275 GHz has the allocation of Fixed and Mobile services. ITU-R has already initiated a study question for exploring the possibilities of coexistence of Fixed and Land mobile radio with existing satellite services in the frequency bands 275-1000 GHz. However, ITU-R has not yet initiated the identification of frequency bands for IMT services in the frequency bands beyond 100 GHz nor floated any study question on this subject.

Moreover, there is no agenda item for WRC-23 or WRC-27 on this subject. No frequency bands beyond 100 GHz could be possible for IMT services till the completion of WRC-27. Identification of frequency bands is possible during the WRC-31 cycle, subject to agenda items that would be set during WRC-23 & WRC-27. Without spectrum allocation, beyond the 5G network cannot be commercially available before 2031.

References

[1] International Telecom Union. www.itu.int

[2] 1G. https://en.wikipedia.org/wiki/1G

[3] Salih et.al. (2020). Evolution of Mobile Wireless Communication to 5G Revolution. Technology Reports of Kansai University. 62. 2139-2151.

[4] T. Mshvidobadze, "Evolution mobile wireless communication and LTE networks," *2012 6th International Conference on Application of Information and Communication Technologies (AICT)*, 2012, pp. 1-7, doi: 10.1109/ICAICT.2012.6398495.

[5] Simon Johansen.1G, 2G, 3G, 4G, 5G. https://its-wiki.no/images/c/c8/From_1G_to_5G_Simon.pdf

[6] About mobile technology and IMT-2000. https://www.itu.int/osg/spu/imt-2000/technology.html

[7] IMT-Advanced. https://www.itu.int/en/ITU-R/study-groups/rsg5/rwp5d/imt-adv/Pages/default.aspx

[8] 3GPP Release 8. https://www.3gpp.org/specifications/releases/72-release-8

[9] A. Pandharipande, "Principles of OFDM," in *IEEE Potentials*, vol. 21, no. 2, pp. 16-19, April-May 2002, doi: 10.1109/45.997971.

[10] 3GPP Release 15, https://www.3gpp.org/release-15

[11] Introducing Radio Spectrum. https://www.gsma.com/spectrum/wp-content/uploads/2017/04/Introducing-Radio-Spectrum.pdf
[12] Ericsson Mobility Report. https://www.ericsson.com/en/mobility-report/reports
[13] https://news.itu.int/wrc-19-agrees-to-identify-new-frequency-bands-for-5g/
[14] Toward 5G Deployment in 2020 and Beyond. https://www.ieee-jp.org/section/tokyo/chapter/SSC-37/ISSCC2016%20NTT%20DOCOMO%205G.pdf
[15] ITU-R M.2370. https://www.itu.int/dms_pub/itu-r/opb/rep/R-REP-M.2370-2015-PDF-E.pdf
[16] Tripathi, P. S. M. (2014). Radio Spectrum Management for Future Wireless Communication Services.
[17] History of ITU. https://www.itu.int/itudoc/about/itu/history/history.txt
[18] ITU-R Study Groups. https://www.itu.int/dms_pub/itu-r/opb/gen/R-GEN-SGB-2020-PDF-E.pdf
[19] About ITU. https://www.itu.int/en/about/Pages/default.aspx
[20] Welcome to ITU-R. https://www.itu.int/en/ITU-R/information/Pages/default.aspx
[21] ITU, Radiocommunication sector (ITU-R), http://www.itu.int/en/ITU-R/Pages/default.aspx,
[22] World Radiocommunication Conference. https://www.itu.int/en/ITU-R/conferences/wrc/Pages/default.aspx
[23] ITU Radio Regulations," https://www.itu.int/pub/R-REG-RR
[24] WARC-92. https://www.itu.int/en/history/Pages/RadioConferences.aspx?conf=4.122
[25] WRC-2000. https://www.itu.int/net/ITU-R/index.asp?category=conferences&rlink=wrc-00&lang=en
[26] WRC-07. https://www.itu.int/net/ITU-R/index.asp?category=conferences&rlink=wrc-07&lang=en
[27] ITU Radio Regulations — 110 years of success. https://news.itu.int/itu-radio-regulations-110-years-of-success/
[28] WRC-12. https://www.itu.int/net/ITU-R/index.asp?category=conferences&rlink=wrc-12&lang=en
[29] WRC-15. https://www.itu.int/en/ITU-R/conferences/wrc/2015/Pages/default.aspx
[30] WRC-19. https://www.itu.int/en/ITU-R/conferences/wrc/2019/Pages/default.aspx

[31] Resolution 238 (WRC-15) https://www.itu.int/dms_pub/itu-r/oth/0c/0a/R0C0A00000C0014PDFE.pdf

[32] Task Group 5/1. https://www.itu.int/en/ITU-R/study-groups/rsg5/tg5-1/Pages/default.aspx

[33] ITU-R Preparatory Studies for WRC-23. https://www.itu.int/en/ITU-R/study-groups/rcpm/Pages/wrc-23-studies.aspx

[34] S. Mumtaz et al., "Terahertz communication for vehicular networks. https://doi.org/10.1109/TVT.2017.2712878

[35] M. Z. Chowdhury, M. Shahjalal, S. Ahmed and Y. M. Jang, "6G Wireless Communication Systems: Applications, Requirements, Technologies, Challenges, and Research Directions," in *IEEE Open Journal of the Communications Society*, vol. 1, pp. 957-975, 2020, doi: 10.1109/OJCOMS.2020.3010270.

[36] I. F. Akyildiz, C. Han and S. Nie, "Combating the Distance Problem in the Millimeter Wave and Terahertz Frequency Bands," in *IEEE Communications Magazine*, vol. 56, no. 6, pp. 102-108, June 2018, doi: 10.1109/MCOM.2018.1700928.

[37] Marco Chiani, et.al. Open issues and beyond 5G. https://www.5gitaly.eu/2018/wp-content/uploads/2019/01/5G-Italy-White-eBook-Beyond-5G.pdf

[38] Milda Tamosiunaite et. al. Atmospheric Attenuation of the Terahertz Wireless Networks. http://dx.doi.org/10.5772/intechopen.72205

[39] H. Sarieddeen, N. Saeed, T. Y. Al-Naffouri and M. Alouini, "Next Generation Terahertz Communications: A Rendezvous of Sensing, Imaging, and Localization," in *IEEE Communications Magazine*, vol. 58, no. 5, pp. 69-75, May 2020, doi: 10.1109/MCOM.001.1900698.

[40] FCC Docket 18-21 "Spectrum Horizons". https://docs.fcc.gov/public/attachments/FCC-18-17A1.pdf

[41] Y. Xing and T. S. Rappaport, "Propagation Measurement System and Approach at 140 GHz-Moving to 6G and Above 100 GHz," *2018 IEEE Global Communications Conference (GLOBECOM)*, 2018, pp. 1-6, doi: 10.1109/GLOCOM.2018.8647921.

[42] Vitaly Petrov, Thomas Kürner, and Iwao Hosako. IEEE 802.15.3d: First Standardization Efforts for Sub-Terahertz Band Communications towards 6G. https://arxiv.org/pdf/2011.01683.pdf

[43] FCC. Rulemaking to Allow Unlicensed Operation in the 95-1,000 GHz Band. ET Docket No. 18-21 Released March 21, 2019

[44] T. Nagatsuma, "Breakthroughs in photonics 2013: Thz communications based on photonics," *IEEE Photonics Journal*, vol. 6, no. 2, pp. 1–5, April 2014

[45] Hadeel Elayan et. al. Terahertz Band: The Last Piece of RF SpectrumPuzzle for Communication Systems. https://arxiv.org/pdf/1907.05043.pdf

[46] ITU-R Study Question 256-1/5. https://www.itu.int/pub/R-QUE-SG05.256

Biographies

P. S. M. Tripathi is currently Deputy Wireless Adviser in Wireless Planning & Coordination (WPC) Wing of Department of Telecommunications, Ministry of Communications, Government of India. Presently, he has been posted as Regional Licensing Officer, Guwahati. Earlier, he was associated with radio spectrum management and planning and engineering activities for telecom services in India, especially in 900 MHz, 1800 MHz and 2100 MHz bands and also associated with spectrum auctions held in India from 2010 onwards. He has more than 20 years of experience in management, and strategy in the Radio Spectrum Management and Radio Spectrum Monitoring Sector, including implementing a prestigious World Bank Assisted Project on "National Radio Spectrum Management and Monitoring System (NRSMMS)" in the WPC Wing.

He was graduated from M M M Technical University, Gorakhpur (India). He worked as a Research fellow at the Center for Tele-infrastructure (CTIF), Italy, Department of Electronics, University of Tor Vergata, Rome, Italy. He was selected in the year 2010 under the Erasmus Mundus "Mobility for Life" scholarship programme of the European Commission for doing PhD at the Department of Electronic Systems, Aalborg University, Denmark. The Degree of Doctor of Philosophy was awarded in 2014.

Ambuj Kumar has hands-on experience of more than 20 years in the Telecom Industry and Academia. He received a Bachelor of Engineering (BE) in Electronics & Communications Engineering (ECE) from Birla Institute of Technology (BIT), Ranchi, India, in the year 2000. As a part of the Bachelor's programme, he underwent an extensive Internship Training in 1999 at the Institut für Hochfrequenztechnik (IHF), Technical University (RWTH), Aachen (Germany). After graduation, Ambuj Kumar worked at Lucent Technologies Hindustan Private Limited, a vendor company, from 200-to 2004. His major responsibilities were the Mobile Radio Network Design, Macro and Microcell planning, and Optimization for the GSM-and the CDMA-based Mobile Communication

Networks. There, he was involved in pan-India planning and optimizing the MCNs of various service providers for both the green field and the incumbent deployments. During the period 2004 to 2007, he worked with Hutchison Mobile Services Limited (now Vodafone-Idea Limited), a service provider company, where he was involved in planning, deployment, and optimization of Hutch's rapidly expanding GSM-and Edge networks across India. Afterwards, he joined Alcatel- Lucent, New Delhi, in 2007 and continued until 2009. He worked for three months in 2009 as Research Associate at the Centre for TeleInFrastruktur (CTIF), Department of Electronic Systems, Aalborg University, and his research area was on 'Identification Optimization parameters for Routing in Cognitive Radio'. Ambuj Kumar was awarded a research scholarship under European Commission -Erasmus Mundus "Mobility for Life" programme for doing a PhD and joined CTIF, Department of Electronic Systems, Aalborg University Aalborg, (Denmark), in the year 2010. Ambuj Kumar has also worked as a Collaborative Researcher at the Vihaan Networks Limited (VNL), India. The work of PhD research was conceptualized at VNL; there, he developed test-bed facilities for experimental studies on 'Advanced Alternative Networks'. He had worked as Research Assistant in the eWall Project, funded by the European Commission, at the Faculty of Science and Engineering (Department of Electronic Systems) during 2015-2016. He was awarded a Doctor of Philosophy (PhD) in 2016 by the Aalborg University (Denmark) on his thesis titled "Active Probing Feedback Based Self Configurable Intelligent Distributed Antenna System for Relative and Intuitive Coverage and Capacity Predictions for Proactive Spectrum Sensing and Management". Currently, Dr Ambuj Kumar has been working as a PostDoc in the Department of Business Development and Technology, School of Business & Social Sciences, Aarhus University (Denmark), since February 2017. Currently, he is playing a pivotal role in three European projects. He is a key person in the EU's project titled "Capacity building and ExchaNge towards attaining Technological Research and modernizing Academic Learning (CENTRAL)". He has more than 30 research publications, including four book chapters in these thematic areas. As the first author, he has contributed more than 12 research papers, including journals such as Wireless Personal Communications (WPC), and International Telecommunication Union (ITU), scientific magazines such as IEEE AESS, and conferences such as Wireless Personal Multimedia Communications (WPMC), Global Wireless Summit (GWS), and in Wireless World Research Forum (WWRF). He is a reviewer of several journals, conferences, and scientific publishers, namely Wireless Personal Communications, IEEE Communications Magazine, IEEE AESS,

CONASENSE, River Publishers, Mesford publishers, Global Wireless Summit (GWS), Wireless Personal Multimedia Communications. His research interests are radio wave propagation, cognitive radio, visible light communications, radio resource management, drones, IoT, AI, distributed antenna systems, business innovation modelling etc.

Dr Ashok Chandra has his PhD in Electronics and a Doctorate of Science (D.Sc.) in Radio Mobile Communications. He joined the Ministry of Communications & Information Technology, Government of India, in 1977. He has worked as Guest Scientist on DAAD Fellowship at the Institute of High-Frequency Technology, Technical University (RWTH), Aachen, Germany and at Bremen University, Bremen (Germany), where he undertook a series of research studies in the area of mobile radio communications. Dr Ashok Chandra has technical experience of over 35 years in Radio Communications/Radio Spectrum Management, including about seven years of experience dealing with the Institutes of Higher Learning in Technical Education. He has contributed over 30 research papers at various International Conferences/Journals in EMI, Radio Propagation, etc. He played a vital role in establishing new Indian Institutes of Technology, the Indian Institute of Management and the Indian Institutes of Information Technology. Dr Chandra was instrumental in implementing the Government of India's "Technology Development Mission" scheme. Dr Chandra is registered with ITU as an Expert in "Radio Spectrum Management".

He had visited various technical Institutions and Universities abroad and took several lectures in mobile radio communications. He has chaired different Technical Sessions at the International Conferences on Wireless Communications.

During his tenure as the Wireless Adviser, he played a key role in preparing necessary documents for the auction of radio spectrum for 3 G and BWA applications in 2010. This spectrum auction, is one of the most successful spectrum auctions globally. Further, he was also involved in finalising auction modalities and coordinating with other Ministries, including the release of spectrum for 2G, 3G & BWA, etc.

Dr Chandra superannuated from the post of Wireless Adviser to the Government of India and was responsible for spectrum management activities, including spectrum planning and engineering, frequency assignment, frequency coordination, spectrum monitoring, and policy regarding regulatory affairs for new technologies and related research & development activities, etc. He was Project Director of a prestigious World Bank Project on "National Radio Spectrum Management and Monitoring System (NRSMMS)".

He served as a Vice-Chairman, Study Group 5 of the International Telecommunications Union (ITU)-Radio Sector. He has represented India at many ITU meetings, including World Radio Conferences (WRC). He served as Councillor from Indian Administration in the ITU Council.

Dr Chandra was Adjunct Professor for three years from February 2013 at the Indian Institute of Technology (IIT), Bombay. He was TPC Executive Chair of Global Wireless Summit (GWS) 2015. Dr Chandra was a Guest Professor at the National Institute of Technology (VNIT), Nagpur (India). He had organized several 'Short-term Training Programmes on Radio Spectrum Management.

12

Deployment of Terahertz Spectrum Band For 6G

Tilak Raj Dua

Tower and Infrastructure Providers Association, 7, Bhai Vir Singh Marg, Gole Market, New Delhi, India
Vice-Chairman, Global ICT Standardization Forum for India (GISFI), and Chairman, ITU-APT Foundation of India
E-mail: tr.dua@taipa.in

Digital infrastructure and services have emerged as key enablers and critical determinants of a country's growth and well-being. India's digital footprint is one of the fastest-growing in the world. India's mobile data consumption is already the highest globally, with over a billion mobile phones and digital identities and more than 700 million internet users.

The continuous evolution and exponential growth in user demands have led to many emerging use cases and technologies. These new use cases require much higher data rates and way lower latency than the current 5G systems can offer. For example, the next generation of Virtual and Augmented Reality (VAR), i.e., holographic teleportation, requires terabits per second (Tbps)-level data rates and microsecond-level latency. Therefore, the evolution to 6G would be necessitated due to evolving expectations and technology dependence of the masses.

6G is in the inception stage, and several candidate technologies are being considered to realise 6G requirements. Though the newer and newer technologies will continue to emerge, technologies such as Terahertz (THz) band, Novel antenna technologies, Metamaterial based Antenna and RF front end etc., are in the race. Among emerging research and development trends in wireless communications, THz band (0.1-10 THz) communications have been envisioned as one of the key enabling technologies for the next decade. The THz band is available in abundance, and due to the ultra-wide spectrum resources, the THz band can provide Tbps links for a plethora of applications. In this chapter, the advantages of the THz band for the evolution of 6G, applications and use cases, likely challenges and mitigation techniques have been elaborated.

12.1 Introduction

The telecom revolution in India has facilitated a multiplier effect on the Indian economy and its population. The innovations and the rapid expansion in telecommunications are boosting the Indian economy by creating a plethora of opportunities. For instance, the broadband internet facilitated the officegoers and employees to work from home. Today internet is playing a pivotal role in modernizing agriculture, developing e-health, e-commerce, e-governance and e-education. India has one of the lowest data tariffs, resulting in astronomical data usage. A graph depicting data tariff v/s data usage during the last six years is illustrated in Figure 12.1.

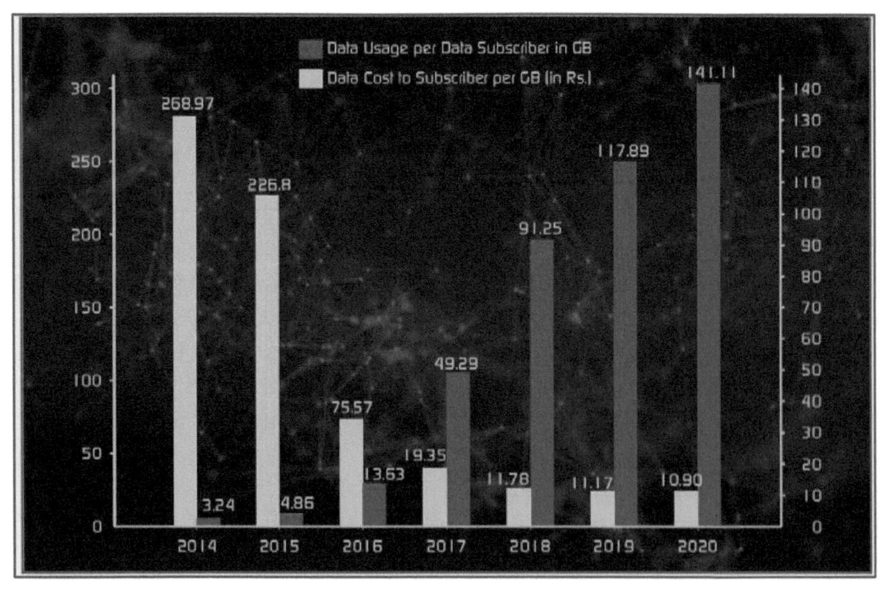

Figure 12.1 With the decrease in per GB data cost, there has been an enormous surge in data usage per subscriber.
Source: TRAI

The high-speed internet and multimedia journey started with 2G in the early 90s. Eventually, we have seen new technology emerge every ten years in the wireless world. The successive decades saw the emergence of 3G, 4G and now 5G. 5G offered a high peak speed of 1 Gbps (1,000 Mbps) compared to 20-100 Mbps on 4G. However, 6G holds the promise of 1 Tbps (1,000 Gbps).

6G will satisfy unprecedented requirements and expectations that 5G cannot meet. It will provide the ultimate experience through hyper-connectivity involving humans and everything. Further, increasing industrial automation and the emergence of Industry X.0 will push connectivity density well beyond 5G capacities. 5G designs are crucial to support high energy efficiency with an increased connection density. Consequently, the research community has shifted its efforts toward addressing the significant challenges. The focus of the ongoing research is shifted more towards exploiting THz band communications, intelligent surfaces and environments, and network automation.

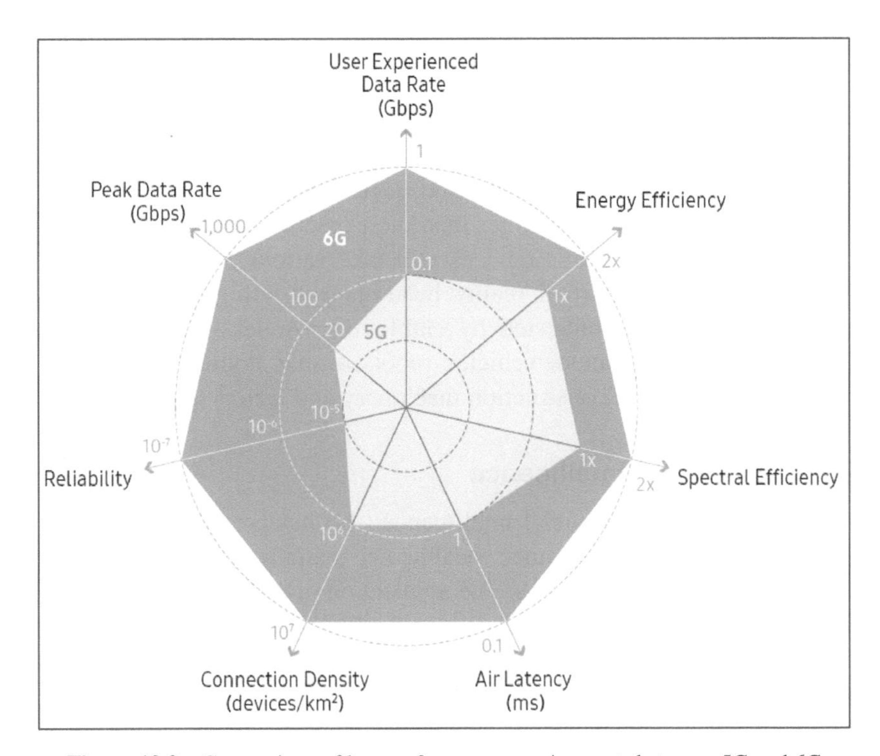

Figure 12.2 Comparison of key performance requirements between 5G and 6G. Source: Samsung 6G white paper

In addition to the high-speed requirement in the range of 1 Tbps in the form of enhanced Mobile Broadband, there would be needed ultra-reliable and low-latency communication (URLLC) and massive machine-type communications (mMTC). Also, with almost every device being connected and sporting an internal antenna, one square kilometre area may connect 10 million devices using 6G compared to a million in the case of 5G.

12.2 6G – Emerging Trends

Applications that take advantage of wireless communications are expanding from connecting humans to connecting various things. Wireless communication is becoming an essential part of social infrastructure. In addition, tremendous growth in advanced technologies such as Artificial Intelligence, robotics, IoT and automation will bring an unprecedented shift in wireless communications.

The four major emerging trends advancing towards 6G are - connected machines, the use of AI, the openness of mobile communications and increased contribution to achieving social goals.

12.2.1 Connected Machines

It is estimated that the number of connected devices will reach 500 billion by 2030, which is about 59 times larger than the expected world population (8.5 billion) by that time. Mobile devices will take various forms, such as augmented reality glasses, virtual reality headsets and hologram devices. Machines will need to be connected by employing wireless communications. Connected machines include vehicles, robots, drones, home appliances, displays, intelligent sensors, construction machinery and factory equipment.

12.2.2 Use of Artificial Intelligence

The recent development in artificial intelligence (AI) has proved its usefulness in various areas such as finance, healthcare, manufacturing, industry, and wireless communication systems. The application of AI in wireless communication can improve performance, reduce CAPEX and OPEX, improve performance of handover operation, optimise network planning, including base station location determination, and reduce network energy consumption.

The 6G network designs will have embedded AI components to support various use cases and services. Adopting 6G during the design stage of network conceptualization can create numerous opportunities to improve performance, cost, and capability to provide new services. 5G ensured the mainstream adoption of mm Wave spectrum, but with 6G, the need for higher data rates and, consequently, larger channel bandwidths will necessitate the incorporation of THz and sub-THz spectrum. At the same time, the opening up of new spectrum bands will also require novel radio designs that can simultaneously sense and communicate over the entire Electro-Magnetic spectrum.

12.2.3 Increased Contribution Towards Achieving Social Goals

The amalgamation of 5G and 6G is expected to reduce greenhouse emissions by 15% by 2030. 6G with inherited features of 5G can successfully address the social challenges such as hunger, climate change and education inequality. Hyperconnectivity will improve and enable access to information. It will provide alternatives to rural exodus, mass urbanization and the related problems. 6G would thus play a significant role in the achievement of UN SDGs and tremendously contribute to the quality and opportunities of human life. Additional services in 6G, which are not possible for 5G, are indeed immersive extended reality (XR), high fidelity mobile hologram, digital replica, etc.

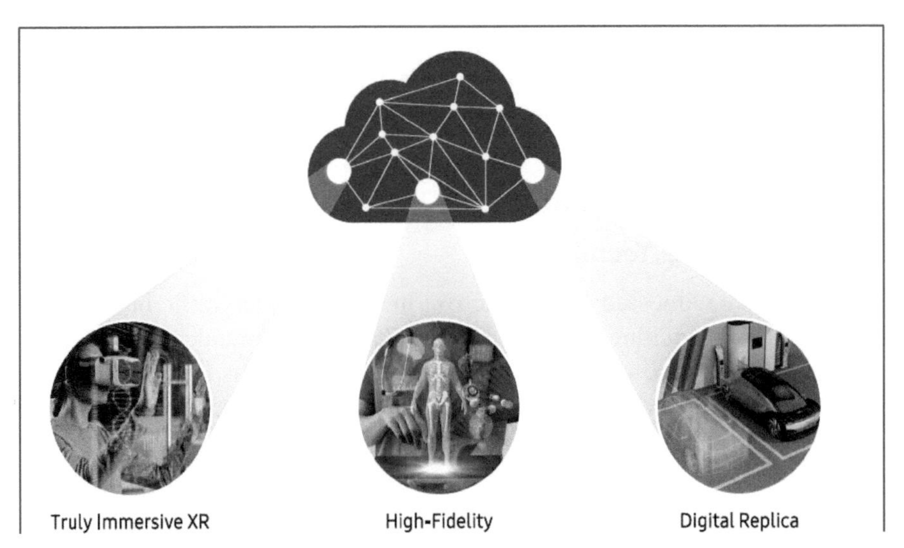

Figure 12.3 Three key 6G services: truly immersive XR, high-fidelity mobile hologram and digital replica.
Source: Samsung 6G white paper

XR is a new term that combines VR, AR, and mixed reality. It finds use cases in entertainment, medicine, science, education, and manufacturing industries. To provide XR media streaming, 0.9 Gbps throughput would be required. The current user experienced data rate of 5G is not sufficient for seamless streaming. The hologram is a next-generation media technology that can present gestures and facial expressions employing holographic display. A hologram display over a mobile device will require at least 0.58 Tbps.

A Digital replica would make it possible to observe changes or detect problems remotely through the representation offered by the digital twin. It will deploy advanced sensors, AI, and communication technologies. Interaction with the digital twin will happen through VR or holographic display. To replicate a 1x1 M area, 0.8 Tbps would be required assuming synchronisation of 100 ms and a compression ratio of 1/300.

12.3 6G Requirements

An increase in mobile computing requirements and battery size would be the first and foremost. While 5G was designed to achieve 20 Gbps peak data, 6G would have peak data of 1000 Gbps and a user experience rate of 1Gbps. Spectral efficiency would be twice that of 5G. Latency targets set up in 6G would be less than1 ms end to end, with extremely low delay jitter in terms of microseconds. Reliability will jump 100 times to support emergency response, remote surgery, etc. The bit error rate would be only 10^{-7}. 6G would support 10 million devices/km^2, ten times more than 5G.

12.4 6G Technologies and Use of THz Band

12.4.1 Emerging 6G Technologies

Recently, we have witnessed a dramatic rise in wireless data traffic brought forth by numerous exciting technologies in wireless communications and assisted by the pandemic impact. This exponential growth has accompanied the demand for higher data rates and better coverage. The pandemic has shrunk the digitalization period by at least 2-3 years.

Figure 12.4 Spectrum usage for different generations.

The services and requirements pose various challenges to developing future wireless systems. Several potential technologies are being considered to realize 6G requirements. Though newer technologies will continue to emerge, the following technologies are in the race.

- Terahertz technologies
- Novel antenna technologies
- Metamaterial based Antenna and RF front end
- Orbital Angular momentum
- Evolution of duplex technology
- Evolution of network topology
- Spectrum sharing
- Comprehensive AI
- Split computing

Among emerging research and development trends in wireless communications, THz band (0.1-10 THz) communications have been envisioned as one of the key enabling technologies for the coming decade because of their high availability. The THz band can provide terabits per second (Tbps) links for many applications, ranging from ultra-fast massive data transfer among nearby devices in Terabit Wireless Personal and Local Area Networks to high-definition video-conferencing among mobile devices in small cells.

12.4.2 Allocation by FCC

Federal communications commission (FCC) has released frequency bands above 95 GHz to 3000 GHz for experimental use and unlicensed applications to encourage the development of new wireless communication technologies. Following this trend, it is inevitable that mobile communications will utilise the terahertz band (0.1-10 THz) in future wireless systems.

The THz band includes an enormous amount of available bandwidth, which will enable highly wideband channels with tens of GHz wide bandwidth and could be utilized up to 3000 GHz. This could potentially provide a means to meet the 6G requirements of the Tbps data rate.

12.4.3 Advantages for THz band Technologies

While the availability of a wideband spectrum is the main driver for THz communications, other benefits can also be realised. Communication in the THz band can provide high precision positioning capability due to the following:

- Extremely wideband waveforms in the THz band would enable accurate ranging between transmitter and receiver.
- LOS link between transmitter and receiver. This will improve the accuracy of distance-based positioning systems.
- Using pencil-point sharp beams will significantly improve angular resolution and triangulation accuracy of 3D position estimation.

The experience of telecom operators in deployment of mm waveband for 5G has not yielded the desired speed of 100 Gbps v/s attainment of 1 Gbps in practical. The THz bands have been applied in imaging and object detection and THz radiation spectroscopy in astronomical research, and their use cases in wireless communications are still under investigation.

12.5 Challenges: Use of THz for 6G

Several wireless summits and workshops have been held during the last year, where researchers and academia have shared their vision about the future of wireless communications. The first two years of every new cycle/ technology are critical to gathering ideas and brainstorming the broad direction for the research efforts. Some fundamental and technical challenges are there in realizing stable THz communications. The key challenges are as under:

12.5.1 Severe Path Loss and Atmospheric Absorption

Free space path loss is proportional to the square of the signal frequency. The severe path-loss in the THz band can be overcome by utilizing massive antenna arrays at the base station, i.e., ultra-massive multiple-input multiple-output (MIMO). In addition, the effect of atmospheric absorption (i.e., absorption by molecules in air) in the THz band is in general severer than in lower frequencies as the absorption lines for oxygen and water are primarily located in the THz band. This will also have to be tackled using technical solutions.

12.5.2 RF Front-end, Photonics and Data Conversion

The THz band is often referred to as the THz gap due to the lack of efficient devices which can generate and detect signals in these frequencies. The device dimensions are significantly oversized relative to the wavelength in these bands, resulting in high power loss or equivalently low efficiency. However, researchers put great efforts into developing chip-scale

THz technologies during the last decade. In addition to improving the output power of the THz signal, the following also need to be addressed.

- Transporting the signal within the integrated system and to the antenna with low loss
- Packaging of the integrated system without significant loss and maintaining proper heat dissipation
- Lowering the mixer phase-noise
- Low power multi-Giga-samples-per-second analogue-to-digital converters (ADCs)and digital-to-analogue converters (DACs)
- Low power digital input/output (IO) to DACs and ADCs to transfer data at Tbps data rate with acceptable power consumption.

Lying between the mm Wave spectrum and infrared light spectrum, THz bands, with their abundant spectrum resources, have been previously deemed a no man's land". However, significant progress in transceiver and antenna design has seen THz links become a promising option for realizing indoor communications networks. More recently, there has been considerable progress in learning wireless network on chip (WNoC) using THz bands.

12.6 6G Timelines

Mobile communication systems have evolved over multiple generations from 2G to 5G approximately every ten years. However, the time spent defining

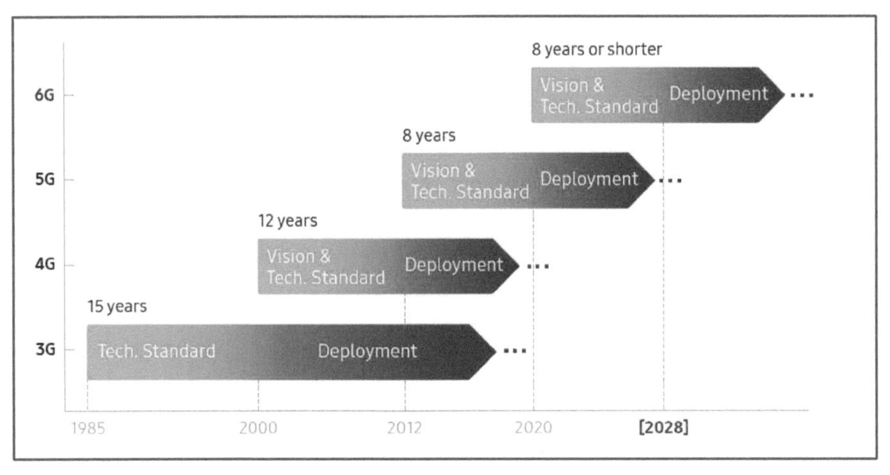

Figure 12.5 6G roadmap.
Source: Samsung 6G white paper

vision and developing technical standards for each generation has shortened from 15 years in 3G to 8 years in 5G.

It is expected that for 6G, ITU-R will begin their work to define the 6G vision in 2021. As depicted in the chart above, the commercialization of 6G could likely happen as early as 2028, while massive commercialisation would start by 2030.

12.7 Conclusions

This study has presented an overview of the standardization process and emphasized 6G standardization requirements. From a technology perspective, 6G will provide a vast opportunity to the users and MNOs to explore different services and use cases. We presented new features of the 6G networks' unique characteristics and associated services, enabling technologies and challenges. This paper also discussed various 6G initiatives going on across the world. Set of new evolved KPIs and requirements. The scope of standards and technology development needs to be broadened to support future ecosystems. The product needs to comply with the standards to keep the full capabilities of upcoming networks. The higher layers are available for research, bringing new challenges and enormous opportunities. Different organizations draft the specifications and standardization requirements for 6G in various use case scenarios.

6G outlines, 100x data throughput, and sub-millisecond latency compared to the 5G networks. The path toward developing future wireless standards and technologies will be compelling. Following are some critical considerations for developing the next generation of standards and technologies. The SDOs need to increase their scope to bring new horizons to the communication ecosystem. There is a need for New Standards Evolution Paradigm to support reusability and scalability. To overcome shortcomings in the SDOs, we need Universal global standards to prevent duplication of the standards. The industry and market should encourage healthy and competitive development of the new and existing solution providers to ensure a stable global supply chain without any significant disruptions.

References

[1] Chowdhury, Mostafa Zaman, Md Shahjalal, Shakil Ahmed, and Yeong Min Jang. "6G wireless communication systems: Applications, requirements, technologies, challenges, and research directions." IEEE Open Journal of the Communications Society 1 (2020): 957-975.

[2] Communications in the 6G Era by Harish Viswanathan1, (fellow, IEEE), and Preben Mogensen2, (member, IEEE)
[3] 6G and Beyond: The Future of Wireless Communications Systems by Ian F. Akyildiz, (fellow, IEEE), Ahan Kak, and Shuai Nie
[4] 6G – The Next Hyper-Connected Experience For All by Samsung

Biography

Tilak Raj Dua is the Director General, TAIPA and Chairman ITU-APT. An engineering graduate with a diploma in Business Management and export Marketing, he has over 35 years of experience in the telecom sector. His experience includes all facets of telecom, be it Product Development, Business Development, Telecom Licensing, Regulatory issues concerning interconnection / roaming /unified licensing and infrastructure sharing / Mobile Number Portability, Spectrum Management, Spectrum Related Issues Like Spectrum Pricing, Efficient Utilization and Spectrum reframing, finalization of joint ventures, Technical collaboration, Introduction of the new product, Launch of cellular services in India and finalization of licence agreements / interconnect agreements etc.

Served as Deputy Director IAF - from 1967 to 1989, Director, Shyam Telecom Ltd. (Regulatory) – from1994 to 2000, Director, Bharti Airtel Ltd. (Corporate Affairs & Regulatory) – from 2000 to 2005, Deputy Director-General: COAI – from 2006 to 2011, Executive Director Augere Mobile BroadBand Wireless Pvt. Ltd. – 2011, Director General Tower and Infrastructure Providers Association: since Jan 2015 till date. Works with institutions like ITU, APT, and WWRF for spectrum/regulatory matters.

13

Economic Challenges for 6G Deployments

Berend W.M. Kuipers[1], Paulo S. Rufino Henrique[2] and Ramjee Prasad[3]

[1]INOV-INESC Inovação / Lusíada University of Lisbon, Lisbon, Portugal
[2]Paulo S. Rufino Henrique, CTIF Global Capsule (CGC) / Aarhus University / Spideo, Paris, France
[3]CTIF Global Capsule (CGC) / Aarhus University - Department of Business Development and Technology, Herning, Denmark
E-mail: martijn.kuipers@gmail.com; rufino@spideo.tv; ramjee@btech.au.dk

Since the release of 5G, investors, business technologists, and board of directors have been looking for methods to interpret and translate the traditional models of network investments to the virtualized networks that began in the 5G era to maximize their ROI (return over investments) in all areas. This chapter investigates a comprehensive technical methodology for the 6G cost modelling measuring the capital expenditure (Capex), and operational expenditure (Opex), including the total cost of ownership (TCO) for 6G for virtualized and nonvirtualized networks. In this aspect, the first part of this chapter evaluates the 6G costs based on the infrastructure deployment from the Terahertz spectrum acquisition, spectrum sharing, energy consumption to 6G virtualized networks, also considering the additional services such as artificial intelligence (AI) and potentially quantum machine learning (QML) to mention a few entities 6G Network Architecture. The second part focuses on identifying different costs for differentiating traffic demands depending on the geographic location versus population distribution to define the 6G rollout strategy considering the cell-sizes distributions. Posterior, there is an analysis of 6G services like Private Networks, 6G mobile virtual networks operators (6GMVNO), and Network Slicing are also considered from this research perspective. The final objective is to create a generic and adaptable knowledge performance index (KPI) for optimal 6G total value of opportunity (TVO). Thus, there is a proposition for 6G business models to define some possible optimal approaches to help decision-makers select the best route to maximize the 6G TVO.

13.1 Introduction

The sixth generation of mobile networks, also known as 6G, is planned to initiate its commercial operations in 2030. Consequently, as with many novel technologies, many expectations exist, which look to better comprehend the 6G outlook from a commercial to a technical standpoint. This novel Network is already being investigated from a network design viewpoint to specific use cases which will benefit from it. Thus, this future wireless research and development (R&D) are already ongoing in several scientific fields worldwide. Nowadays, academic institutions, governments, and standardization bodies assemble to work on future wireless standards. Hence, important 6G books [1] are also being published by the researcher communities, including recently released articles and companies' white papers. Following the initial 5G rollout in the different global markets and the sharp growth of the mobile economy [2] during the coronavirus (COVID-19) pandemic, people are becoming more interested in the intrinsic of knowing the scientific innovations and social-economy development in the next 9 to 10 years.

In this chapter, the discussion centres on investigating 6G cost modelling that enables the discussion amongst Telco's decision-makers in their boarding rooms. C-level executives represented by chief executive officers (CEOs), chief technology officers (CTOs), chief information officers (CIOs), and chief finance officers (CFOs) are going to be in the quest to balance out short, medium and long-term the return of investments (ROI) for 6G. Therefore, the future network infrastructure utilizing existing network infrastructure invested in 5G and its predecessors will be considered, without excluding the challenges of deploying 6G over the brownfields network, if it proves feasible to support 6G terabits per second data rate. These perspectives will be considered to maximize the cost efficiency of 6G deployment and control risks. As a starting point, the analyses here presented will begin gauging the capital expenditure (CAPEX) to create the innovative network infrastructure and the operational expenditure (OPEX) to run it for at least ten years. Additionally, a cost modelling for future wireless network technology will be reviewed to pave the way for an initial investigation to allow an early analysis of the economic model strategy that will support the 6G technical rollout plan. The preceding discussion will look at the total cost of ownership (TCO) and link it to the total value of opportunity (TVO).

What is known so far is that 6G will continue the digital transformation of all vertical and horizontal sectors beyond 2030. Also, it will be a crucial enabler for a human-centric cellular network underpinned by the Society 5.0 framework [3] and defined by the United Nations (UN) 17 Societal Development Goals (SDGs) [4]. Then, it is imperative to evaluate some critical aspects defined in Society 5.0 as a novel framework conceptualized by the Japanese Cabinet Office and cross-relate them to the SDGs before delving into the 6G cost models approach presented in this chapter.

Society 5.0 concept is an essential innovative framework based on Science, Technology, and Innovation (STI) policy, in which UNESCO [5] emphasized that the ICT industry should aim to support a fairer society. It is embedded in this policy the necessity of using technologies to converge the cyberworld and the physical realm addressing societal challenges intertwined to achieve economic improvements beyond 2030 [6]. Like its predecessors, Society 5.0 is the next generation of society to focus on the societal and environmental challenges with the aid of technologies and hyperconnected society, the future quantum internet era. In this new society, technology is entirely used to promote quality of life (QoL) for all. Additionally, it is propelled by Artificial Intelligence (AI), Machine Learning (ML), distributed via Cloud Computing Systems, and hyper wireless connectivity, linking everything, everyone and responding to social needs proactively. For instance, technology will be a concierge and oracle to serve the ageing population better, support future job allocation for youths, and proactively control the waste and resources in the Smart Cities. Besides, protecting the environment with autonomous surveillance to preserve its ecosystem and reduce the future impact of climate change. Having thought that, Society 5.0 will need a robust and intelligent wireless network to enable such idealized objectives to come to fruition, which will similarly trigger the rapid implementation of the integration of Communication, Navigation, Sensing and Services (CONASENSE) [7] and Knowledge Human Bond Communication Beyond 2050 (Knowledge Home). Knowledge Home is a technological concept aiming to integrate the five human senses via technology and wirelessly connectivity that offers services for humans in a personal area network (PAN) [8]. Combined and buttressed by 6G networks, these strategic technological frameworks and concepts will inaugurate the era of the Internet of Beings (IoB) [9].

On the other hand, the International Telecommunication Union (ITU) has also initiated an agenda to promote the SDGs [10] and outlined the main areas that the Information Communication Technologies (ICT) services and the future wireless Network can focus on before and beyond 2030. These SDGs areas outlined by ITU in the Connected Agenda 2030 [11] bolstered by ICT, which are:

- **Growth** (Increase access to Telecom Systems/ICT supporting Digital Economy)
- **Inclusiveness** (bridge digital divide/broadband access for all)
- **Sustainability** (managing emerging risks/challenges and opportunities due to Digital Transformation)
- **Innovation & Partnership** (enable innovation in ICT for social digital transformation)
- **Partnership**(Strengthen ITU partnership to support all ITU strategic goals)

Consequently, in this study, the primary focus of the 6G cost modelling will be based on the densely populated areas governed by the Smart Cities. Such a decision is rooted in the statement provided in the UN project United for Smart Sustainable Cities (U4SSC) [12] initiative that states that by 2050, cities will host 70% of the world population, which is an increase of 20% on top of how the world population is distributed and along with this remark many challenges are envisaged. As a response to these challenges, a newly published report by U4SSC entitled Key Performance Indicators: A Key Element for cities wishing to achieve the Sustainable development Goals [13] defines ninety-one key performance indicators (KPIs), allowing cities to implement policies to converge them into very effective SSC, which utilizes ICT technologies and digital technologies to improve citizens QoL, managing energy consumption efficiently, delivering dynamic and intelligent smart transportations, reducing waste, creating perpetual life-cycle of efficient recycling products, reducing carbon emissions, etc.

Thus, evaluating the 6G rollout investment and the opportunities in a Smart City densely populated will be readily justifiable from the commercial perspective. Furthermore, it will balance out the ROI expectations.

Figure 13.1 shows the interrelationship between Society 5.0 and the UN SDGs that sets out the roadmap discussions for 6G.

Based on the evidence available of how the 5G rollout is planned, it seems fair to suggest beginning the cost modelling for the end to end 6G networks for Smart Cities the need to investigate CAPEX and OPEX models for:

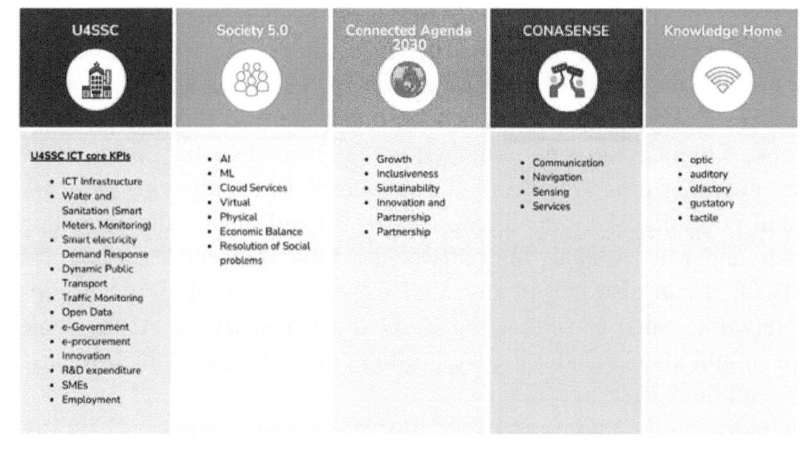

Figure 13.1 6G for society 5.0, SDGs, CONASENSE and knowledge home.

- 6G Spectrum Acquisition/ Spectrum sharing
- Fronthaul Infrastructure Strategy
- MidHaul Infrastructure Strategy
- Backhaul Infrastructure Strategy
- Traffic Demand
- Energy Consumption and Energy Savings
- Total Cost of Ownership (TCO)
- Total Value of Opportunity (TVO)

Finally, a study of the universal 6G costs of implementation, costs of maintenance, and the return of investment (ROI) for the future 6G investors, Telco, within ten years minimum.

13.2 5G Economical Modelling

As 6G is the successor to 5G, it makes sense to look at the 5G cost modelling and see if that can be adapted to 6G. It is also interesting to see where the great challenges lie in implementing and maintaining 5G mobile networks, as this can also impact the upcoming 6G networks.

Networks have seen exponential growth in data over the years. However, the revenues do not match this growth [14]. The rise in traffic demands on the network infrastructure of the Mobile Network Operators(MNO) has been a challenge to keep up with, and the capacity needs to be countered by profitability, which has become increasingly difficult [15]. According to many in the industry, the deployment of 5G has become more of an economical challenge rather than a technical one [16].

This means the Total Cost of Ownership of the 5G (and 6G) network will drive the deployment of the networks, much more so than for previous generations of mobile networks. This requires cost models that optimise the Capex by ensuring efficient use of the resources while allowing network growth and controllable Opex. These models should aid in understanding how, where and when virtualizing parts of the network is more economical for the TCO. It can also help understand the advantages of Cloud Radios Access Networks, such as centralized Baseband Units (BBU). At the same time, one should also consider the supporting core networks, which require an end-to-end model.

Most cost modelling papers on 4g and/or 5G networks, such as [17] [18], look into rural, urban and dense urban scenarios and simulate the deployment of the number of macros, micro and picocells driven by traffic demands and population density.

5G access networks have three spectrum bands that can be used, which are located around 700MHz (lower), 3.5GHz (middle) and 26GHz (higher). The use of the higher frequencies, with their reduced range, has become imperative for small cell deployment and high traffic.

Another approach for cost-saving can be obtained in certain scenarios, where the traffic is produced and consumed close to each other. There would be no need to send the data over a backhaul network for such a scenario, introducing extra delays and costs, but rather process it directly at the edge, hence the name Multi-Access Edge Computing (MEC).

The backhaul of the mobile operators' network must handle the peak-data rates, which implies high-speed interconnections using a myriad of technologies. At the same time, it must be resilient to temporary failures of some links.

The Core Network (CN) is responsible for aggregation, switching, authentication, switching and generally controlling the entire network. The CN also includes pivotal functionalities, such as Access and Mobility Management, Session Management and Authentication. Already in 5G, new concepts for the CN are being introduced, such as Software Defined Networking (SDN) and Network Function Virtualization (NFV) [15]. In SDN, the control plane of the equipment, which has the local forwarding functionality, is separated into a new entity. The idea is that this entity has the information from a single component in the network and can make the best decision based on data from many components. This allows for a much more dynamic network that can adapt to changes and even work around failures of some components in a manageable and cost-effective manner [19]. In NFV, the network functions are implemented in generic computing environments, which allows the use of custom the shelf (COTS) solutions to reduce the cost [15] further. It is expected that most of the NFV functionalities will run in virtualized machines or hypervisors, able to scale more quickly when demand requires.

The cost model used in 4G and 5G networks is mainly based on each area's capacity planning estimates. After which, the TCO can be calculated as the sum of all Capex values for each network part and their annual Opex expenditures times the number of years. The Capex and Opex expenditure components must be obtained from the access, backhaul and core network portions. So, the Capex is based on the acquisition costs of fibre and radio components, such as the OLTs, ONTs and splitters, and the cost of the antennas and switches used. The Capex also includes the cost of deploying cables and the installation cost of these components.

The Opex are the operational costs, such as power consumption, maintenance, fault management and reparation costs. It is expected that the main Opex costs for 5G are involved with power consumption.

However, modelling is more complex as the network or part of the network can be virtualized, creating multiple deployment strategies to be investigated.

13.3 6G Network Architectures

The future cellular Network will be more advanced and powerful than its predecessor, 5G. Looking at the 5G Network architecture as a standing point of analysis for the 6G future infrastructure, it is clear that this latter will encompass the existing extended services. For instance, 5G is the Network that brought into the market the possibility of having network slicing, in which the traffic can be prioritized by different types of applications and devices, offering to 5G user's equipment (UE) different Service Level Agreement (SLA). The 6G network will not shy away from this perspective. It will go above and beyond. In its current R&D process, 6G is planned to provide continuity to the network slicing embedded in 5G release 17. As shown in Figure 13.2,

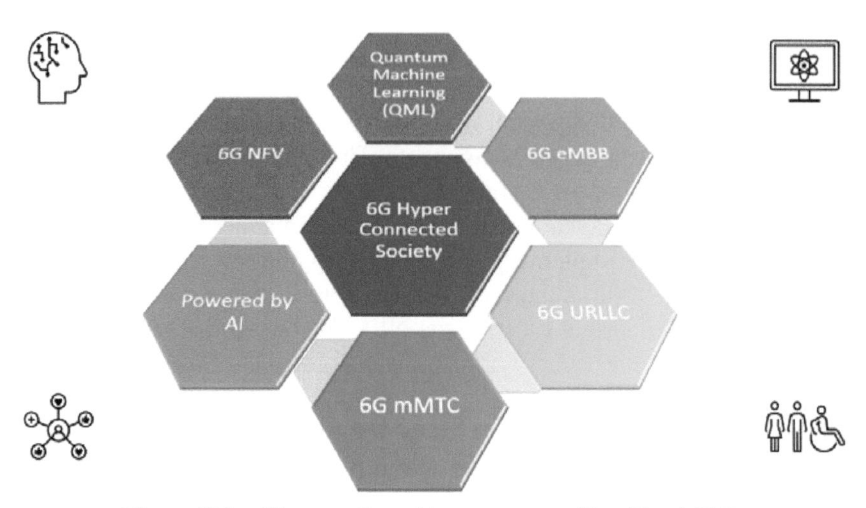

Figure 13.2 6G new radio architecture powered by AI and QML.

the 6G Network will also outreach the intrinsic capacity that exists in the 5G Radio New Radio [20] regarding the features of massive Machine Type Communications (mMTC), enhanced Mobile Broadband (eMBB), and the ultra-Reliable Low Latency Communications (uRLLC) features.

The figure above shows that 6G will enable a more intelligent and agile network slicing. For this intelligence, 6G Cloud-RAN will require an AI to be integrated into the Multi-Access Computing Network (MEC) [21], aligned with a Quantum Machine Learning (QML) [22] to overcome the be challenges of Big Data generated on the end-to-end of network architecture and Additionally, it will optimize the network traffic prioritization more efficiently, in much higher speeds that can achieve the nominal velocity of terabits per second for its premium services. It will significantly increase 100% over the data rate transmission envisaged on the 5G.

More importantly, 6G will continue utilizing the millimetre and centimetre wave spectrum. Therefore, it is imperative to add the benefits of Ultra-Massive MiMO (UM-MiMO) antennas, possibly embedded in a Graphene [23] surface, as an innovative solution to exploit further the electromagnetic phenomenon of multipath signal propagation [24] to rally and boost the antenna's gains. As the radio frequency (RF) is becoming scarce and moving up to the Terahertz domain, it is necessary to explore other options like the Optical Wireless Communications (OWC) to offer omnipresence to the 6G architecture [25].

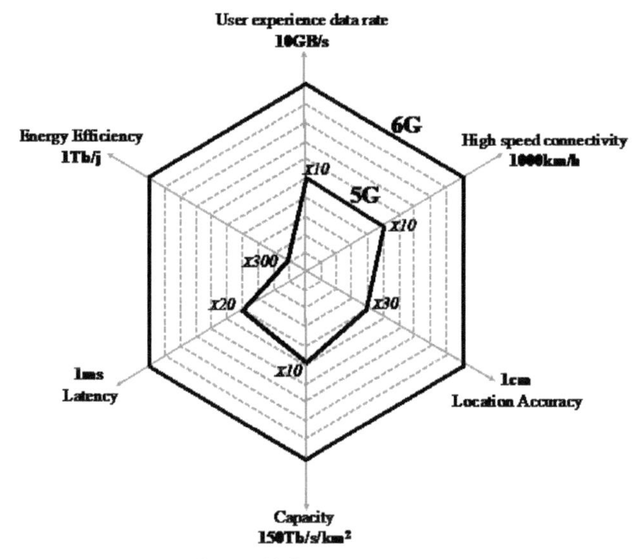

Figure 13.3 Improving 5G.

Some KPIs commonly agreed for the next generation of cellular networks will overpass the existing KPIs currently delivered at the 5G., see Figure 13.3.

As shown in Table 13.1, these are the KPIs envisaged for 6G and are designed to deliver new use cases that cannot be covered on the 5G NR. Hence, to achieve Terabit's data throughput, the 6G Fronthaul, MidHaul and Backhaul must be connected to artificial intelligence operating full-time aligned to quantum computing (QC) to prioritize the data traffic to automatically control the network resources based on specificities of the SLAs. All this technology is required for the 6G network slicing (6GNS), 6G Private Networks, and 6G MVNOs, including all types of the network and application layers. In terms of the medium to interconnect the Network physically and to support such a great data speed and high QoS and QoE, there is a need to invest in super-fibre optics rollouts and OWC to anchor the Fronthaul to the Backhaul to offload traffic congestions and offer wider bandwidth.

It is important to note that 6G architecture will be costly for those Mobile Network Operators (MNO) that are not entirely covered by a well-equipped core of fibre networks, then it seems to be prudent for MNO to invest part of 5G ROI in the CAPEX for the future 6G infrastructure. However, it is important to observe that the domain of hyper-scale [26] will continue, as

Table 13.1 6G KPIs envisaged for the cellular network architecture.

6G KPIs	
Data Throughput	Terabits/s
Power Consumption	1Terabit/Joule
SLA Network Availability	99.9999%
Latency	1ms
RF Spectrum	mm-wave / cm-wave light spectrum
High-Speed Data Connectivity	1000 km/h

they will deliver important NFV to virtualize network nodes services via Cloud-as-a-Service (CaaS) to deliver the 6G networks components of the architecture. Therefore, in the next chapter, there is an evaluation of the 5G technical aspects and network costs followed by the principles envisaged here to roll out an economical 6G network with the basic settings required.

13.4 6G Economical Modelling Challenges

The forecast predicts that by 2035, the mobile communication sector [28] suggests that by 2035 the mobile communication sector has an economic output of $13 T. This requires an investment from the industry of $235 B annually and generates 22 M jobs.

The 6G Architecture is still at the very beginning of its definition. What is certain is that 6G will bring innovation to the wireless networks and that innovation will exceed the reach of the known 5G Networks. However, it is essential to note that 6G will be based on network function virtualization as presented in 5G. Thus, 6G Networks will be presented in many flavours based on cell sizes and to serve different purposes. The important part to consider in the 6G CAPEX is the Spectrum acquisition. 6G will combine the terahertz spectrum and the light spectrum as it is known. Therefore, the first strategic CAPEX to consider is the RF Spectrum acquisition acquired by Telecom Operators. For this part, Spectrum acquisition will be considered as $\mathbf{CAPEX}_1$. Possibly, a new trend possible to happen is that the Mobile Service Providers will reutilize the 4G spectrum to deliver new 6G uses cases and to increase their capabilities of reach. This opportunity of reutilizing old RF Spectrum can strategically save CAPEX and boost ROI for Telecom investors. This behaviour is currently being observed by the Telecom industry [29] globally, which is turning off 2G and 3G networks for rolling out 5G use cases. Based on this assumption, the phase-out of 4G networks to benefit 6G use cases will be characterized as $\mathbf{CAPEX_2}$.

In light of 6G offering a terabits data throughput, as previously presented, it will require that Telecom Operators start preparing to increase their super-fibre networks footprint from now to alleviate the future costs of the 6G fibre network. Then, it is vital to say that part of the 6G success in the Physical Layer of the Open System Interconnection (OSI protocol) is related to the availability of fibre optics networks. Therefore, this cost can be defined as $\mathbf{OPEX_1}$ or $\mathbf{CAPEX_3}$ if the Telecom Operator starts investing in fibre networks before or after the 6G spectrum acquisition. Some well-established telecom operators can also offer solutions to future 6G Mobile Operators that do not have an excellent fibre optics footprint. The solution to be offered by Hyperscalers or Telecom Operators to 6G Mobile Operators without enough fibre networks infrastructure is the novel technological concept defined as Network as a Service (NaaS) [29]. In this case, this capital expenditure will be considered $\mathbf{CAPEX_4}$ in case of the additional need for extra fibre network resources availability in the 6G Core Network.

Energy efficiency and low carbon strategies have attracted much concern. The goal for green 6G networks drives the Information Communication Technologies (ICT) sector to strategies that incorporate modern designs for low carbon and sustainable growth. Moreover, an essential part of the OPEX is due to the electricity demands. This chapter looks into the areas impacting the CAPEX and OPEX values of 6G networks by having a standing point of analysis and comparison with the previous generations.

6G is expected to have at least double the high-band spectrum, which will also work within the terahertz frequency range. Experimenting with the range from 95GHz to 3THz is already ongoing, but many hurdles still have to be overcome. For example, the signal attenuation above 95GHz is enormous, making this communication only suitable for very short distances (a few meters at best) but with extremely high data rates. This will require a new network infrastructure to process and distribute all the data from these access points. With their reduced wavelength, these higher frequencies seem to have advantages for massive MIMO gains [30].

Short ranges are also an exciting scenario for Cloud Radio Access Networks (C-RAN). C-RAN employs the latest Common Public Radio Interface (CPRI) standards to allow the transmission of a baseband signal over long distances to a Remote Radio Head (RRH). The radio signal can be created in a virtualized Baseband Unit (BBU), reducing the overall cost while obtaining high reliability, low latency, and a less high bandwidth interconnect network

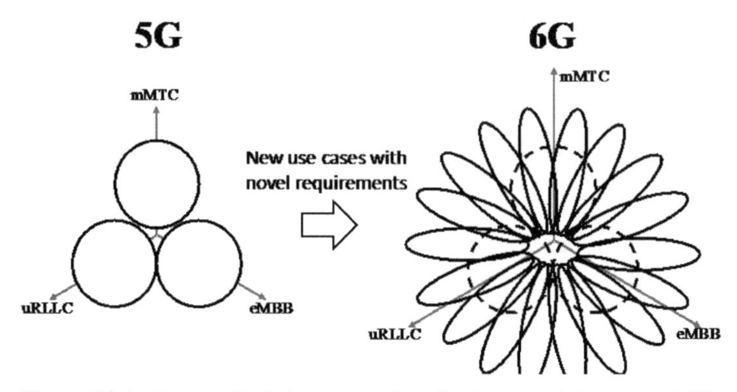

Figure 13.4 Image of wireless network technology evolution toward 6G.

nodes. According to [31], the C-RAN approach can reduce the Capex to 50% while lowering the Opex by 15%.

Ever since the earliest cellular communications systems, a change can be noticed in the location of the base stations and/or access points. Firstly, base stations were affixed to high buildings, but the frequency reuse and the rise in frequencies require a more dense cellular infrastructure. It has often been suggested that every lamppost can be an access point for the network.

The use cases for 5G networks can be roughly divided into three areas: eMBB, mMTC and URLLC. However, 6G is much more complex, where different requirements are needed based on the specific use case, as shown in Figure 13.4. It will not be economically viable for the operators to build products around each use case, where many will not be successful, but maybe one or two will be.

Therefore, 6G will create opportunities for new technologies and business innovations by addressing vertical markets. Each vertical market, such as automotive, energy, government, healthcare and city management, has different requirements from ultra-low latency, high bandwidth and mobility, and high reliability.

Already in 5G networks, we see the rise of MVNOs, but these are merely competing with the MNOs on branding and pricing. In 6G, the market will be operated by the current MNOs, but will also see the rise of niche market MVNOs, creating new customer segments [32]. This trend will only be fortified by the rapid deployment of IoT related devices, which IoT MVNOs will run.

13.5 Conclusions

In 6G, the communication architecture is moving away from a pure cellular deployment and will see the formation of networks as a service (NAAS). This requires a complex technology [32], drastically changing their operation. MNOs and MVNOs must devise new business models and deployment strategies to have any chance of recuperating their investments as the classical "overprovisioning" method is no longer a viable solution in 6G.

6G relies on virtualization of the network, which will use resource provisioning mechanisms based on a self-managed reliable and trustworthy AI algorithm. Networks need to adapt dynamically to the required demand and requirements and must do so without interrupting the service.

This creates new opportunities for existing players and opens the market for new operators. Even though 6G is still being defined, it is of the utmost importance to understand the new dynamics, models and opportunities that it will bring us.

References

[1] P. S. Rufino Henrique and R. Prasad, "*6G The Road to the Future Wireless Technologies 2030*," River Publishers: Professional Books, 31-Mar-2021. [Online]. Available: https://www.riverpublishers.com/book_details.php?book_id=920. [Accessed: 03-Jun-2021].

[2] M. Warwick, "*Mobile broadband costs keep developing nations on the fringes of the global Internet economy*," TelecomTV, 05-Mar-2021. [Online]. Available: https://www.telecomtv.com/content/sustainability/mobile-broadband-costs-are-keeping-developing-nations-on-the-fringes-of-the-global-internet-economy-40993/. [Accessed: 03-Jun-2021].

[3] Cabinet Office, Society 5.0. [Online]. Available: https://www8.cao.go.jp/cstp/english/society5_0/index.html. [Accessed: 03-Jun-2021].

[4] UN, "*THE 17 GOALS |Sustainable Development*," United Nations. [Online]. Available: https://sdgs.un.org/goals. [Accessed: 03-Jun-2021].

[5] UNESCO, "*Japan is pushing ahead with Society 5.0 to overcome chronic social challenges*," UNESCO, 09-Sep-2020. [Online]. Available: https://en.unesco.org/news/japan-pushing-ahead-society-50-overcome-chronic-social-challenges. [Accessed: 03-Jun-2021].

[6] A. G. Kravets, A. A. Bolshakov, and M. Shcherbakov, "*Society 5.0: Cyberspace for Advanced Human-Centered Society*," Springer. [Online]. Available: https://www.springer.com/gp/book/9783030635626. [Accessed: 03-Jun-2021].

[7] E. Cianca, M. D. Sanctis, A. Mihovska, and R. Prasad, "*CONASENSE:Vision, Motivation and Scope*," Journal of Communication, Navigation, Sensing and Services (CONASENSE). [Online]. Available: https://www.riverpublishers.com/journal_read_html_article.php?j=JCONASENSE%2F1%2F1%2F1. [Accessed: 06-Jun-2021].
[8] R. Prasad, "*Knowledge home*," 2016 International Conference on Advanced Computer Science and Information Systems (ICACSIS), 2016, pp. 33-38, doi: 10.1109/ICACSIS.2016.7872717.
[9] Rahimi M, Prasad R. *An introduction to the business model for human bond communications*. In 2017 Global Wireless Summit, GWS 2017. Vol. 2018-January. Institute of Electrical and Electronics Engineers Inc. 2018. p. 230-234 https://doi.org/10.1109/GWS.2017.8300501
[10] ITU, "*ITU Activities & Sustainable Development Goals*," ITU. [Online]. Available: https://www.itu.int/en/action/environment-and-climate-change/Pages/ITU-in-the-UN-Environmental-Agenda.aspx. [Accessed: 06-Jun-2021].
[11] ITU, "*Connect 2030 – An agenda to connect all to a better world*," ITU. [Online]. Available: https://www.itu.int/en/mediacentre/backgrounders/Pages/connect-2030-agenda.aspx. [Accessed: 06-Jun-2021].
[12] UN, "*United 4 Smart Sustainable Cities*," ITU. [Online]. Available: https://www.itu.int/en/ITU-T/ssc/united/Pages/default.aspx. [Accessed: 06-Jun-2021].
[13] U. N. ITU, "*Key Performance Indicators: A Key Element for cities wishing to achieve the Sustainable development Goals* ." [Online]. Available: https://www.itu.int/en/ITU-T/ssc/united/Documents/U4SSC%20Publications/KPIs-for-SSC-concept-note-General-June2020.pdf. [Accessed: 06-Jun-2021].
[14] Frank H. et al. (2020) *Resource Analysis and Cost Modeling for End-to-End 5G Mobile Networks*. In: Tzanakaki A. et al. (eds) Optical Network Design and Modeling. ONDM 2019. Lecture Notes in Computer Science, vol 11616. Springer, Cham. https://doi.org/10.1007/978-3-030-38085-4_42
[15] "5G Infrastructure Requirements in the UK," Tech. Rep., 2017
[16] R. Charni and M. Maier, "*Total cost of ownership and risk analysis of collaborative implementation models for integrated fiber-wireless smart grid communications infrastructures*," IEEE Transactions on Smart Grid, vol. 5, no. 5, pp. 2264–2272, 2014.

[17] M. Mahloo, P. Monti, J. Chen, and L. Wosinska, "*Cost modeling of backhaul for mobile networks*," in IEEE International Conference on Communications Workshops (ICC), 2014, pp. 397–402.

[18] M. De Andrade, M. Tornatore, A. Pattavina, A. Hamidian, and K. Grobe, "*Cost models for baseband unit (bbu) hotelling: From local to cloud,*" in IEEE International Conference on Cloud Networking (CloudNet), 2015, pp. 201–204.

[19] M. Jaber, M. A. Imran, R. Tafazolli, and A. Tukmanov, "*5G backhaul challenges and emerging research directions: A survey*," IEEE access, vol. 4, pp. 1743–1766, 2016.

[20] NTT Docomo, "*White Paper 5G Evolution 6G - NTT Docomo.*" [Online]. Available: https://www.nttdocomo.co.jp/english/binary/pdf/corporate/technology/whitepaper_6g/DOCOMO_6G_White_Paper EN_20200124.pdf. [Accessanded: 06-Jun-2021].

[21] P. S. Rufino Henrique and R. Prasad, *6G: the road to future wireless technologies 2030*, 1st ed. S.l.: RIVER PUBLISHERS, 2021. Chapter 5, page 53

[22] J. Biamonte, P. Wittek, N. Pancotti, P. Rebentrost, N. Wiebe, and S. Lloyd, "*Quantum machine learning*," Nature News, 14-Sep-2017. [Online]. Available: https://www.nature.com/articles/nature23474/. [Accessed: 06-Jun-2021].

[23] D. P. Harrop, "*Graphene for 6G Communications*," IDTechEx, 14-May-2021. [Online]. Available: https://www.idtechex.com/en/research-article/graphene-for-6g-communications/23776. [Accessed: 06-Jun-2021].

[24] R. Henrique, Paulo Sergio. "*TV Everywhere and the Streaming of Ultra High Definition TV over 5G Wireless Networks* - Performance Analysis". Brunel University London. 2016.

[25] H. Harald, J. Elmirghani, and I. White, "*Optical wireless communication,*" *Philosophical Transactions of the Royal Society*, 02-Mar-2020. [Online]. Available: https://royalsocietypublishing.org/doi/10.1098/rsta.2020.0051. [Accessed: 06-Jun-2021].

[26] "*Hyperscale your cloud journey*," Accenture, 08-Dec-2020. [Online]. Available: https://www.accenture.com/us-en/insights/cloud/hyperscale-cloud-journey. [Accessed: 06-Jun-2021].

[27] Telecompaper, "*Cosmote to shut down 3G network from September*," Telecompaper. [Online]. Available: https://www.telecompaper.com/news/cosmote-to-shut-down-3g-network-from-september--1379564. [Accessed: 06-Jun-2021].

[28] https://www.qualcomm.com/media/documents/files/ihs-5g-economic-impact-study-2019.pdf

[29] "*What Is Network as a Service (NaaS)?*," Cisco, 26-Apr-2021. [Online]. Available: https://www.cisco.com/c/en/us/solutions/enterprise-networks/network-as-service-naas.html#~{ }q-a. [Accessed: 06-Jun-2021].

[30] H. Frank, "*Interference mitigation for femto deployment in next generation mobile networks*," in Proceedings of the International MultiConference of Engineers and Computer Scientists, vol. 2, 2016.

[31] A. Checko, H. L. Christiansen, Y. Yan, L. Scolari, G. Kardaras, M. S. Berger, and L. Dittmann, "*Cloud RAN for mobile networks - A technology overview*," IEEE Communications surveys & tutorials, vol. 17, no. 1, pp. 405–426, 2015.

[32] https://telecoms.com/opinion/the-future-of-mvno-market-from-a-single-niche-to-multiple-niches-of-opportunities/

[33] C. Sergiou, M. Lestas, P. Antoniou, C. Liaskos, A. Pitsillided, 2020, *Complex systems: a communication networks perspective towards 6G*, IEEE Access, May 2020.B.G. Buchanan and E.H. Short life. *Rule-Based Expert Systems: The MYCIN Experiments of the Stanford Heuristic Programming Project.* Addison-Wesley Publishing Company, 1984.,

Biographies

Berend Willem Martijn Kuipers received a B.Sc. from the Rijswijk University of Technology, the Netherlands, in computer science in 1996. In 1999, he received his M.Sc. in telecommunications from the Delft University of Technology. He received his PhD in telecommunications from Aalborg University, Denmark, in 2005. He developed a novel multicarrier access scheme for 4G systems during his PhD. Currently, he is employed by INOV-INESC Inovação in Lisbon, where he is involved in applying artificial intelligence algorithms for data analysis, such as clustering algorithms, seasonal ARIMA forecasting, and machine learning. He has supervised more than 30 M.Sc. students and was involved with courses on telecommunications and computer networks, artificial intelligence and data structures. He has taken part in many national and European projects. He is also a professor and coordinator for the bachelor's degree in Computer Science and Engineering at the Lusíada University of Lisbon.

Paulo Sergio Rufino Henrique (Spideo-Paris, France) CTIF Global Capsule, Department of Business Development and Technology, Aarhus University, Herning, Denmark. Paulo S. Rufino Henrique holds more than 20 years of experience working in telecommunications. His career began as a field engineer at UNISYS in Brazil, where he was born. There, Paulo worked for almost nine years in the Service Operations, repairing and installing corporative servers and networks before joining British Telecom (BT) Brazil. At BT Brazil, Paulo worked for five years managing MPLS networks, satellites (V-SAT), and IP-Telephony for Tier 1 network operations. He became the Global Service Operations Manager overseeing BT operations in EMEA, Americas, India, South Korea, South Africa, and China. After a successful career in Brazil, Paulo got transferred to the BT headquarters in London, where he worked for six and a half years as a service manager for Consumers Broadband in the UK and IPTV manager for BT TV Sports channel.

Additionally, during his tenure as IPTV Ops manager for BT, Paulo also participated in the BT project of launching the first UHD (4K) TV channel in the UK. He then joined Vodafone UK as a quality manager for Home Broadband Services and OTT platforms and worked for almost two years. During his stay in London, Paulo completed a Post-graduation Degree at Brunel London University. His thesis was entitled 'TV Everywhere and the Streaming of UHD TV over 5G Networks & Performance Analysis'. Presently, Paulo Henrique holds the Head of Delivery and Operations position at Spideo, Paris, France. He is responsible for integrating the Spideo recommendation platform on the OTT and IPTV providers. He is also a PhD student under the supervision of Professor Ramjee Prasad at Global CTIF Capsule, Department of Business at Aarhus University, Denmark. His research field is 6G Networks - Performance Analysis for Mobile Multimedia Services for Future Wireless Technologies.

Dr Ramjee Prasad is the Founder and President of CTIF Global Capsule (CGC) and the Founding Chairman of the Global ICT Standardization Forum for India. Dr Prasad is also a Fellow of IEEE, USA; IETE India; IET, UK; a member of the Netherlands Electronics and Radio Society (NERG); and the Danish Engineering Society (IDA). He is a Professor of Future Technologies for Business Ecosystem Innovation (FT4BI) in the Department of Business Development and Technology, Aarhus University, Herning, Denmark. He was honoured by the University of Rome "Tor Vergata", Italy as a Distinguished Professor of the Department of Clinical Sciences and Translational Medicine on March 15, 2016. He is an Honorary Professor at the University of Cape Town, South Africa, and KwaZulu-Natal, South Africa. He received the Ridderkorset of Dannebrogordenen (Knight of the Dannebrog) in 2010 from the Danish Queen for the internationalization of top-class telecommunication research and education. He has received several international awards, such as the IEEE Communications Society Wireless Communications Technical Committee Recognition Award in 2003 for contributing to the field of "Personal, Wireless and Mobile Systems and Networks", Telenor's Research Award in 2005 for outstanding merits, both academic and organizational within the field of wireless and personal communication, 2014 IEEE AESS Outstanding Organizational Leadership Award for: "Organizational Leadership in developing and globalizing the CTIF (Center for TeleInFrastruktur) Research Network", and so on. He has been the Project Coordinator of several EC projects, namely, MAGNET, MAGNET Beyond, and eWALL. He has published more than 50 books, 1000 plus journal and conference publications, more than 15 patents, and over 150 PhD. Graduates and a more significant number of Masters (over 250). Several of his students are today worldwide telecommunication leaders themselves.

Ramjee Prasad is a member of the Steering, Advisory, and Technical Program committees of many renowned annual international conferences, e.g., Wireless Personal Multimedia Communications Symposium (WPMC); Wireless VITAE, etc.

14

6G and Green Business Model Innovation

Peter Lindgren

CGC - Aarhus University, Birk Centerpark 40, DK-7400 Herning Denmark
E-mail: peterli@btech.au.dk

The green business model is one of the trending areas among the Businesses, societies and academia. It amalgamates creation, visualization, delivery, reception and consumption techniques with various reconfiguration techniques that focus on transforming existing business models to become greener with the development of the tactical parameters, tools and actions. Green business model parameters are measurement objects which will benefit from the advancing 6G technologies leading to the evolution of the Green Business Model innovation and development.

6G technologies combined with Artificial Intelligence and Artificial Reality will take Green Business Model innovation into the next generation of green transformation. 6G technologies will enable businesses and societies globally to fulfil their vision of becoming green fully.

The chapter presents work and thoughts on 6G technologies related to the green business model definition, green business model tools, green business model innovation and development. It reports on challenges to green business model innovation, growth and measuring and discusses innovation of green business model dashboard - enabling measurements of how green business models are anywhere and anytime – with anybody and anything. It builds on the multi business model approach and theory and analyses different green business model challenges from diverse business model ecosystems. The goal is to understand the challenges of green business model innovation and development and how to build absolute trust in green business model transformation supported by 6G technologies.

14.1 Introduction

More and more businesses and societies have been caught in greenwashing [1]. Some are due to a lack of knowledge of "green" related to business operation and innovation, and others because of a chosen greenwashing strategy to signal and brand green to users, customers and society. It seems as if there is a big gap in literature and practice on how to work ethically and trustworthy with green business model innovation. Maybe this is because the technologies available still are not advanced enough, and 6G technologies could be a solution.

Sustainable and circular business models might surprisingly not be green business models [2], although sustainable and circular certificates are given by well known and highly recommended certification businesses. Research shows that Green business model Innovation (GBMI) and development have a big challenge in measuring green business models (GBM) over a lifetime and in real-time. This challenges users, customers, networks, employees, businesses and business model ecosystems to become, demand and buy green business models. It challenges this trust in the global agenda's green transformation and could potentially build barriers and resistance to the worldwide society's vision to become green and create a green economy.

Both businesses and societies have to take up this challenge seriously, and 6G technology providers could have great potential to capitalize on and support this mission.

GBMI and development measured in real-time seem today to be a significant challenge because of at least two critical challenges:

- Lack of clear definitions of a Green Business Model and What is Green Business Model Innovation.
- Lack of technologies and tools can measure Green Parameters of Business Models in real-time anywhere with anybody and any business.

No, doubt it is challenging to develop such Green business model measurement tools. However, numerous green business model certificates [3, 4] and databases exist and could be used as the baseline. Several consultancies try to lead and capitalize on the billion-dollar green business model certificate ecosystem to measure and validate the transformation of Business model ecosystems [4], businesses and their business models into the green. Their experience could be valuable knowledge to innovate and develop GBM and GBMI measurement tools.

However, unfortunately, most GBM and GBMI definitions and certificates seem to be very static and include numerous other terms and approaches.

Those measuring practices and certifications that we investigated seem to be measuring green within a specific date and on behalf of certain standards – auditory dates and embedded with Sustainable, circular and UN 17 goal terms and objectives. None of the terms that we investigated was based on or defined pure green business model definitions. Further, the terms did not have any guarantee or systems that measured the business and its model after the day, week and minute the certificate was given. Sustainability, circular and UN 17 goal certificates also seem to be built on and related to old iso measurement standards and tools. They seem not to be mainly focused on green business model parameters and, in this sense, able yet to measure green business models – and in real-time. Most certificates, certification practices and standards in this field seem to be based on and include former iso standards pitfalls, pains and potential to "jump the fence" – greenwashing. Most green business model approaches take the point of entry from the technical and/or the sustainable, circular and UN 17 goals [6] view and do not delimit their measurement to the green business model parameters [2]. Finally, the efficiency of green business models, business model dimensions, and business model innovation levels view are not considered, although many call their business models green or green business models.

Therefore, Green Business Model Innovation, Development and Measurement calls for a new and much more advanced green business model innovation approach, measurement, tool and technology innovation and development. Herein could future 6G technology potentially play a significant role.

The current situation shows the importance of innovating faster digital technologies and solutions to the Green Business Model Innovation Ecosystem to continue the successful operation of all spheres of green life, green education, green business and green social interaction with Green Business Models – here under support and increase the green transformation of the current state (AS IS) BM and innovation of future state (TO BE) GBM.

This gives a special flare and hopes for the efficiency of future 6G systems, technologies and business models – opposite to previous 5G systems and technologies. It is envisioned that 6G technologies will be able to support the further enhancement of the green transformation. With the expected full-broadband (FeMBB) services to all types of users, customers, networks and business models, it is envisioned that 6G technologies will be a universal technology and infrastructure – that will be able to take green business model innovation and development into a new era – a more advanced, trusted, ethical and sustainable business level of green business model innovation.

The current trends of green business model digitalization and the business models stakeholder's increasing requirements for access and investment to trustful green business models lay pressure on the transmission of AS IS Business Models into green or greener business models. Reconfiguring AS IS BM's is undoubtedly the largest BMES and objective to green transformation. It requires innovation of trust, complete definition of green business models and advanced measurement of green business model data while on the move - and in real-time. For networking and intelligence in all spheres of the green business models lifecycle – cradle to cradle [8, 9] – it is necessary to use and embed more advanced technologies – hereunder 6G technologies and related new technologies. The demand for 6G as the accelerator and umbrella [10] of transformation and innovation on a global scale and deep penetration of green business model innovation is apparent. Current application trends in the green business model innovation business model ecosystem can be observed as the emergence of services based on Augmented Reality (AR), Virtual Reality (VR), Mixed Reality (MR), wireless green business model brain-computer interaction, green cities, tactile green business model communications, and holographic green business model communications.

These developments challenge the current capabilities of the enabling and existing wireless communication systems from various aspects, such as delay, rate, degree of intelligence, coverage, reliability, ability, and capacity. The ambition of green business model innovation cannot be achieved just by evolutionary research and classical linear business model practice – neither existing circular and sustainable business model [8–12] approach and practice. Green Business Model Innovation and development research will need to seek breakthroughs from the current network architecture and communication theory to provide novel concepts that can be key for designing a radically new green business model innovation system supported by and embedded with 6G. At the same time, it is essential to enable such revolutionary GBMI technology developments to stay 'green' and take into account major environmental concerns, such as climate change, which can be achieved by novel, 'green' digitalized business models.

Undoubtedly, many businesses are motivated and heavily pushed by society and other stakeholders (investors) to design, reconfigure, and develop Green business models (GBM). However, it is well known that reconfiguring existing business models (AS IS BM) is complex, but it is not that known that reconfiguration of AS IS BM's to become efficient GBM's is more complex and resource-consuming than expected. Reconfiguring AS IS BM into GBM's is the primary task and covers an estimated 80 – 90 % of the potential of GBMI. Many BM from the past were created "black" or "half-black" business models and now have to be turned into green or greener. Further balancing monetary and non-monetary value formulas of GBM's and developing indispensable GBM's strategies to commit the business to the green economy and vision embedded with 6G is an enormous task. It includes 6G GBMI on all business model dimensions and levels.

More and more businesses are aiming at transforming into the green. The term "Green Business Model" (GBM), as used in our framework, terminology and vocabulary, relates purely to the environmental dimension of green (e.g. Use of Material and Resources, Consumption of Material and Resources, Waste reduction, Pollution, Recycling of BM's waste and pollution and at the same time the green business model economic perspective (e.g. turnover – cost = earnings, the efficiency of GBM's and GBMI). The 6G GBM construction perspective in our terminology focus on the input, operation and output of GBM's that become operative 6G GBM that are both environmentally (Green Business Model Parameters) friendly and economically sustainable (Economic Business Model Parameters) to the business and its Business Model Ecosystem (BMES) throughout the entire lifecycle of the 6G GBM and its related BM and BMES's.

Businesses today are globally involved in executing their business role to minimize environmental damage effects – but at the same time, they are also responsible in a world of capitalism to try to capitalize or seek other values on the green economy and their GBM's. 6G and related technologies are expected to play an essential role in this green transformation in businesses and their BMES. To act upon the climate change and balance their related BMES's complemented with 6G technologies and 6G GBM's the "fight" on environmental challenges as, e.g. higher temperatures, Co2 emission, numerous frequencies of floods and storms, lack of food and water, pressure on the health care system, biodiversity both in Atmosphere, soil and in the water a new use of wireless and related technologies is needed.

The green transformation includes, according to literature, among other green parameters, reduced use of energy and non-renewable materials and resources, increased delivery of reliable renewable energy supplies called green energy and increased focus on GBMI. It includes heavy investment in GBM through the digitalization of the GBMs and the GBMI processes. However, scientists and researchers from both businesses and academia discuss how to deploy and measure GBMI and strategies from an efficiency perspective and the entire GBMI Process, as indicated in Figure 14.1.

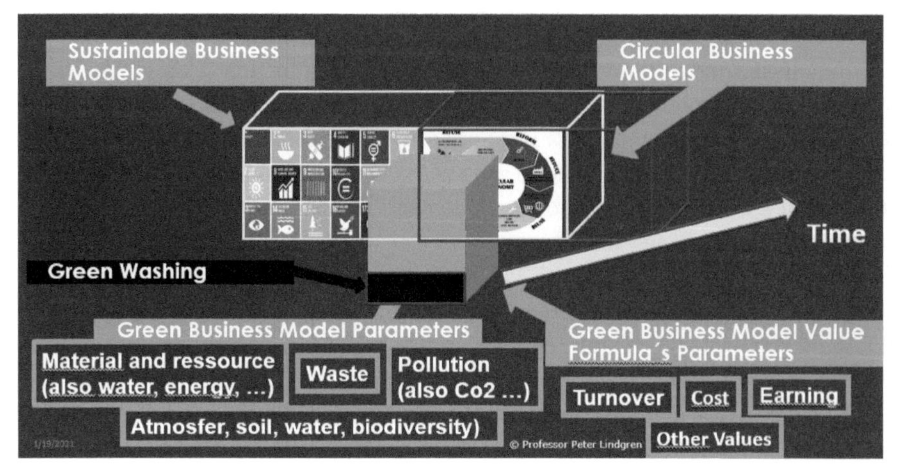

Figure 14.1 Green business model parameters and process.

Our long term approach to the GreenBizz project [19] relates to the measurement of GBM and GBMI. With 6G and complementing advanced technology - 6G as an umbrella – the aim is to measure all green business model parameters and the Green Business Model Value formula's in all phases of the green business model innovation, as shown in Figures 14.1. Further, the aim is to measure the individual BM, Business and BMES, but the entire value network of the BM, Business and BMES that the GBM and GBMI process is a part of.

14.2 Measuring Green Business Models

At this point, we begin the work and the measurement of Green Business Models – or Business Models in general - from the "inside in" perspective [11] of the GBM, Green Business and Green BMES. We intend to commence measuring green on the BM Components, Dimensions, Portfolio, Business and BMES level and then increase the measurement to the "outside-in" and "outside out" of the GBM [11] – covering green through the relations and valuing to and from other BM's, Businesses and BMES in the value network.

The GBM parameters can be divided into some main measurement areas following the value proposition and value chain function flow from:

As can be seen, inf Figure 14.1, all green parameters are related to the term green, green economy, green business, GBM's hereunder green technology – the environmental part of the green. However, sustainable business models [12] and Circular business models [13] have tried to embed the GBM into their terms and vocabularies with great difficulties. It has been challenging for these to "cut" the green part of their terms and give a clear definition and answer: What is green? Especially when sustainable and circular business models turn out not to be green – but greenwashing [4] - when deeper investigated both in short and long term perspective – the green parameters and definitions of these terms are questioned [14, 15].

All topics shown in Figure 14.1 are seen as solutions to fulfil the Global Society's vision, mission and goals to become a green economy. However, none have organized the green discussion on the business model framework relating and defining the different business model dimensions and innovation levels to both the term green and the terms economic-related to BM. Few have tried to bridge the Green Parameters to the Economics Parameters of GBM's. In our literature study, it was impossible to precisely find a measurement of GBMI and Green Business Model Development (GBMD) related to the GBM and the green parameters. These gaps in literature and practice are the visions of using "the 6G umbrella" to measure green related to business models possible, visual and operative shortly.

14.2.1 Relating Measurement of Green Parameters to Business Model Dimensions and Innovation

Previous literature study on GBM's and GBMI shows no generally accepted, clear, and precise definition of GBM and GBM parameters nor GBMI [12, 14, 15]. Many "talks" about GBM and GBMI, e.g. as related to 100% Co2 neutral business model, circular business models [12], sustainable business models [13] and even just GBM's [12, 14, 15]. The big question is if a business, business models and BMES ever can and will be 100% green on different GBM parameters and stages (lifetime of BM) and levels of GBMI?

As sketched in Figure 14.2, our observations are that many BMES, Businesses and BM's will relate to GBMI and **try to do GBM reconfiguration** – focusing on changing or reconfiguration [16] of their operative AS IS BMES, Businesses and BM's into Greener BMES, Businesses and BM.

Fewer businesses design [16] Green Businesses (GB) and GBM's that are "**born green**", and **fewer design new green businesses** - Startup businesses - that are **"born pure green"** – and **pure green from beginning to the end of their business and BM's life**. This would be equal to a startup business built with "**green gens**" or "**green Business Model Components**" in its core business model – the strategic highest level of a business and following all

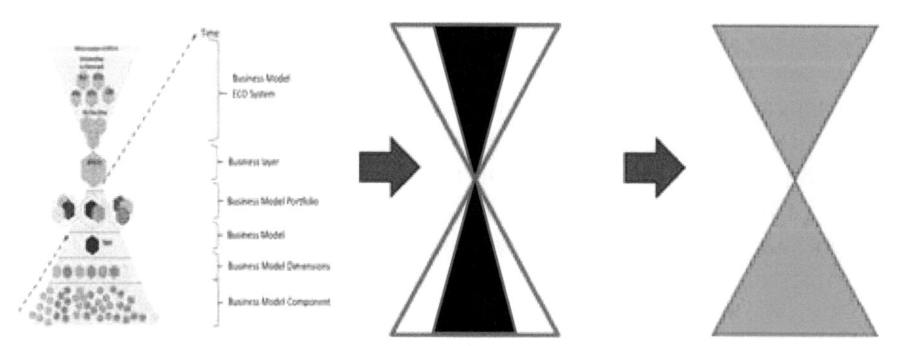

Figure 14.2 Different stages and levels of green business model innovation inspired by Lindgren 2021 [17].

related BM's. Both to AS IS and TO BE businesses, we claim this level of GB and GBM is radical and, in some cases, disruptive - extremely difficult and complex to fulfil, operate and reach.

Observations from our research showed that most BMES, Businesses and BM's experience "the greening" of a business and its "greening" of related BM's as **"a Green Business Model Innovation Journey and Process"** - **"a Green Business Model Innovation transformation"** - from being "black" or "half-black" to becoming incrementally more green - but maybe seldom and never completely pure green as a sketch in Figure 14.3 Our research shows that pure green seems very difficult to become to any BMES, Business and BM – or near to impossible. To achieve this stage would include all

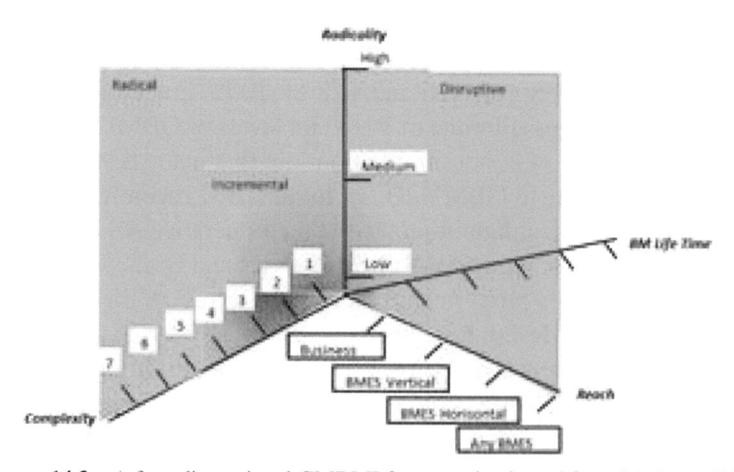

Figure 14.3 A four-dimensional GMBMI framework adapted from Lindgren [10].

related BMES, Businesses, BM's at all BMI levels and all GBM parameters to be and become green – and thereby green in their BM's entire value network of BM's related to any BMES. 6G and related technologies could help measure and visualise this green transformation of the AS IS BM and design of TO BE BM's.

We propose that businesses BM can be designed or reconfigured into the green on seven levels: BM component, BM dimensions, BM, BM portfolio, Business, BMES layer and BM/BMES Process level. GBMI can be designed or reconfigured on any of the 7 Green Multi Business Model Innovation (GMBMI) levels and can be measured related to the

- **Radically of green** – defined as the **degree that the BM's dimensions are changed** into green (incremental or radical)
- **The complexity of green** – is defined as how many **dimensions of each BM are changed** into green
- **Reach of green** – defined as **the impact of the change of green** the BM has on the business, vertical- and horizontal BMES [12] or the world
- **Time** – defined as **the degree of green of the BM through its entire life cycle**

Related to measuring a BM's green complexity, if all BM dimensions are changed to Green, including all BM dimensions components, the BM could be classified as a Radical GBM. However, this was not found in any case in our investigation of 106 SME businesses GBMI projects and processes as reported in previous articles [18] seen in Table 14.1 enclosure 1.

The businesses in our investigation primarily **focused on "greening" or innovating green on the Business Models BM competence dimension and on the component Layer at the technology part of the BM Competence dimension,** as seen in Tables 14.1 and 14.2 inclosure 1. The BM competence dimension consists of 4 component groups [8] – **1. Technology** (product- and service technology, production technology and processes technology), **2. HR**, **3. Organisational Systems,** and **4. Culture.** According to our research, the last three competence component groups were hardly touched upon. In general, a very small part of the BM competence dimension technology group was that businesses changed and wanted to change when doing GBMI, as indicated in Figure 14.4.

It is primarily product technology – energy, water, material – that is changed. This we marked with the grey arrow in Figure 14.5. However, the businesses investigated in our research also tried to change the **BM value proposition dimension product component group** as they tried to innovate

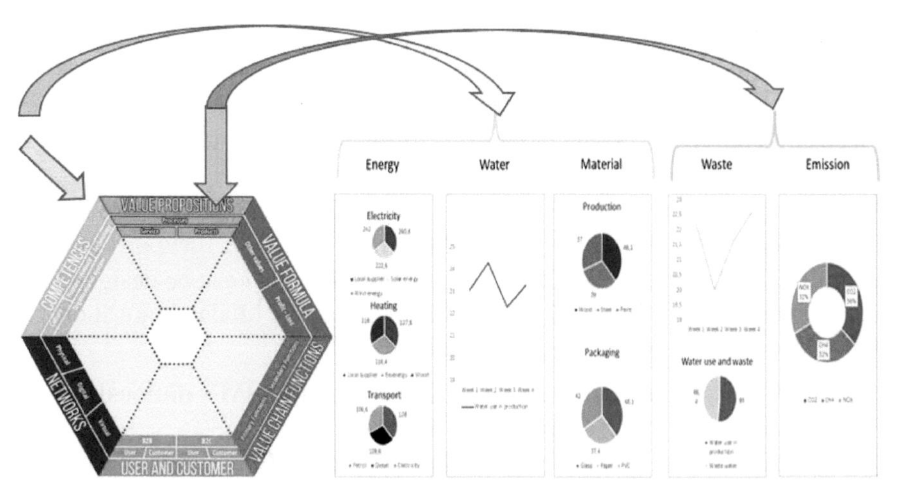

Figure 14.4 Example of a green business modelling dashboard focused on competence BM dimensions and technical components.

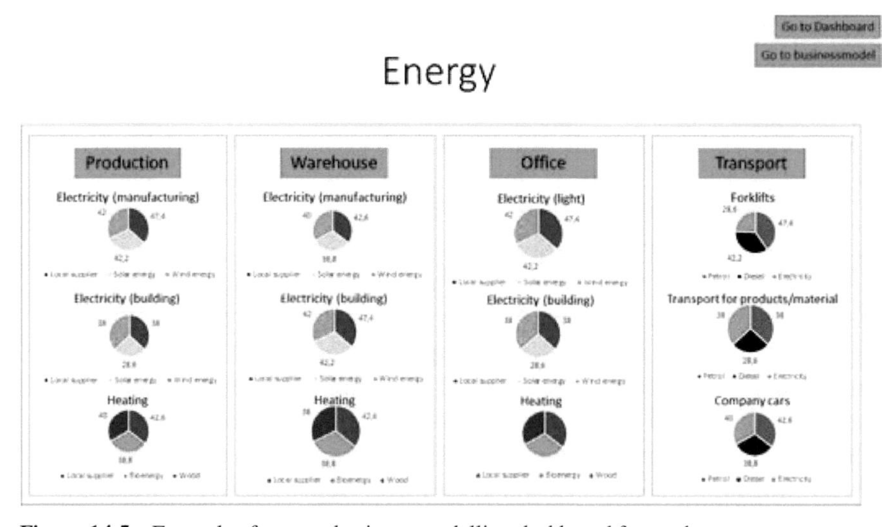

Figure 14.5 Example of a green business modelling dashboard focused on energy consumption – black and green energy competence BM dimension and technology component group measured on different business model areas.

and, at this moment, reduce waste and pollution – CO_2. This we marked with the pink arrow in Figure 14.5.

We found in our research that there is a clear overweight to focus on innovating, investing and implementing green on the BM competence component technology – mainly **reducing the use of energy, water, material, consumption and changing energy type to more green energy and/or greener material/resource types.** Changing into more **green production technologies** was also seen in some business cases – e.g., investment in solar energy systems for renewable energy production to supply the business with more green energy. Further, in some business cases, investment in convex energy production system was seen to reuse heat from building to supply heat to the business. **Waste reduction and pollution, especially CO_2 emissions,** were further seen as focus areas for GBMI. There was also a focus on **recycling waste** either internal the business by valuing other business models or by valuing network or customers' business models**.** These last-mentioned **Green Business Model Parameters (GBMP)** [2, 19] have been shown to reduce pollution CO2 positively. The complete and real-time calculation and measurement of resources used to recycle were not seen calculated in the businesses and cases of the research.

In some cases, the impact on CO_2 focusing on "greening" material, resources, waste and pollution reduction showed a much higher impact on CO_2 than reducing energy consumption and changing **from black to green energy**. However, the measurement of energy consumption is sketched in Figure 14.5. was observed in our investigation as more accessible and less critical to the business to innovate green and implement in the businesses than innovating on other GBM Parameters.

However, very few of the businesses investigated, as shown above in Figures 14.5 and 14.6, had an overview of the Competence BM dimensions of energy consumption related to different departments serving or engaging in the operation of other BM's. Further, the energy consumption was not split out to the different BM's and devices – production technologies - internal the business, so it would be possible to see energy consumption pr. BM and per devices. As a result, it was not possible in most cases to see and analyse the efficiency, Return of Investment (ROI) and progress of the investment in green production technology and relate these to different BM's. Technical: There is not much challenge to measure this as advanced software technologies are already available and offered by many technology providers but are exponentially proposed to the Green Business Model Ecosystem. Many businesses want to join the green transformation. If implemented, it could help businesses measure energy consumption and other green business model parameters even close to real-time [20, 21, 2]. We found examples of energy consumption tools and software that could measure daily, hourly and second, as seen in Figure 14.6.

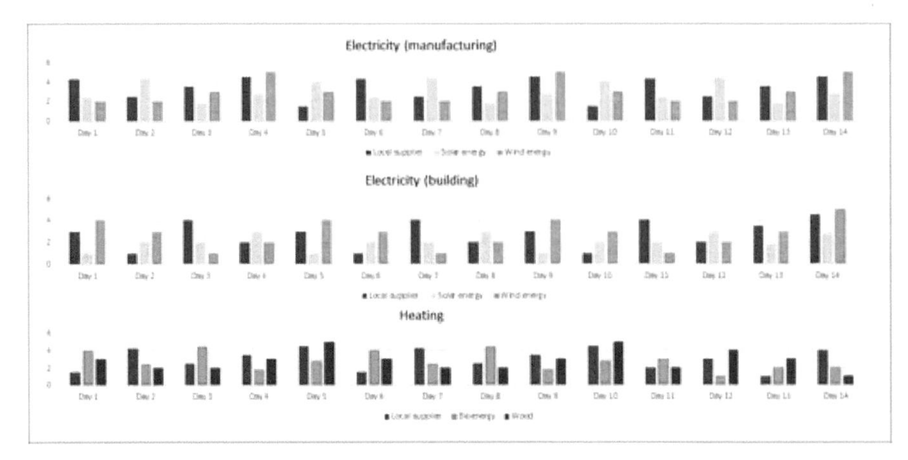

Figure 14.6 Example of a green business modelling dashboard focused on daily energy consumption and energy source types – separated on supplier, black and green energy in different business model areas.

It was observed that most accountant systems at the businesses were not prepared for splitting out these Green Parameter measurements on different BMs. Measuring the green parameters - material and resources consumption – use of black and green energy, use of water, use of production technology, waste, recycling of BM and waste, pollution and relating them to different BM's and economic terms is illustrated and sketched in Figure 14.7. as an example.

This should be possible with present technology to do. However, there are some challenges with existing technology.

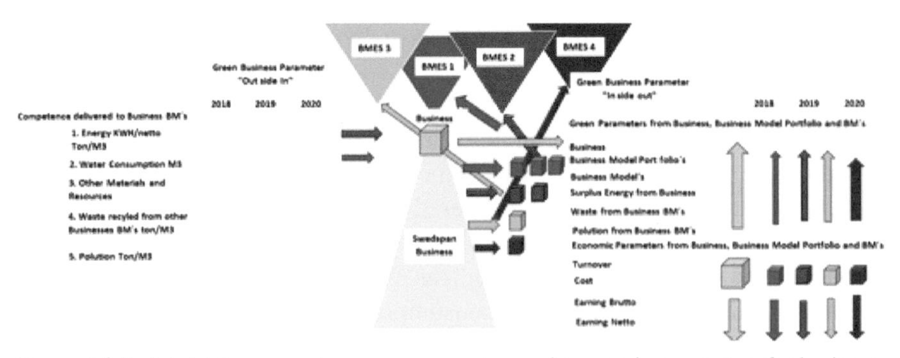

Figure 14.7 Model for measuring green parameters and economic parameters for business, business portfolio, business models.

14.3 Challenges Measuring Green on Technology Dimension in Business Models

Our research found some significant challenges in measuring green, particularly in the technology dimension of business models.

14.3.1 Product and Service Technology - Material and Resources

Concerning the supply of black and green energy to a business and its business models, there is much confusion, doubt, and actual paradox. Although several energy businesses sell and guarantee the supply of green energy to businesses today, the reality is that e.g. green electricity and green gas (Biogas) - supplied and sold as Green Energy to a business - often are and cannot be guaranteed as the supply is most often a mix of "black and green electricity" or "black and green gas". Even though businesses get subsidies for buying and consuming green energy, it is a fact that no technology yet and in reality can measure a mix of green and black energy as no technology and energy business yet can measure a mixture of black and green energy and thereby measure and guaranty pure, e.g. green electricity or green gas. Only when offered in separate energy systems can it be done, but these systems are few and expensive to establish and give other challenges related to the societies and business's energy systems. Therefore, it is impossible for businesses today to verify and claim that they are pure green on this green parameter and GBM dimension. The national energy system in Denmark and more countries confirms that pure green electricity and pure green gas were supplied to the energy system and region around the specific business – but there is no actual proof of this estimate and claim. It is still an estimate and not based on the actual energy supply measurement. Here we see great potential for 6G technology and research to change this situation so it will be possible to measure mixtures of energy in the future.

14.3.2 Volatile Green Energy Production and Demand

Today, most existing energy systems cannot adapt and store a surplus of renewable energy, e.g. electricity from windmills on a stormy day [23]. Further, the energy demand is extremely volatile. This means that Denmark's windmills were "taken out of the wind" – stopped and taken out of production estimated 50 days in 2020. It resulted in Danish wind energy production businesses paying German and Norwegian businesses money to eliminate the overproduction of green energy. But it also works opposite as, e.g. in 2018, German Windmill businesses paid Danish Windmill owners 190 mill DKK to stop the Danish Wind Energy production because the German Businesses would receive a fine from the German State if they quit their windmill production in Germany. In other words, a GBM with a negative earning is dependent on which GREEN BMES the businesses are related to.

The green energy BMES is well known as being volatile, and the GBM's around green energy are very difficult to manage both for energy businesses and society. Therefore, several businesses are working on this GBMI challenge [24] to find a better storage system, e.g. Power2X systems [25, 26, 27], able to use the surplus of wind energy to power energy heavy production storable energy. Innovation of new battery systems and new and better energy forecast systems are also invested in. Several businesses profit enormously by capitalising on these fluctuations, volatility and different regulations in different BMES in the production and supply of green energy. Advanced wireless technology can play a significant role in this GBMI and the development of Green Energy BMES.

14.3.3 Tracking And Measuring Green on Product-, Service-, Production- and Process Technology

A significant challenge in greening business models is to classify product-, service-, production and process technologies' impact on the environment previously, now, and in the future. We found that most GBMI investments and projects are today taking place on BM reconfiguration level and technology component as seen in Tables 1 and 2 in enclosure 1. Most GBMI investments are made on existing AS IS BMs, but it is last. This will be a significant focus in future GBM design because new BM will be pushed to be born green so that "repair" of AS IS BM will be diminished dramatically. The EU is already beginning to introduce restrictions on finance business's possible to borrow money for businesses' investments in, e.g. BMI, as these have to be proved green [28, 29].

In this process, the measurement of the technologies on the GBM parameters becomes more and more essential to secure high-quality estimation and verification of the effects and progress of greening of different GBMI projects and investments. We found that these measurement tools – LCA software [30] is still very complicated to use, expensive, lack user-friendliness, and are not implemented in most businesses, although they should be easy to implement as they are heavily needed. 6G technologies have a great potential to fill out this gap. In our Greenbizz project, we studied an auto recycling business Salling Autogenbrug. One of the most significant challenges to the business was classifying what kind of product technologies existed in the old car – e.g. a Peugeot 202. Although several databases were available with the specification of the product technologies, it was challenging and time-consuming to verify the product technologies embedded in the car. Therefore it was also challenging to take decisions on further recycling, and humans and human judgement made all classifications. This could, via 6G, AI and AR with preference, be carried out automatically, and as a result, the business model could be scaled up.

Another challenge was that different product technologies – materials – were mixed in the car's different parts – ferrous and non-ferrous metals were mixed in the motor and other parts of the car. Even non-ferrous metals – plastic parts, auto bags, and seats were created as mixed product technology components. This made it extremely difficult, expensive and sometimes unhealthy, and inefficient to recycle the parts. Advanced technologies could support and help in this case.

Many business models are embedded with a mix of product and service technologies and produced with production and process technologies. 6G and supporting advanced technologies could help verify these technologies in AS IS BM's and help secure that TO BE BM's are designed green.

14.3.4 Single and Multi Green Business Modelling

The majority of the businesses we studied were still limiting their GBMI to greening at a single business model innovation reconfiguration level as seen in Table 14.2 enclosure 1 – and still not in particular designing GBM's. Businesses were generally uncertain of GBMI investments, and they have not yet fully adapted the GBM approach and GBMI to the entire and higher levels of the company – and the new BMI area. In other words, GBMI seems still in the very early days - strategically, GBM and GBMI have not yet been embedded into critical and more significant parts/levels of the businesses. One primary concern in the businesses is the lack of understanding and verifications technologies to show the efficiency of GBM's and GBMI. 6G and – related advanced technologies could help businesses to overcome this challenge.

As can be seen in both Tables 14.1 and 14.2, enclosure one and as we observed in many of the GreenBizz projects [19], there is much more potential for greening the businesses BM's when focusing not just on the

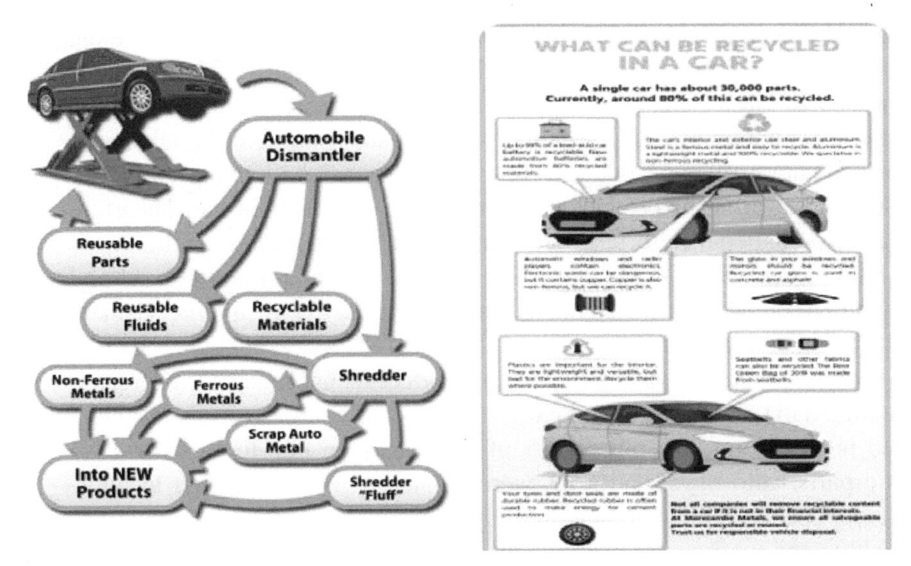

Figure 14.8 Automobile recycling [31][32].

business BM's competence dimensions – technology. We found in our investigation that the technical part of the competence dimension only releases some part of the green business model potential – but when extended to other components in the competence dimension – HR, Organisational systems, and culture more potential can be released. When the greening is extended to the value proposition and other parts of the BM, the value to the business, its surrounded BMES and society has a large impact. Very few of the businesses included in our research focused on these components and dimensions and could therefore not be classified as 100% green or pure green to the business, business portfolios, business models and business model dimensions. 6G Technology could initially help measure other BM dimensions and other BM Innovation levels. Hereby the greening process could be increased to more than just now. In most cases, the competence technology component.

14.3.5 Measuring Incremental, Radical and Disruptive Green Business Models

The degree of green of the GBM can become even " more green", or what we classify as radical and disruptive green, when GBM is related to the degree of "external impact" of greening. The GBM's impact on vertical, horizontal and/or any BMES can transform GBMs into radical and disruptive

GBMI. Then the greening process can potentially take an exponential speed. However, this cannot be realized without measuring the individual business's GBM's interaction with other BM's – the Green Multi Business Model approach. If GBM relations are established to other BM's outside the BM, green value propositions will flow out of the BM into other BM's and enable receiving and consuming **GBM Value Propositions** from other BM's and hereby make the BM greener and enable other BM's to become green. Radicality and complexity, however, today, in many businesses and GBM projects take the view- the point from the business side – inside out perspective [11]. However, if all BM dimensions are changed green, the BM is changed green – meaning it could be classified as being in the disruptive green zone – close to pure green. Dependent on its impact on vertical, horizontal and any BMES – it could be classified as **disruptive green**. The **green impact** on the **reach** axis in figure 14-4 – green to whom - becomes hereby related to defining and measuring how green the BM is and the degree of impact that the green BM and GMBMI in the business has – seen **"from outside – in"** and **"outside – out" viewpoint** and hereby the effect of green of GBM's on other BM's. It measures the change in existing BM's – "AS IS BM" and "TO BE BM" – related to the **green to the business**, **green to vertical BMES**, **green to horizontal BMES** and **green to any BMES** – **green to the world**.

However, this transformation and measurement of green will require substantial investments to transform, operate, measure, and validate the degree of green. 6G, blockchain and beyond technologies could support this measurement, but businesses and society always have the final strategic say on How Green they want their business and their related BM and GBM to be.

GBMI seems to be a long journey to businesses and society - with a beginning but probably a long end - if ever. There will always be components, dimensions, business models, business model portfolios, businesses, business model ecosystems and business model processes that can be innovated green or green [2]. Society and Businesses will continuously learn new technics and approaches to become greener and measure GBMI more precisely. GBMI is strongly linked to continuous improvement, continuous innovation and, not least, learning. **Learning** will always be the raw material for any Green Multi Business Model Innovation (GMBMI). Learning to become greener and building green competencies in technology, HR, organisational systems and culture of the BM's and businesses to be able to innovate BM's to become greener - we expect - will take businesses and society several years - through several iterations, "learning loops" including many "fails and bugs". However theoretical, it should be possible to measure GBM and GMBMI on a scale on all BM Dimensions and MBMI levels – not just the BM Competence and Value Proposition dimensions.

14.4 Towards Measurement of Green Business Models

As more and more businesses face significant strategic challenges related to choosing a strategy or road for the business green transformation, it becomes even more interesting to measure GBM and GBMI on a scale of "green parameters" together with the "economic parameters" of each BM's. The green parameters are indicated as discussed above in the BM competence technical area:

- BM Competence Product Technical area – Material and Resources (e.g. all types of materials and resources, Energy, Black and Green Energy, Water and even waste from other BM's if used as raw material in the particular BM.
- BM Value proposition area – Waste, Pollution (including, e.g. CO_2 and other types of pollution)

However, other BM Competence Technical areas, such as production and process technology, are just beginning to be measured and implemented as business measurement objectives. Typically this measurement is done by LCA measurement tools and calculation systems [33].

Table 14.2 in enclosure one shows, based on our preliminary investigations in 2020/2021 of 106 Danish Small and Medium size businesses, GBMI projects and processes [34] their **single green business model strategies**. A screening questionnaire screened all businesses, and all data from this screening were carefully analysed, grouped and scaled into different green categories and strategies, as seen in Tables 14.1 and 14.2. Our investigation showed that most GBMI strategies in these SMEs were focused on limited numbers of Business Model dimensions Table 14.1 enclosure one and mostly on Green Business Model reconfiguration Tables 14.1 and 14.2 enclosure 1. In GBMI research, we distinguish between Green Business Model Innovation (Green Business Model Design and Green Business Model Reconfiguration) [10, 25] and Green Business Model Development. Green Business model development focus on the implementation and introduction, growth, maturity and decline phase of the GBM with classical BM development tactical parameters, tools and actions. Green Business Model Development also covers the continuous improvement of GBMs. We found very few GBM cases that had entered the GBM development phase [8 incidences], often because the businesses could not find enough efficiency in investing green and did not have the numbers available on green parameters and economy measured on individual BM's – both on AS IS BM's and To BE BM's. Those we found that had invested in GBM's were primarily into the introduction phase and mostly focused on initial promotion and improvement of GBMs and focusing on branding their business or BM on being green. However, they did not have the exact numbers separated on the individual claimed GBM's. In Figure 14.6, we show an example on GBM's from one anonymous business called Swedspan, that had the numbers available on elected Green parameters and some BM's but only at an overall business level for economy.

In Figure 14.9, we divided the green parameters technology (material and resources) - coming into the business (Outside in) and those value propositions green parameters offered by the different business models – going out from the business (inside out) to different BMES. As Swedspan uses more energy types (e.g. Electricity, Gas, Diesel), the energy input to the business is calculated in KWH and net ton, as seen in Figure 14.9. Hereby we relate the Green parameters to the business model relation axiom quadrants 2 and 3 [11]. As Swedspan has no Green Parameter measurement e.g. LCA, device energy consumption, LCA on production and process technology and other GBM Dimensions and GBMI levels inside the business – yellow lower triangle in Figure 14.9 – it is not possible to comment on green parameter measurement inside the BM's related to quadrant 1 in the relations axiom. If 6G and more advanced technology had been embedded in the GBM, this would have been possible.

We also found very few GBM in the GBM Design phase [12 incidences], indicating that most GBMI is not focused on creating BM's as GBM when they are "born". None of the GBMI projects and GBM could be related to

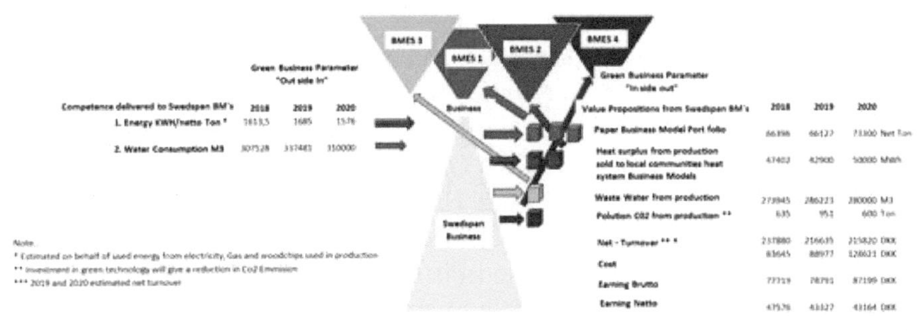

Figure 14.9 Green parameters "outside-in" and "inside out" together with Swedspan's total turnover, cost and earning measured at the business level.

radical or disruptive GBMI. We would expect that 6G and the introduction of more advanced ICT technologies would push to more radical and disruptive GBM and GBMI.

There is still not much validated research that documents the efficiency of these GBMI projects and developments. However, some of the GBMI and Development projects were carried out due to subsidies [35] and arguments of energy cost savings, but indications show that material and resource savings could give even higher cost savings and value adds to the businesses. In this case, advanced Life Cycle analysis (LCA) [32, 33] supported with 6G technologies of BM's becomes essential and a new measurement area for GBM's. LCA studies of a BM is a "snapshot in time" measure of burdens. The lower the burdens across the LCA of a BM or a network of BM's, the smaller the footprint – CO2 impact. An LCA measures burdens—what goes in (how much energy and raw materials it takes to make a BM or a network of BM) and what goes out (how much waste, water pollution, and emissions to air) across the BM's life cycle. 6G and related more advanced technologies are expected to make this measurement more detailed, visual, and real-time.

14.5 Discussion

Green Strategic Business Model Innovation is a very new strategic BMI game and tool that can potentially be used in many businesses globally. Our study showed how SME businesses strategically handled single GBMI from early 2020 – to early 2021. The research showed clearly that GBMI by SMEs is in the very beginning of "the roll-out phase" – an maybe in more cases not even with a very strategic focus – but more as a necessity or tricked by a "push" by society, politically [35, 36, 37, 38] and from different BMES's [39, 40, 41, 42]. The SME business focuses on competence BM dimension innovation and value proposition dimension at a BM component level – mostly changing product and production technology into greener technology. We expect that this will expand into other BM dimensions and BMI levels as soon as businesses can penetrate the potential of the BM competence technology area via 6G technology. Then GBMI will move into more "soft" and qualitative BMI areas and levels, e.g. Human Ressource, organisational systems, culture and the business level. This will mean that **businesses will begin to "be green"** – radical and disruptive green. This will mean that companies will demand and be able to check related business models, networks and even users and customers' innovation and action on the green agenda. We expect GBMI and GBMD to spread out to the entire business value network and BMES. We are still in the process of finding measurements tools for other BM Dimension areas than the competence dimension, but 6G technologies will enable us to move faster on this innovation and development track. We expect measuring these upcoming areas will be more complex and challenging and require other measurement competencies than purely technical and quantitative engineering skills. 6G technologies will be able to provide businesses with these today's lacking facilities and services.

14.6 Conclusions

The chapter reports on an investigation of How 6G and related advanced technologies can support measuring Green Business Models and Green Business Model Innovation in the future. The research was carried out in 2 Green Business Model Innovation projects – The danish ECSMV project – www.ecsmv.dk . and the Nordic EU Interreg Kask Greenbizz project – www.Greenbizz.eu. These projects were used to find the needs and challenges in green business model innovation to address where 6G and related advanced technologies can contribute and make the greening of business models more detailed, precise, visual, and real-time.

In the ECSMV Green Business Model (GBM) project, we investigated 106 SME businesses' green business model innovation (GBMI) projects. The purpose was to find out How green the businesses were and how to measure GBM and GBMI. The investigation showed that most businesses' strategic approach to GBM is at a single GBMI approach and a very low, narrow and bottom level of the GBMI potential. Most businesses investigated focused on the BM competence dimension, especially green technology – product technology and, to some extent, production technology related. Many other Business Model dimensions and components were left behind, and Green Business Model Innovation was not taking place at higher levels of Business Model Innovation. Green Business Model Innovation in the SME businesses was found mainly taking place at a very small and limited GBMI level. Businesses seemed not yet to have adapted the GBM approach and GBMI to the higher levels of their businesses. Very few businesses had spread out their GBMI to their entire value network – maybe because technologies that can support this demand are not available.

In the Greenbizz project, we were able to investigate and innovate deeper into different GBM and GBMI challenges and measurement tools with the aim shortly to get learning How to innovate a new range of GBM and a new GBMI dashboard that can measure How Green a BM, Business and BMES really are – and in realtime and hopefully on all Green Business Parameters. However, several challenges are present still, and there is some way to go. In our investigation, we found some of the challenges to be dealt with.

References

[1] Delmas, M. A., Burbano, V. C. (2011). The drivers of greenwashing. *California Management Review,* 54, 64–87.

[2] Lindgren P. (2021) A Scoping Review and Framework of Green Business Models Related to Future Wireless Technology: Bridging Green Business Models to Future Wireless Technology Journal of NBICT, Vol. 1, 1-34. doi: 10.13052/nbjict1902-097X.2020.016

[3] https://www.gbci.org/#certification

[4] https://geca.eco/?gclid=CjwKCAjwn6GGBhADEiwAruUcKuKSBlGMM9FY3ZyI56Oql_ufODJqvhA8VIHFEZ-dkfcYMpi2iH-lehoCzDQQAvD_BwE

[5] Lindgren, P. (2016) The business model ecosystem. Journal of Multi Business Model Innovation and Technology, 4, no. 2, 61-110.

[6] UN 17 goals - https://sdgs.un.org/goals

[7] https://www.c2ccertified.org/get-certified/product-certification

[8] Toxopeusa M.E., B.L.A de Koeijera, A.G.G.H. Meijb (2015) Cradle to Cradle: Effective Vision vs. Efficient Practice? May 2015 Procedia CIRP 29DOI: 10.1016/j.procir.2015.02.068 Elsevier Science Direct

[9] Bocken, N.M.P et all (2014) A literature and practice review to develop sustainable business model archetypes - Journal of Cleaner Production Volume 65, 15 February 2014, Pages 42-56

[10] Henrique Paulo Sergio Rufino and Ramjee Prasad (2021) 6G: The Road to the Future Wireless Technologies 2030 (River Publishers Series in Communications River Publishers ISBN 8770224390

[11] Lindgren, P., and Rasmussen, O. H. (2018) The business model relation axiom. In: River Publishers Series in Multi Business Model Innovation, Technologies and Sustainable Business (eds), The Multi Business Model Innovation Approach Part 1, pp 119-147.

[12] Marina P.P. Pieroni*, Tim C. McAloone, Daniela C.A. Pigosso (2017) *Comparison of scopes of CE-oriented and sustainability-oriented business model innovation approaches*. Technical University of Denmark (DTU), Department of Mechanical Engineering, Nils Koppels Alle 404 / Room 229, DK- 2800 Kgs, Lyngby, Denmark

[13] Bocken, Nancy; Schuit, Cheyenne; Kraaijenhagen, Christiaan (2018) Experimenting with a circular business model: Lessons from eight cases *Environmental Innovation and Societal Transitions* DOI: 10.1016/j.eist.2018.02.001 2018
[14] Sommer, A (2012) *Managing Green Business Model Transformations* Springer Verlag ISBN 978-3-642-28848-7
[15] Lindgren P., Susanne Durst, Xingyu Zhu, Gitte Kingo, Rita (2021) Green Business Models literature study – What did we learn? Journal of Nordic Baltic ICT River publishers In press
[16] Lorenzo, M., Christopher L. Tucci, Allan Afuah (2017) a critical assessment of business model research *Academy of Management Annals* 2017, Vol. 11, No. 1, 73–104. https://doi.org/10.5465/annals.2014.0072
[17] Lindgren P., (2018) *The Multi Business Model Innovation Approach - Part 1* River Publishers
[18] Lindgren, Peter et all (2021) "Green Multi Business Models" How to measure Green Business Models and Green Business Model Innovation?
[19] ESMV - projektet - https://ecsmv.dk/
[20] The Greenbizz project – EU Interreg. Kask - https://www.greenbizz.eu/
[21] Nordlys/Niras - https://norlys.dk/?gclid=CjwKCAjwvMqDBhB8EiwA2iSmPPB8_y15CYEZ1lszzNJahNaKi_0lKxmLdt2wpiMYnK3LMtDmXupbKBoCwRIQAvD_BwE
[22] Tracknamic - https://www.tracknamic.com/functionalities/#fuel-consumption-based-on-probe
[23] Kamstrup Metro - https://www.kamstrup.com/en-en
[24] https://www.wind-watch.org/news/2019/12/12/tyskere-betaler-for-at-fa-slukket-danske-vindmoller-nar-det-blaeser/
[25] Energinet Danmark 2021 -
[26] https://en.wikipedia.org/wiki/Power-to-X
[27] https://www.danskenergi.dk/nyheder/pressemeddelelse/17-virksomheder-power-to-x-partnerskab-med-dansk-energi
[28] https://www.mm.dk/misc/202024.pdf
[29] EU CoE Transmission - https://ec.europa.eu/info/business-economy-euro/banking-and-finance/sustainable-finance/eu-taxonomy-sustainable-activities_en
[30] EU taxonomy for sustainable activities https://ec.europa.eu/info/business-economy-euro/banking-and-finance/sustainable-finance/eu-taxonomy-sustainable-activities_en#what
[31] https://simapro.com/customers/

[32] https://www.todaystopquestions.com/how-does-car-recycling-work/

[33] https://www.pinterest.com/pin/811210951616390782/

[34] Guinée, Jeroen B. , Reinout Heijungs, Gjalt Huppes, Alessandra Zamagni, Paolo Masoni, Roberto Buonamici, Tomas Ekvall, and Tomas Rydberg Life Cycle Assessment: Past, Present, and Future Environ. Sci. Technol. 2011, 45, 1, 90–96 Publication Date:September 2, 2010 https://doi.org/10.1021/es101316v

[35] ESMV – projektet - https://ecsmv.dk/

[36] smv:digital https://smvdigital.dk/groen-tilskudspulje/tilskud-til-privat-raadgivning-paa-100-000-kr

[37] EU Green Deal Call https://ec.europa.eu/commission/presscorner/detail/en/ip_20_1669

[38] EU CoE Transmission – https://ec.europa.eu/info/business-economy-euro/banking-and-finance/sustainable-finance/eu-taxonomy-sustainable-activities_en

[39] EU taxonomy for sustainable activities https://ec.europa.eu/info/business-economy-euro/banking-and-finance/sustainable-finance/eu-taxonomy-sustainable-activities_en#what https://www.globaltimes.cn/content/1209841.shtml

[40] *Edie* https://www.edie.net/news/9/China-s-carbon-neutral-target-for-2060–What-does-it-mean-for-global-climate-action-/

[41] Ministry of Finance Green Climate Deal Danish Government and Industry (2020) https://fm.dk/nyheder/nyhedsarkiv/2020/juni/bred-klimaaftale-bringer-danmark-tilbage-i-den-groenne-foerertroeje/

[42] Greenest City Vancouver 2020 - https://vancouver.ca/files/cov/Greenest-city-action-plan.pdf

Enclosures

Enclosure: 1

Table 14.1 Green business model innovation related to green business model components and dimensions in 106 SME businesses 2020/2021.

Green Business Model Innovation related to Business Model Dimensions and components					
	Business Model Innovation				
Business Model Dimensions	**Business Model Design**		**Business Model Reconfiguration**		**Business Model Development**
Value Proposition - Products - Services - Processes	Green Value Proposition Innovation		Green Value Proposition Reconfiguration	19	Green Value Proposition Development
User and Customer - User - Customer	Green User and Customer Innovation	1	Green User and Customer Reconfiguration	31	Green User and Customer Development
Value Chain Function - Primary - Secondary	Green Value Chain Function Innovation		Green Value Chain Function Reconfiguration	37	Green Value Chain Function Development
Competence - Technologies - HR - Organisational Systems - Culture	Green Competence Innovation	11	Green Competence Reconfiguration	104	Green Competence Development

(Continued)

Table 14.1 Continued.

⬡ **Network** - Physical ⬡ - Digital ⬢ - Virtual ⬢	Green Network Innovation	⬢	Green Network reconfiguration	⬡ 30	⬢ Green Network Development
⬡ **Value Formula** - Monetary ⬢ - Other values ⬡	Green Value Formula Innovation	⬢	Green Value Formula Reconfiguration	⬡ 18	⬢ Green Network Development
⬢ **Relations** - Tangible ⬢ - Intangible ⬢	Green Relation Innovation	⬢	Green Relation Reconfiguration	⬡ 1	⬢ Green Relation Development
Total incidents in 106 businesses		**12**		**240**	**8**

⬢ = **No GBMI activity and investment**
⬡ = **Some GBMI activity and investment**

Table 14.2 Single green business model innovation related to business model innovation levels [29].

Single Green Business Model Innovation related to Business Model Innovation Levels					
	Green Business Model Innovation				
Business Model Innovation Levels	⬡ **Business Model Design**		**Business Model Reconfiguration**		**Green Business Model Development**
⬡ **Business Model Component –**	Green Business Model Component Innovation	⬡ 11	Green Business Model Component Reconfiguration	⬡ 104	⬢ Green Business Model Component Development

Table 14.2 Continued.

Business Model Dimension –	Green Business Model Dimension Innovation		Green Business Model Dimension Reconfiguration	1	Green Business Model Dimension Development
Single Business Model	Green Business Model Innovation	2	Green Business Model Reconfiguration	46	Green Business Model Development
Business Model Portfolio	Green Business Model Portfolio Innovation		Green Business Model Portfolio Reconfiguration	3	Green Business Model Portfolio Development
Business	Green Business Innovation	3	Green Business Reconfiguration	43	Green Business Development
Business Model Ecosystem	Green Business Model Ecosystem Innovation	1	Green Business Model Ecosystem Reconfiguration	42	Green Business Model Ecosystem Development
Business Model Innovation Process	Green Business Model Process Innovation		Green Business Model Process Reconfiguration	2	Green Business Model Process Development
Total incidents in 106 businesses		**241**		**17**	8

Biography

Peter Lindgren holds a full Professorship in Multi business model and Technology innovation at Aarhus University, Denmark – Business development and technology innovation and is Vice President of CTIF Global Capsule (CGC). He is Director of CTIF Global Capsule/MBIT Research Center at Aarhus University–Business Development and Technology and is a member of the Research Committee at Aarhus University – BSS. He has researched and worked with network-based high-speed innovation since 2000. He has been head of Studies for Master in Engineering – Business Development and Technology at Aarhus University from 2014–2016 and a member of the management group at Aarhus University Btech 2014–2018. He has been a researcher at Politecnico di Milano in Italy (2002/03), Stanford University, USA (2010/11), University Tor Vergata, Italy (2016/2017) and has in the period 2007–2011. He has been the founder and Center Manager of International Center for Innovation at www.ici.aau.dk at Aalborg University, founder of the MBIT research group and lab http://btech.au.dk/forskning/mbit/ and is cofounder of CTIF Global Capsule www.ctifglobalcapsule.org. He has worked as a researcher in much different multi-business model and technology innovations projects and knowledge networks among others E100 http://www.entovation.com/kleadmap/ Stanford University project Peace Innovation Lab http://captology.stanford.edu/projects/peace-innovation.html. The Nordic Women in the business project–www.womeninbusiness.dk/ The Center for TeleInFrastruktur (CTIF), FP7 project about "multi busi- ness model innovation in the clouds" –www.Neffics.eu, EU Kask project – www.Biogas2020.se, Central Project, Motor5G, Recombine, Greenbizz. He is cofounder of five startup businesses amongst others–www.thebeebusiness.com, www.thedigibusiness.com, www.vdmbee.com. He is author of several articles and books about multi-business model innovation in networks and Emerging Business Models. He has an entrepreneurial and interdisciplinary approach to research. His research interests are multi business model and technology innovation in interdisciplinary networks, multi business model typologies, sensing-, persuasive- and virtual- business models.

15

An Introduction to Privacy Preservation in 6G

Aaloka Anant[1] and Ramjee Prasad[2]

CGC - Aarhus University, Birk Centerpark 10, DK-7400 Herning Denmark
E-mail: aaloka@anantprayas.org; ramjee@btech.au.dk

Quality real-world data is the foundation on which valuable information and intelligent insights can be constructed. Therefore, the availability of real-world data is imperative for developing solutions that rely on intelligent insights based on historical or contemporary data. Individuals generate data and always contain information that, which if misused, may cause harm to the individual. A significant development in the current times is the development of applications based on machine learning and artificial intelligence, which need lots of data. Data is generated and stored in various forms by government bodies, corporates and individuals. Data may contain either public or personal information. With technology advancements like 6G, sharing information between different parties becomes more accessible, real-time, and voluminous. The challenge to preserve the individual's privacy and retain the utility of data becomes more prominent. It can help governments and organizations better serve humankind by evolving new approaches and innovations. At the same time, it would create new challenges for protecting the privacy of individuals. Privacy preservation of data can help the two ends meet by keeping the utility of data while protecting the privacy of individual data subjects.

15.1 Introduction

Any data exchange or information processing must adhere to privacy preservation regulations to respect the privacy of persons whose data is concerned. The data is usually stored with consent from the persons for specified purposes only, primarily guided by terms of use in a contract. The simplest way to preserve the privacy of individuals is to use anonymization, but true anonymization depletes the information content of data and compromises the utility of data. It's a challenge to ensure that data-carrying personal information is anonymized without losing its utility for a given purpose. Organization managing the data has to ensure that any shared data is fully anonymized but still has the utility for usages such as collaboration in clinical trials, financial fraud detection and training of machine learning software for different purposes.

This chapter presents the topic of privacy preservation with details on the legal, technical and business aspects. Few privacy preservation methods like Homomorphic Encryption and Differential privacy are briefly described to explain the techniques with examples. The particular focus highlights the importance of privacy in the emerging 6G world. Several techniques for data anonymization and processes for privacy preservation are outlined through the implementation of these techniques that are not covered in this chapter.

The chapter is divided into five parts, followed by a Summary at the end. The first part is an introduction followed by a discussion of the global laws and the rationale. This is followed by the section on significant techniques for privacy preservation in use today with industry acceptance, highlighting their shortcomings and advantages. Next section deals with the significant challenges in this area. A brief view on Named Entity Recognition (NER) is presented as to how it can help in the privacy preservation of data. The following section relates the enhanced challenges and relevance of privacy preservation topic with the rise of 6G technology-based applications and highlights that privacy topic can be a "do or die" subject for achieving a higher adoption among individuals and industry.

15.2 Privacy Laws and Global Awareness

Across the globe, there are different privacy laws. As of May 2021, there are no global standards for privacy preservation. The background and culture have importance in the rules formulated in a country for privacy preservation. The Universal Declaration of Human Rights in the United Nations was adopted in 1948, "No one shall be subjected to arbitrary interference with his privacy, family, home or correspondence, nor attacks upon his honour and reputation. Everyone has the right to protect the law against such interference or attacks". Countries with more freedom for individuals have more substantial protection than countries where the government has more dominating influence over its citizens. There are several countries where there are no laws on data privacy. However, most countries are gradually adopting and enacting laws on data privacy preservation. In the last ten years, more than 60 countries have enacted laws on data privacy protection [1]. The topic is gaining importance with digitization touching every aspect of human life. Information security is becoming more critical than ever. Many countries have strict data retention laws, even if they do not have data privacy laws. Data retention laws restrict the digital data from being carried out of the physical geography of a country. Privacy preservation laws, wherever they exist, also have provisions for dealing with data retention.

There are two primary principles in driving the legal structures across the globe. One is to empower the individuals and leave the decision to the individual for sharing their data with or without privacy preservation as the individual seems fit. The second is to empower the system and governance structure to protect the individual, considering they are vulnerable. The authorities in America mainly seem to be relying on the first principle to empower the individual with information and let them choose what to do with their data. At the same time, authorities in Europe seem to be more protective of laws like GDPR. GDPR has emerged as the most impactful of the privacy laws globally, with the default design being to protect the most vulnerable individual. Contrary to the two principles, there are some regimes like China, where the government has much higher control over data and data security for all the data subjects [2]. Cultural differences and regimes in a country affect the individual country and impact the globe [3], with technology exports seeding surveillance societies globally.

Figure 15.1 shows a few prevalent privacy laws on a political world map. These are mainly GDPR – General Data Protection Law enacted in major countries across the European Union; IPDP represents the Indian Personal Data Protection Bill; CCPA – represents the California Consumer Privacy Protection Act; HIPAA – represents the Health Insurance Portability and Accountability Act; LGPD – represents the Brazilian General Data Protection Law; CPB – represents the Chile Privacy Bill Initiative; NZPB – represents the New Zealand Privacy Bill. These all are not enacted laws but a good representation of the laws across different parts of the globe enacted or planned to be enacted by 2021.

The privacy laws represented in Figure 15.1 also come with a financial penalty to the defaulter, prescribed in the law with certain limits. In general, even in the absence of any specific law on privacy preservation, a country's

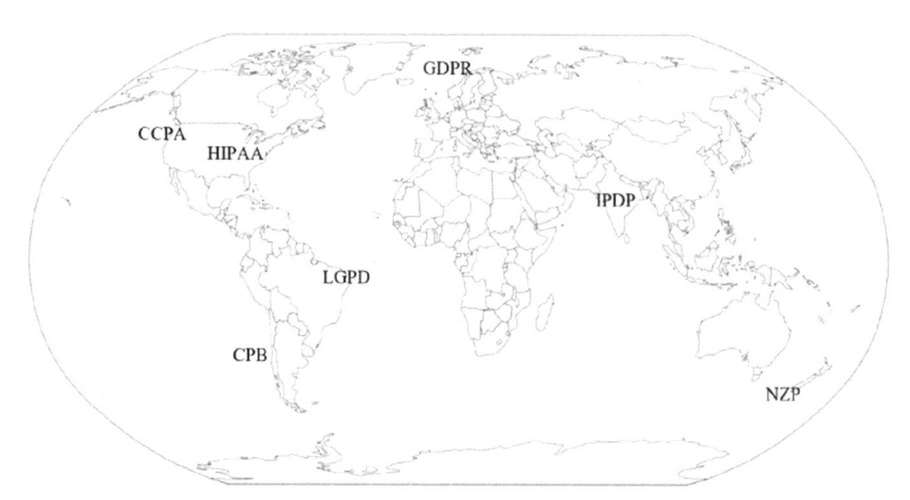

Figure 15.1 Some of the privacy related laws across the globe.

constitution provides rights to its citizens to protect their privacy. However, laws make it easy to enforce such rights of individuals rather than individuals going to the Judiciary to claim their rights. Once a law is enacted like GDPR, there are specific authorities and provisions set out, which must be adhered to by anyone capturing personal information as a proactive step. Any company doing business or activity in the areas governed by the law must report mandatorily to the authorities. Most of these laws, like GDPR, mandate an accountable role inside a company or any organization to ensure that the company takes privacy protection-related measures.

Some of the aspects covered in the law are outlined below:

- Lawful grounds for data processing: The organization collecting data must ensure that the data is being processed only for specific reasons as permitted by the law.
- Consent for processing data: The company or the organization storing and analyzing personal information must have valid consent from the data subject whose data is being stored and processed.
- Data Subject Rights: The individual or the entity whose data is stored or processed must have the required information and rights. For example, to know the purposes for storing the data. In most laws, the data subject must have the right to know what an organization stores data about the individual. The data subject must be provided with the option to revoke consent and must be provided with the knowledge on how to withdraw

consent. Laws like GDPR give a timeline of one-month maximum for the data subject rights to be fulfilled. Laws like LGPD keep this timeline to 15 days.

- Data breach notification: The data subject must be notified if there is any breach of data known to the party storing the data, and remediation action must be taken.
- Data Security: Data stored must be secure and protected for access. They are also protected against breaches and hacks by using appropriate technology for encryption, pseudonymization, or anonymization.
- Accountability and governance of privacy protection processes: The company or organization storing and processing personal data must have a defined process and designated accountable people to govern the process.
- Data Transfers restrictions: Data can be stored or processed only within certain physical boundaries specified in the law. Also, companies or organizations dealing with other companies as partners in processing the data have to ensure that the data, if in any way leaves the specified region, the personal data is adequately protected for access. Also, the data subject's rights can be successfully exercised across the data processing process. In many cases, there are restrictions, and no data may be exchanged outside of the defined boundaries by law.
- Automated decision making: Some laws restrict companies from using automated decision-making based on the personal data of a data subject, except if the subject has explicitly agreed to the same.
- Privacy by design and default: The organization storing and processing personal information must design any step related to data processing with full consideration of privacy. In addition, there must be default options to provide the best privacy preservation, with data subjects having the right to choose to allow further data processing.

15.3 Privacy Preservation Techniques: A Preview

In this section, different methods in use are outlined. These methods are mainly used for privacy preservation. Several other techniques are used for data security, which is not covered in this section. Also, this section does not cover the privacy preservation topic and state of the art in privacy preservation of Enterprise data [4]. A few popular methods used for data privacy preservation are presented below.

15.3.1 Anonymized Data Using Homomorphic Encryption

Encryption is a widely accepted method for data security and communication between multiple parties and hybrid landscapes (cloud and non-cloud environments). Encryption for data transmission keeps the data safe during transit as the data cannot be decrypted (converted back to original data) without the required key. Technically, it can be possible to decrypt an encrypted text using sizeable computational power, though the intended recipient can easily interpret the original data using the key.

The use of encryption with keys to scramble data for privacy is not considered safe. The data subject can be identified back using keys available or stolen. This makes key-based encryption only a data security method and not an anonymization method.

Homomorphic encryption is needed to operate on encrypted data for privacy preservation. Computations performed on anonymised data using homomorphic encryption can result in the same result as computations on original data. The computation mechanisms and the available technology to deal with the same are expensive and beyond reach for dealing with enormous data sets. One possible and simplistic definition of Homomorphic encryption is below [5].

An encryption algorithm *E* is called homomorphic over an operation *f* on two messages, *m1* and *m2* if it supports the following equation

$$f(E(m1), E(m2)) = \mathrm{E}(f(m1, m2)), \tag{15.1}$$

where *m1*, *m2* are a subset of all the possible messages set *M*

An example in Figure 15.2 below explains the equation above with a simple demonstration. The two messages are numbers *m1* being ten and *m2* being 20. The encryption algorithm converts them to 100 and 200, respectively. The results obtained from operating on 100 + 200 would be the same as after encryption of the sum of *m1* and *m2*. Meaning 30, encrypted using the same algorithm *E* would become 300, the same as the sum of encrypted values 100 and 200. Such an arrangement makes operations on the data easy to move outside of the organization holding individuals' private data, preserving the data subject's privacy.

Figure 15.2 presents the example for numeric data, though homomorphic encryption can also be done for alphanumeric data [6]. Homomorphic encryption gives an excellent use case for an organization to store and process data anonymized using homomorphic encryption. No original data needs to be stored. Operations on the encrypted data can be used to perform calculations,

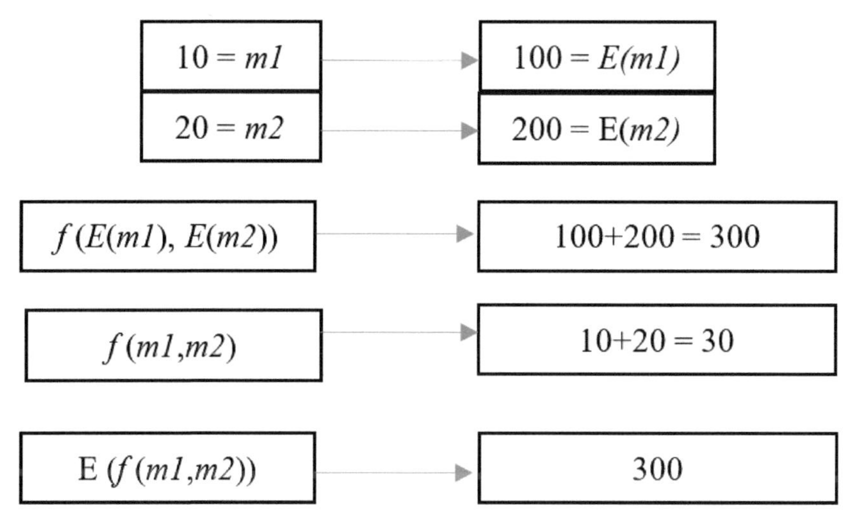

Figure 15.2 Demonstration of homomorphic encryption using a simple example.

and results can be provided for interpretation. In case of a breach of data or any other accidents, which lead to exposure of the homomorphically encrypted personal data, the data loss would not be significant and would not lead back to the data subject. The example presented may reflect simplicity. The computing power needed to achieve homomorphic encryption is enormous and has kept this practice expensive enough to be commonplace in mainstream applications. This method is used only for a few cases, where very high compute power is available for performing encryption and for processing significant functions on the encrypted data. There are several public libraries for using the fully homomorphic encryption provided by IBM and other leading software companies like Microsoft in the open-source space.

15.3.2 Anonymization using Differential Privacy

Methods like differential privacy add noise to the dataset to make it anonymous [7]. It can be explained in simple terms with an example of smokers in a group of people if there is a dataset with n data subjects, where the information if they smoke or not is captured. The dataset is anonymized using differential privacy. After the anonymization, the individuals would not be identifiable, though the dataset would still represent the same statistical distribution of smokers and non-smokers in the group. If one of the group members requests to remove their data from the anonymized dataset, the overall distribution of the anonymized dataset would remain the same. There would be no way to identify who is the member in the group whose data is represented.

There is a balance between utility and anonymity which must be maintained using parameters epsilon and sensitivity in case of differential privacy. Paper [8] presents the most advanced method, though it still has shortcomings and issues where these methods may not have the data utility retained. One of the definitions for differential privacy is given below [9].

Differential privacy is a mechanism M with parameters $(\varepsilon,\ \delta)$, where $\varepsilon = 0$ and $\delta = 0$ if for all neighbouring databases D0 and D1, databases differing in only one record, and for all sets S$\subseteq$ [M], where [M] is the range of M, the following in-equation holds:

$$\mathrm{Pr}[\mathrm{M(D0)} \in \mathrm{S}] = \exp(\varepsilon) * \mathrm{Pr}[\mathrm{M(D1)} \in \mathrm{S}] + \delta. \tag{15.2}$$

If a mechanism satisfies pure differential privacy, it satisfies (ε,0) differential privacy. And it satisfies approximate differential privacy (ADP) if it satisfies (ε, δ) differential privacy for $\delta > 0$. Differential privacy is quite helpful for preserving privacy, remarkably since it is preserved under any post-processing step. No matter which functions are applied to the output of a differentially private mechanism, the results do not reduce the privacy guarantee. In other representation, if M is $(\varepsilon,\ \delta)$ differentially private and T is any randomized algorithm, then T (M), defined as T(M)(x) = T(M(x)), is also (ε, δ) differentially private. ε – Epsilon relates to the probability of an individual record contributing to the overall result though δ – sensitivity steers the magnitude of the noise.

Differential privacy is the most widely accepted mechanism for privacy preservation. However, the utility of data varies with different implementations. There has been widespread use of this mechanism in the data for medical research, statistical calculations like smart meters and others. Google claims to use differential privacy [10] for data stored for location from Google Maps users. Apple claims to protect the privacy of its users by using differential privacy in its devices while processing user inputs for spelling suggestions in its product announcement since the launch of the iPhone 10.

15.3.3 Privacy Preservation Using Pseudonymization Methods

Pseudonymization is accepted as a valid method for privacy preservation in different countries. With GDPR in Europe, there has been broad discussion on this topic and guidelines published by various countries on which circumstances and how the method of pseudonymization can be considered a valid form of anonymization [11].

Pseudonymization presents a unique method in practice since time is unknown to replace an original value with another value in a dataset. The replacements are made consistently in the dataset to have the possibility to revert the original data if needed in some cases. The replaced values may have meaning or may not have any purpose and only represent a tag, for example. The practice of pseudonymization retains a dictionary or mapping of the original value and the replaced value. This practice makes this method a high-risk method. The pseudonymized data can be converted to the original data [12].

Table 15.1 below demonstrates an example of the pseudonymized dataset. The first record represents original data, and the next row represents pseudonymized data. The pseudonymization has been shown with nonrealistic values. For example, the name John is replaced with J123. Though using pseudonymization, it may be replaced with meaningful values like another name Jill. Depending upon the usage of data post privacy preservation, the dictionary or mapping between original and replaced values can be prepared. Only Name, Employee ID and Designation have been pseudonymized in this example table below.

Pseudonymization standalone is a weak method for privacy preservation and not even an accepted method for anonymization. When combined with other methods, it presents a promising approach for preserving privacy, though it does not prove to be an anonymization method as the existence of a dictionary always leaves the vulnerability for the original data. This method enhances data security in an organization and permits several use cases by granting access to such data to a broader audience. This method remains a popular practise due to its simplicity and implementation.

Large data sets can be processed quickly with machine learning and massive computation power. Hence, methods like pseudonymization that retain

Table 15.1 Example of an original data and pseudonymized data.

Original/ Pseudo	Name	Employee ID	Address	Age	Designation	Salary
Original	John	E0001	Port St., NY	25	User	50000
Pseudo	J123	PS01	Port St., NY	25	L1	50000
Original	Jacob	E0010	Dublin, IE	35	Lead	80000
Pseudo	J125	PS02	Dublin, IE	35	L2	80000

the original structure of data and a one-to-one mapping between original and pseudonymized records prove pretty ineffective for complete privacy preservation.

15.3.4 Combination of Multiple Methods of Anonymization

Increasingly accepted is the practice of a combination of more than one anonymisation method to protect the private data of data subjects. This stream of research focuses on innovations like federated learning, multi-party deep learning [13] and others and combines methods like differential privacy with homomorphic encryption to achieve privacy preservation. Considering the increasing requirements of data and increasing use of technology to touch different aspects of human life and Enterprise businesses, the practice of combining methods for privacy preservation seems to be the only hope.

In addition to a combination of privacy preservation methods in the same dataset, there are also practices evolving to apply different methods at different storage locations of the data and consume the data. For example, evolving practices with edge computing and massive computing power in edge devices like mobile phones enable the use of federated learning and hence promote anonymization of data in the source itself before using machine learning algorithms. Also, extracting information from the data where it exists and deriving intelligent insights help protect the individual's privacy as the individual's data is not duplicated, eliminating the need for privacy preservation in many cases.

Table 15.2 below presents the same table as in example 1, but with more methods used on the same record. The dataset presented here is only for simplicity and does not represent a real-world example as having such a small dataset as the subject of privacy preservation is unseen. Large datasets are the subject in general and present more complex challenges, as covered in the later sections of this chapter. In Table 15.2, Address has also been changed from the original. Age has been anonymized. The age, in this case,

Table 15.2 Example of an original data and anonymized data.

Original/ Anonymized	Name	Employee ID	Address	Age	Designation	Salary
Original	Johny	E0001	Port St., NY	25	User	40000
Anon	N123	PE01	Street A, PA	20	Employee	50000
Original	Jill	E0010	Dublin, IE	35	Lead	70000
Anon	N125	PE02	Street Z, IE	30	Employee	60000

can be anonymized using differential privacy principles. Salary can also be anonymized with the addition of noise. The designation has been generalized in the below example, presenting another form of privacy preservation method called K-anonymity, which is not covered in this chapter.

15.4 Challenges in Privacy Preservation

Privacy preservation topic is evolving rapidly, with new challenges surfacing with new technology and new approaches to doing the same thing. Personally Identifiable Information, generally called PII, is one of the main concepts in privacy regulations. It defines privacy statutes and regulations [14]. Most privacy regulations focus on collecting, using, and disclosing PII and leave non-PII unregulated. One of the biggest challenges is in identifying the PII. Personal data can be found in places where it may never be expected, for example, comments on a delivery note of a postal package, which says Happy Anniversary with the name of the couple, address, and date. This information alone, which the sender of the postal packet very casually gives to ensure delivery, passes several hands and several systems in digital format. The message may be stored in free text. Free text in a digital notation can store any information. Free text is not supposed or expected to have personal information, different from a field, which is supposed to store personal information like the name or address of a person. If the data is to be anonymized, the fields like Name, and Address, are things that readily come to attention. In contrast, something like free text does not come to notice unless the contents of free text can be analyzed to find that it contains PII.

Named Entity Recognition (NER) is widely used in machine learning practice to identify an entity type in a free text. Such NER performed for free text intending to identify personal information serves as a vital input for privacy preservation activity where free text is relevant. The field of machine learning with Natural Language processing has made several advances in identifying the PII as NER. Microsoft with an open-source project, Presidio, Google with DLP and Spacey are some of the most advanced software packages in NER. Though NER is designed not only for PII identification, it serves a great purpose in finding PII.

Any information which can reveal a unique individual may be considered a PII. Consider the statement, “the first man to step his foot on the moon won a lottery for 1 Million Euros”. This statement does not sound to have any personal information. Though the report in the statement very clearly identifies a unique person. Neil Armstrong may or may not like to reveal the information that they have won a lottery, but the information presented in the above statement would identify him. To retain the privacy of the lottery winner, the statement may be written using other attributes of the person, “a resident of Unites States of America won a lottery for 1 Million Euros”. Any single or combination of PII attributes of a person may reveal the individual uniquely. Some of the Personally Identifiable information attributes in the GDPR, CCPA and other laws are outlined below. Though it may never be exhaustive, such a list of attributes is for guidance.

- Name of person
- Home address
- Email address
- Telephone / mobile number
- Personal Identification numbers like Passport number, driving license etc.
- Social Security Number or another identification number in different countries
- Income of person
- Cultural profile of a person
- Data held by a doctor or hospital (which uniquely identifies a person for health purposes)
- Internet protocol (IP) address
- Racial or ethnic origin
- Sexual orientation
- Political opinions
- Religious or philosophical beliefs
- Memberships of different bodies like trade unions etc.
- Personal data related to criminal convictions and offences
- Hometown / birthplace
- Relationship status
- Date of Birth
- Job details

15.5 Enhanced Challenges with 6G

New technologies bring new opportunities, a new way of doing things and new challenges to solve. People would never have thought of talking far distances until the telephone was invented. More an invention simplifies our lives. More is the adoption of such an invention. What people can do using 6G would lead to an increase in the adoption of 6G. Emerging applications such as the Internet of Everything, Holographic Telepresence, collaborative robots, and space and deep-sea tourism highlight the limitations of existing fifth-generation (5G) mobile networks.

6G technologies make possible low latency applications like holographic telepresence [15]. A person can be seen standing next to another using such technology, while these two are physically standing miles apart. It would need massive data sets to be captured using devices like cameras, stored across the network, processed and may be retained in digital form for future usage. Data is stored, which can be used to construct a complete person's identity using the face, eyeballs(retina), fingerprints, to name some of the things that can uniquely identify a person. Such information can be highly sensitive for the person as it can be used/misused if fallen into the wrong hands to impersonate the individual. Many new attributes need to be added to the list of PII, as discussed in the previous section of this chapter. These PII would be particularly challenging and may need more protection than anonymization to safeguard the data subject and pose a different challenge to privacy preservation techniques.

The applications made possible by the advancements in communication technology are numerous. For simplicity, we have outlined some of the applications which became possibly and easily consumable with the different generations of mobile communication technology in Table 15.3 below.

Shifting voice calls to VOIP using apps like WhatsApp transformed the way the telecom industry worked for several decades, giving way to different business models for telecommunication companies. Similarly, increasing experiments to make driverless cars and driverless trucks present a different economic model for societies. With technologies like Telepresence becoming a reality, we will witness a sea change in how communications are perceived in the age of 6G.

The topic can be explained in more detail if we look deeper into what kind of PII and how much it is getting converted into the digital medium,

Table 15.3 Applications that became popular with technology adoption.

Usage/ area	3G	4G	5G	6G
Personal Communication	Voice calls	Video calls		Holographic Telepresence
IoT and device communications	Location parameters	Automated driving	Driverless cars	Automated Vehicles
Human-Computer Interfaces		Virtual Reality	Augmented motion	Human Brain and computer interface

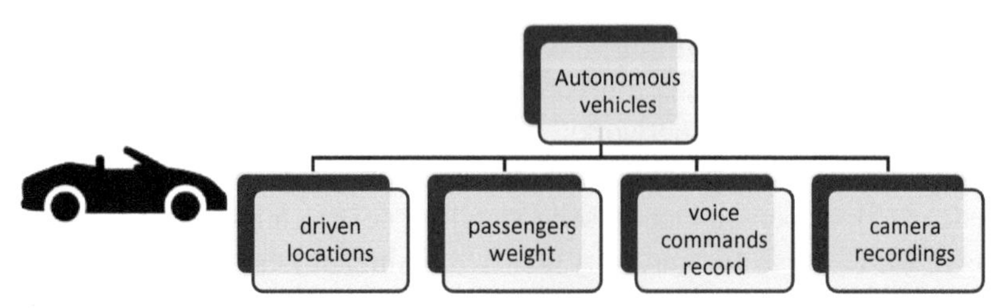

Figure 15.3 Increased PIIs storage, processing, transmission and duplication.

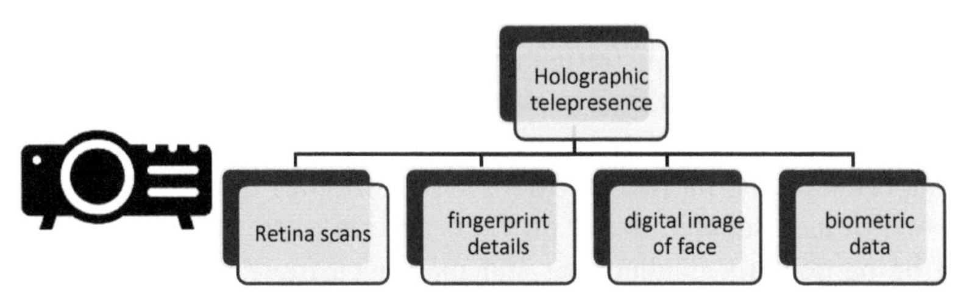

Figure 15.4 Increased PIIs storage, processing, transmission and duplication.

posing new demands for privacy preservation. Figure 15.3 outlines some PII that need to be digitized and relevant for privacy preservation. Autonomous vehicles [16] capture much information on the passengers. The cars also run on voice commands and record voice commands, including driver preferences on music, driving speed etc. Such vehicles are fitted with numerous IoT sensors to measure passengers' weight (estimating their age, need for the car seat, etc.) and equipped with multiple cameras to record video footage, including faces and several PIIs of passengers and passersby.

Another example is Holographic telepresence [17], where many more PIIs like retina images, details of fingers, digital images of the face and other biometric data may be easy to derive from the minute details captured in such a session. It's presented in Figure 15.4 below. Such a communication claims to establish eye contact with the participants of a meeting, and hence the level of details needed on a personal level would create a lot of PII data, which must be protected, and privacy preserved if it is to be stored used further.

15.6 Conclusions

This chapter gives a brief view into the expanding topic of privacy preservation for data exchange. Enormous personal information needs security and new ways of storage, exchange, and utilisation to prevent any harm to the data subject. Several data sanitisation methods are in place to preserve data privacy and retain utility. However, these methods need adoption at different stages of data processing. Laws across the globe are pushing organizations to adopt measures to safeguard the privacy of data subjects. Cultural awareness and government structures across the globe push adoption of privacy preservation technologies. With advancements in technology and applications made possible by the attributes of 6G communication networks, privacy has become a more central topic. If not taken care of, advancements may lead to chaos with misuse of technology, making the data subjects vulnerable to manipulation by profit-making organizations. Such weakness in the development and adoption of privacy preservation technologies can also slow down the adoption of applications enabled by 6G, hence slowing down the adoption of 6G itself. If taken care of, privacy preservation can open up the possibility of many new applications, with the vast data processing capability and other innovations with 6G.

References

[1] G. Greenleaf and B. Cottier, "2020 Ends a Decade of 62 New Data Privacy Laws," Social Science Research Network, Rochester, NY, SSRN Scholarly Paper ID 3572611, Jan. 2020. Accessed: May 11, 2021. [Online]. Available: https://papers.ssrn.com/abstract=3572611

[2] X. Qiang, "The Road to Digital Unfreedom: President Xi's Surveillance State," *J. Democr.*, vol. 30, no. 1, pp. 53–67, Jan. 2019, doi: 10.1353/jod.2019.0004.

[3] S. Romaniuk and T. Burgers, "How China's AI Technology Exports Are Seeding Surveillance Societies Globally," *The Diplomat*, vol. 18, 2018.

[4] A. Anant and R. Prasad, "State-of-the-art in Privacy Preservation for Enterprise Data," in *2020 23rd International Symposium on Wireless Personal Multimedia Communications (WPMC)*, Oct. 2020, pp. 1–6. doi: 10.1109/WPMC50192.2020.9309459.

[5] A. Acar, H. Aksu, A. S. Uluagac, and M. Conti, "A Survey on Homomorphic Encryption Schemes: Theory and Implementation," *ACM Comput. Surv.*, vol. 51, no. 4, pp. 1–35, Sep. 2018, doi: 10.1145/3214303.

[6] M. Ogburn, C. Turner, and P. Dahal, "Homomorphic Encryption," *Procedia Comput. Sci.*, vol. 20, pp. 502–509, Jan. 2013, doi: 10.1016/j.procs.2013.09.310.

[7] C. Dwork, F. McSherry, K. Nissim, and A. Smith, "Calibrating Noise to Sensitivity in Private Data Analysis," in *Theory of Cryptography*, Berlin, Heidelberg, 2006, pp. 265–284. doi: 10.1007/11681878_14.

[8] N. C. Abay, Y. Zhou, M. Kantarcioglu, B. Thuraisingham, and L. Sweeney, "Privacy-Preserving Synthetic Data Release Using Deep Learning," in *Machine Learning and Knowledge Discovery in Databases*, Cham, 2019, pp. 510–526. doi: 10.1007/978-3-030-1092 5-7_31.

[9] S. Meiser, "Approximate and Probabilistic Differential Privacy Definitions.," *IACR Cryptol EPrint Arch*, vol. 2018, p. 277, 2018.

[10] M. E. Andrés, N. E. Bordenabe, K. Chatzikokolakis, and C. Palamidessi, "Geo-Indistinguishability: Differential Privacy for Location-Based Systems," *Proc. 2013 ACM SIGSAC Conf. Comput. Commun. Secur. - CCS 13*, pp. 901–914, 2013, doi: 10.1145/2508859.2516735.

[11] P. Štarchoò and T. Pikulík, "GDPR principles in Data protection encourage pseudonymization through most popular and full-personalized devices - mobile phones," *Procedia Comput. Sci.*, vol. 151, pp. 303–312, Jan. 2019, doi: 10.1016/j.procs.2019.04.043.

[12] P. Barbas, A. Clifford, K. Emanowicz, and P. G. O'Sullivan, "Identification of pseudonymized data within data sources," US10657287B2, May 19, 2020 Accessed: Jun. 29, 2020. [Online]. Available: https: //patents.google.com/patent/US10657287B2/en

[13] M. Gong, J. Feng, and Y. Xie, "Privacy-enhanced multi-party deep learning," *Neural Netw.*, vol. 121, pp. 484–496, Jan. 2020, doi: 10.1 016/j.neunet.2019.10.001.

[14] D. J. Solove and P. M. Schwartz, "The PII Problem: Privacy and a New Concept of Personally Identifiable Information," p. 82, 2011.

[15] E. C. Strinati and S. Barbarossa, "6G Networks: Beyond Shannon Towards Semantic and Goal-Oriented Communications," *ArXiv201114844 Cs Math*, Feb. 2021, Accessed: May 18, 2021. [Online]. Available: http://arxiv.org/abs/2011.14844

[16] M. Wang, T. Zhu, T. Zhang, J. Zhang, S. Yu, and W. Zhou, "Security and privacy in 6G networks: New areas and new challenges," *Digit. Commun. Netw.*, vol. 6, no. 3, pp. 281–291, Aug. 2020, doi: 10.101 6/j.dcan.2020.07.003.

[17] C. D. Alwis et al., "Survey on 6G Frontiers: Trends, Applications, Requirements, Technologies and Future Research," *IEEE Open J. Commun. Soc.*, vol. 2, pp. 836–886, 2021, DOI: 10.1109/OJCOMS.2021.30 71496.

Biographies

Aaloka is the founder of MAYA Data Privacy Limited, based out of Ireland. Company enables organizations to use data in a better way, complying with GDPR and other privacy regulations. He has held leadership and senior positions in SAP and Honeywell since 2004 and also worked with start-ups like Idea Device Technologies, MovidDLX, NGeneR and co-founded a non-profit organization Anant Prayas.

He is a researcher at CTIF Global Capsule, Aarhus University, Denmark since October 2019. He attained his Post Graduate degree in Enterprise Management from Indian Institute of Management Bangalore and B.Sc in Electronics and Communication Engineering from BIT Sindri in India. He teaches students for Masters program in Data Science as Associate Lecturer in National College of Ireland, Dublin and has taught previously at Furtwangen University, Germany. He is actively pursuing research on the topic of Privacy Preservation. His work focusses on new approaches for privacy preservation of Enterprise data and missing technology and structural framework for achieving end-to-end data privacy.

Dr Ramjee Prasad is the Founder and President of CTIF Global Capsule (CGC) and the Founding Chairman of the Global ICT Standardization Forum for India. Dr Prasad is also a Fellow of IEEE, USA; IETE India; IET, UK; a member of the Netherlands Electronics and Radio Society (NERG); and the Danish Engineering Society (IDA). He is a Professor of Future Technologies for Business Ecosystem Innovation (FT4BI) in the Department of Business Development and Technology, Aarhus University, Herning, Denmark. He was honoured by the University of Rome "Tor Vergata", Italy as a Distinguished Professor of the Department of Clinical Sciences and Translational Medicine on March 15, 2016. He is an Honorary Professor at the University of Cape Town, South Africa, and KwaZulu-Natal, South Africa. He received the Ridderkorset of Dannebrogordenen (Knight of the Dannebrog) in 2010 from the Danish Queen for the internationalization of top-class telecommunication research and education.

He has received several international awards, such as the IEEE Communications Society Wireless Communications Technical Committee Recognition Award in 2003 for contributing to the field of "Personal, Wireless and Mobile Systems and Networks", Telenor's Research Award in 2005 for outstanding merits, both academic and organizational within the field of wireless and personal communication, 2014 IEEE AESS Outstanding Organizational Leadership Award for: "Organizational Leadership in developing and globalizing the CTIF (Center for TeleInFrastruktur) Research Network", and so on. He has been the Project Coordinator of several EC projects, namely, MAGNET, MAGNET Beyond, and eWALL. He has published more than 50 books, 1000 plus journal and conference publications, more than 15 patents, and over 150 PhD. Graduates and a more significant number of Masters (over 250). Several of his students are today worldwide telecommunication leaders themselves.

Ramjee Prasad is a member of the Steering, Advisory, and Technical Program committees of many renowned annual international conferences, e.g., Wireless Personal Multimedia Communications Symposium (WPMC); Wireless VITAE, etc.

Index

About the Editors

Ramjee Prasad, Professor-CGC Research Lab, Department of Business Development and Technology, Aarhus University, Denmark. ramjee@btech.au.dk

He is the Founder and President of CTIF Global Capsule (CGC) and the Founding Chairman of the Global ICT Standardization Forum for India. Dr Prasad is also a Fellow of IEEE, USA; IETE India; IET, UK; a member of the Netherlands Electronics and Radio Society (NERG); and the Danish Engineering Society (IDA). He is a Professor of Future Technologies for Business Ecosystem Innovation (FT4BI) in the Department of Business Development and Technology, Aarhus University, Herning, Denmark. He was honoured by the University of Rome "Tor Vergata", Italy as a Distinguished Professor of the Department of Clinical Sciences and Translational Medicine on March 15, 2016. He is an Honorary Professor at the University of Cape Town, South Africa, and KwaZulu-Natal, South Africa. He received the Ridderkorset of Dannebrogordenen (Knight of the Dannebrog) in 2010 from the Danish Queen for the internationalization of top-class telecommunication research and education. He has received several international awards, such as the IEEE Communications Society Wireless Communications Technical Committee Recognition Award in 2003 for contributing to the field of "Personal, Wireless and Mobile Systems and Networks", Telenor's Research Award in 2005 for outstanding merits, both academic and organizational within the field of wireless and personal communication, 2014 IEEE AESS Outstanding Organizational Leadership Award for: "Organizational Leadership in developing and globalizing the CTIF (Center for TeleInFrastruktur)

Research Network", and so on. He has been the Project Coordinator of several EC projects, namely, MAGNET, MAGNET Beyond, and eWALL. He has published more than 50 books, 1000 plus journal and conference publications, more than 15 patents, and over 150 PhD. Graduates and a more significant number of Masters (over 250). Several of his students are today worldwide telecommunication leaders themselves.

Ramjee Prasad is a member of the Steering, Advisory, and Technical Program committees of many renowned annual international conferences, e.g., Wireless Personal Multimedia Communications Symposium (WPMC); Wireless VITAE, etc.

Anand Raghawa Prasad, Partner at Deloitte Tohmatsu Cyber (DTCY), Tokyo, Japan & Board of Director at Digital Nasional Berhad, Malaysia.
anandraghawa.prasad@tohmatsu.co.jp

Anand Raghawa Prasad is a Deloitte Tohmatsu Cyber (DTCY) partner, where he leads connectivity security. He is also a board of directors at Digital Nasional Berhad, Malaysia. Before DTCY, Anand was Founder & CEO, wenovator LLC, which now forms part of Deloitte and Senior Security Advisor, NTT DOCOMO. He was CISO, Board Member of Rakuten Mobile, where he led all aspects of enterprise and mobile network security (4G, 5G, IoT, Cloud, device, IT, SOC, GRC, assurance etc.) from design, deployment to operations. Anand was Chairman of 3GPP SA3, where, among others, he led the standardisation of 5G security.

He is also an advisor to several organizations such as CTIF Global Capsule, GuardRails and German Entrepreneurship Asia. Anand is an innovator with 50+ patents, a recognized keynote speaker (RSA, MWC etc.) and a prolific writer with six books and 50+ publications. He is a Fellow of IET and IETE.

Albena Mihovska, Associate Professor CGC Research Lab, Department of Business Development and Technology, Aarhus University, Denmark. amihovska@btech.au.dk

Albena Mihovska is a Senior Academic and Research Professional, currently an Associate Professor at the Dept of Business Development and Technology (BTECH) at Aarhus University, Herning, Denmark. She leads the 6G Knowledge Research Lab at the CTIF Global Capsule(CGC) research group at BTECH and is the Technical Manager of several EU-funded projects in Beyond 5G networks coordinated by Aarhus University. She is a member of IEEE INFORMS and has more than 150 publications.

Nidhi, PhD Student-CGC Research Lab, Department of Business Development and Technology, Aarhus University, Denmark. nidhi@btech.au.dk

Nidhi is an Early-Stage Researcher in the project TeamUp5G, a European Training Network in the frame of (MSCA ITN) of the European Commission's Horizon 2020 framework. Currently enrolled as a PhD student at Aarhus University in the Department of Business Development and Technology. She received her Bachelor's degree in Electronics and Telecommunication and master's degree in Electronics and Communication (Wireless) from India. Her research interests are small cells, spectrum management, carrier aggregation, etc.